Viral Entry into Host Cells

ADVANCES IN EXPERIMENTAL MEDICINE AND BIOLOGY

Recent Volumes in this Series

Volume 782
PROGRESS IN MOTOR CONTROL
Michael A. Riley, Michael J. Richardson and Kevin D. Shockley

Volume 783
THE NEW PARADIGM OF IMMUNITY TO TUBERCULOSIS
Maziar Divangahi

Volume 784
KISSPEPTIN SIGNAING IN REPRODUCTIVE BIOLOGY
Alexander S. Kauffman and Jeremy T. Smith

Volume 785
CROSSROADS BETWEEN INNATE AND ADAPTIVE IMMUNITY IV
Bali Pulendran, Peter D. Katsikis and Stephen P. Schoenberger

Volume 786
TRANSCRIPTIONAL AND TRANSLATIONAL REGULATION OF STEM CELLS
Gary Hime and Helen Abud

Volume 787
BASIC ASPECTS OF HEARING
Brian CJ Moore

Volume 788
NEUROBIOLOGY OF RESPIRATION
Mieczyslaw Pokorski

Volume 789
OXYGEN TRANSPORT TO TISSUE XXXV
Sabine Van Huffel, Gunnar Nualaers, Alexander Caicedo, Duane F. Bruley and David K. Harrison

Volume 790
VIRAL ENTRY INTO HOST CELLS
Stefan Pöhlmann and Graham Simmons

A Continuation Order Plan is available for this series. A continuation order will bring delivery of each new volume immediately upon publication. Volumes are billed only upon actual shipment. For further information please contact the publisher.

Viral Entry into Host Cells

Edited by

Stefan Pöhlmann, PhD
Institute for Virology, Hannover Medical School, Hannover, Germany, and German Primate Center, Göttingen, Germany

Graham Simmons, PhD
Blood Systems Research Institute, and Department of Laboratory Medicine, University of California San Francisco, San Francisco, California, USA

Springer Science+Business Media, LLC

Landes Bioscience

Springer Science+Business Media, LLC
Landes Bioscience

Springer Science+Business Media, LLC, 233 Spring Street, New York, New York 10013, USA
http://www.springer.com

Please address all inquiries to the publishers:
Landes Bioscience, 1806 Rio Grande, Austin, Texas 78701, USA
Phone: 512/ 637 6050; FAX: 512/ 637 6079
http://www.landesbioscience.com

The chapters in this book are available in the Madame Curie Bioscience Database.
http://www.landesbioscience.com/curie

Viral Entry into Host Cells, edited by Stefan Pöhlmann and Graham Simmons. Landes Bioscience / Springer Science+Business Media, LLC dual imprint / Springer series: Advances in Experimental Medicine and Biology.

ISBN 978-1-4614-7650-4

While the authors, editors and publisher believe that drug selection and dosage and the specifications and usage of equipment and devices, as set forth in this book, are in accord with current recommendations and practice at the time of publication, they make no warranty, expressed or implied, with respect to material described in this book. In view of the ongoing research, equipment development, changes in governmental regulations and the rapid accumulation of information relating to the biomedical sciences, the reader is urged to carefully review and evaluate the information provided herein.

Library of Congress Cataloging-in-Publication Data

Viral entry into host cells / edited by Stefan Poehlmann, Graham Simmons.
p. ; cm. -- (Advances in experimental medicine and biology ; v. 790)
Includes bibliographical references and index.
ISBN 978-1-4614-7650-4 (alk. paper)
I. Poehlmann, Stefan, 1970- II. Simmons, Graham, Ph. D. III. Series: Advances in experimental medicine and biology ; v. 790.
[DNLM: 1. Virus Internalization. 2. Endocytosis. 3. Host-Pathogen Interactions. QW 160 / W1 AD559 v.790 2013]
QR201.V55
571.9'92--dc23

2013013131

PREFACE

For a virus to invade a host cell it needs to penetrate the physical barrier imposed by the plasma membrane. Viruses have evolved specialized surface proteins to meet this challenge. These proteins facilitate delivery of the viral genetic information into the host cell by either fusing the viral envelope with a host cell membrane (enveloped viruses) or by forming membrane pores (non-enveloped viruses). Membrane fusion and pore formation critically depend on the engagement of host cell receptors and receptor choice is a key determinant of viral tropism. The multi-faceted interplay between viral and cellular factors during virus entry is a fascinating field of study, which can provide important insight into viral pathogenesis and define new targets for intervention. This book provides a comprehensive overview of this exciting field of research.

The first step in viral entry is the attachment of virions to target cells. Cellular factors which promote viral attachment and their role in virus infection are reviewed by Jolly and Sattentau. Two chapters review key processes underlying host cell entry of non-enveloped viruses: Bergelson and Coyne discuss the cell biology of picornavirus entry. These viruses exploit various endocytic pathways to invade cells and, by binding to cell surface receptors, activate signaling cascades which prime the cells for infection. Cellular entry of reoviruses is discussed by Danthi and colleagues. Members of the reovirus family depend on the cellular proteases cathepsins B and L for disassembly and viral structures exposed during disassembly can induce signaling cascades, which drive cells into apoptosis.

Enveloped viruses have evolved three related yet distinct solutions to bind and enter target cells: Class I, II and III membrane fusion proteins. Class I membrane fusion proteins are discussed in four chapters. Sun and Whittaker introduce the prototype class I membrane fusion protein, the influenza virus hemagglutinin. Simmons discusses entry of Ebola and Marburg virus, the only members of the filoviridae family. These viruses enter an extremely broad range of cells in a pH-dependent fashion. However, the pH-dependence is indirect: An acidic milieu is required for the activity of cathepsins B and L, which prime the viral glycoprotein for membrane fusion. A particular solution to host cell entry has been evolved by paramyxoviruses, which encode two distinct proteins to accomplish attachment to host cells and membrane fusion. Bossart and Broder describe how these proteins cooperate during host cell entry. Retroviruses comprise important human pathogens and are frequently used to study virus-host interactions during entry. Lindemann and colleagues review how foamy virus and HIV select and enter target cells. Class II membrane fusion proteins are structure- and sequence-wise different from class I fusion proteins but employ related mechanisms to merge the viral and cellular membranes. The key features of class II membrane fusion proteins are discussed by Modis. Class III membrane fusion proteins combine elements of the other classes and are only found in herpes-, rhabdo- and baculoviruses. Regan and Whittaker guide the reader through each step of host cell entry of rhabdoviruses. Entry of herpes viruses into host cells is facilitated by several viral glycoproteins and is regulated by glycoprotein-glycoprotein and glycoprotein-receptor interactions. The respective processes are reviewed by Krummenacher and colleagues, with a particular focus on the structures involved in receptor binding.

Stefan Pöhlmann, PhD
Institute for Virology, Hannover Medical School, Hannover, Germany,
and German Primate Center, Göttingen, Germany
and
Graham Simmons, PhD
Blood Systems Research Institute, and Department of Laboratory Medicine,
University of California San Francisco, San Francisco, California, USA

ABOUT THE EDITORS...

STEFAN PÖHLMANN is Professor for Infection Biology and Head of the Infection Biology Unit of the German Primate Center. His research is focused on virus host cell interactions, particularly on the molecular processes underlying host cell entry of human immunodeficiency virus and emerging viruses. Stefan Pöhlmann received his academic degrees (Dr. rer. nat., venia legendi) from the Friedrich-Alexander-University Erlangen-Nürnberg. Before his appointment at the German Primate Center, he worked as a postdoctoral fellow at the University of Pennsylvania, as a junior research group leader at the Friedrich-Alexander-University Erlangen-Nürnberg and as an assistant professor at Hannover Medical School. He serves on the editorial board of the *Journal of Virology* and *Virology* and serves as academic editor for *PLoS ONE*.

ABOUT THE EDITORS...

GRAHAM SIMMONS is an Associate Investigator at Blood Systems Research Institute and Associate Professor in the Department of Laboratory Medicine at the University of California, San Francisco. He is interested in many aspects of viral entry, particularly for blood transfusion transmissible agents and other emerging and re-emerging viruses that may be a threat to the blood supply. Graham Simmons received his doctoral training from the Institute of Cancer Research in London prior to performing an EMBO-sponsored postdoctoral fellowship at the University of Pennsylvania.

PARTICIPANTS

Jeffrey M. Bergelson
Department of Pediatrics
University of Pennsylvania Perelman School of Medicine
Division of Infectious Diseases
Children's Hospital of Philadelphia
Philadelphia, Pennsylvania
USA

Katharine N. Bossart
CSIRO Livestock Industries
Australian Animal Health Laboratory
Geelong, Victoria
Australia

Christopher C. Broder
Department of Microbiology and Immunology
F. Edward Hébert School of Medicine
Uniformed Services University
Bethesda, Maryland
USA

Andrea Carfí
Protein Biochemistry
Novartis Vaccines and Diagnostics
Cambridge, Massachusetts
USA

Gary H. Cohen
Department of Microbiology
School of Dental Medicine
University of Pennsylvania
Philadelphia, Pennsylvania
USA

Carolyn B. Coyne
Department Microbiology and Molecular Genetics
University of Pittsburgh School of Medicine
Pittsburgh, Pennsylvania
USA

Pranav Danthi
Department of Biology
Indiana University
Bloomington, Indiana
USA

Terence S. Dermody
Lamb Center for Pediatric Research
and
Department of Pediatrics
and
Department of Pathology, Microbiology and Immunology
Vanderbilt University School of Medicine
Nashville, Tennessee
USA

Roselyn J. Eisenberg
Department of Pathobiology
School of Veterinary Medicine
University of Pennsylvania
Philadelphia, Pennsylvania
USA

Geoffrey H. Holm
Department of Biology
Colgate University
Hamilton, New York
USA

Clare L. Jolly
Sir William Dunn School of Pathology
University of Oxford
Oxford
UK

Claude Krummenacher
Department of Pathobiology
School of Veterinary Medicine
University of Pennsylvania
Philadelphia, Pennsylvania
USA

Dirk Lindemann
Institute for Virology
Medical Faculty Carl Gustav Carus
Technische Universität Dresden
Dresden, Germany

Yorgo Modis
Department of Molecular Biophysics
and Biochemistry
Yale University
New Haven, Connecticut
USA

Stefan Pöhlmann
Institute for Virology
Hannover Medical School
Hannover
and
German Primate Center
Göttingen
Germany

Andrew D. Regan
Department of Microbiology
and Immunology
Cornell University
Ithaca, New York
USA

Quentin J. Sattentau
Sir William Dunn School of Pathology
University of Oxford
Oxford
UK

Graham Simmons
Blood Systems Research Institute
and
Department of Laboratory Medicine
University of California San Francisco
San Francisco, California
USA

Imke Steffen
Institute for Virology
Hannover Medical School
Hannover
Germany
and
Blood Systems Research Institute
San Francisco, California
USA

Thilo Stehle
Department of Pediatrics
Vanderbilt University School of Medicine
Nashville, Tennessee
USA
and
Interfakultäres Institut für Biochemie
Eberhard-Karls Universität
Tübingen
Germany

Xiangjie Sun
Department of Microbiology
and Immunology
Cornell University
Ithaca, New York
USA

Gary R. Whittaker
Department of Microbiology
and Immunology
Cornell University
Ithaca, New York
USA

CONTENTS

1. ATTACHMENT FACTORS....................1

Clare L. Jolly and Quentin J. Sattentau

Abstract....................1
Introduction....................2
Nonspecific Virus-Receptor Interactions....................2
Specific Virus-Receptor Interactions....................9
Conclusion....................18

2. PICORNAVIRUS ENTRY....................24

Jeffrey M. Bergelson and Carolyn B. Coyne

Abstract....................24
Introduction....................24
Picornavirus Structure....................25
Attachment to a Receptor....................26
The A Particle, a Likely Uncoating Intermediate....................28
The 80S Empty Capsid....................29
Cell Biology of Virus Entry: Endocytic Mechanisms....................30
Poliovirus: Does "Entry" Involve Internalization of the Virion?....................32
Foot and Mouth Disease Virus (FMDV) Uncoats in an Acidic Endosome....................32
Rhinoviruses Use Both Receptor-Mediated and Endosomal Triggers for Uncoating....................33
Echovirus 1 Needs Caveolae for Entry....................34
Coxsackieviruses: Multiple Receptors Serve Distinct Functions....................34
Virus-Induced Signals are Required for Entry....................36
PVR Signals in Poliovirus Entry....................36
Conclusion and Future Perspectives....................37

3. REOVIRUS RECEPTORS, CELL ENTRY, AND PROAPOPTOTIC SIGNALING 42

Pranav Danthi, Geoffrey H. Holm, Thilo Stehle and Terence S. Dermody

Abstract 42
Introduction 43
Pathogenesis of Reovirus Infection 43
Attachment Receptors: Cell-Surface Sialic Acid and Junctional Adhesion Molecule-A 44
Internalization Receptors: β1 Integrins 46
Stepwise Disassembly in the Endocytic Compartment 47
A Critical Role for Cathepsins in Reovirus Endosomal Disassembly 48
Outer-Capsid Protein σ3, the Major Target for Endosomal Protease Activity 49
Membrane-Penetration Protein μ1 50
Viral Determinants of Apoptosis Induction by Reovirus 51
Signaling Pathways Activated by Reovirus Infection 54
Cellular Gene Expression Profiles following Reovirus Infection 57
Effectors of Reovirus-Induced Apoptosis 59
Reovirus-Induced Apoptosis In Vivo 62
Conclusion and Future Directions 62

4. ENTRY OF INFLUENZA VIRUS 72

Xiangjie Sun and Gary R. Whittaker

Abstract 72
Introduction 72
Influenza Virus Receptors 73
Viral Internalization and Endocytic Trafficking 76
Membrane Fusion 78
Virus Uncoating and Nuclear Import of Viral Components 79
Conclusion and Future Perspectives 79

5. FILOVIRUS ENTRY 83

Graham Simmons

Abstract 83
Introduction 83
Filovirus Glycoprotein 84
Cellular Tropism 85
Receptors for Filovirus Entry 85
Attachment Factors 86
Routes of Entry 88
Mechanisms of Membrane Fusion 89
Conclusion 90

6. PARAMYXOVIRUS ENTRY 95

Katharine N. Bossart and Christopher C. Broder

Abstract 95
Introduction to Paramyxoviruses 96
Paramyxovirus Entry 98
Attachment Glycoproteins and Their Receptors 100

Attachment and Fusion Glycoproteins Work Together to Facilitate Entry 110
Receptor Induced Conformational Changes in Attachment Glycoproteins 113
Fusion Models 118
Conclusion and Remaining Questions 120

7. CELLULAR ENTRY OF RETROVIRUSES 128

Dirk Lindemann, Imke Steffen and Stefan Pöhlmann

Abstract 128
Introduction 128
Cellular Entry of Avian Sarcoma and Leukosis Virus (Alpharetroviruses) 129
Cellular Entry of Foamy Virus (Spumaviruses) 131
Cellular Entry of Human Immunodeficiency Virus (Retrovirus) 135
Conclusion 141

8. CLASS II FUSION PROTEINS 150

Yorgo Modis

Abstract 150
Introduction 150
Overall Architecture 152
Maturation and Priming 154
The Fusogenic Conformational Rearrangement 155
The Fusion Loop 156
Mechanism of Membrane Fusion 158
Strategies for Fusion Inhibition 161
Conclusion 162

9. ENTRY OF RHABDOVIRUSES INTO ANIMAL CELLS 167

Andrew D. Regan and Gary R. Whittaker

Abstract 167
Introduction 167
Viral Attachment 169
Membrane Fusion 170
Conclusion 173

10. ENTRY OF HERPESVIRUSES INTO CELLS: THE ENIGMA VARIATIONS 178

Claude Krummenacher, Andrea Carfí, Roselyn J. Eisenberg and Gary H. Cohen

Abstract 178
Introduction 179
Steps of Herpesvirus Entry 179
Entry Glycoproteins: Specificities of Herpesviruses Subfamilies 181
Entry Pathways: Fusion at the Plasma Membrane and Endocytosis 183
Triggering the Fusion Machinery: HSV gD Binding to Receptors 184
Conclusion and Future Challenges 189

INDEX 197

ACKNOWLEDGMENTS

The editors are very deeply indebted to all chapter authors for their excellent contributions, support and patience. We are particularly grateful to the authors of chapter updates. We are also thankful to Celeste Carlton and Cynthia Conomos at Landes Bioscience for their support and patience.

CHAPTER 1

ATTACHMENT FACTORS

Clare L. Jolly and Quentin J. Sattentau*
Sir William Dunn School of Pathology, University of Oxford, Oxford, UK
**Corresponding Author: Quentin J. Sattentau—Email: quentin.sattentau@path.ox.ac.uk*

Abstract: As obligate intracellular parasites, viruses must bind to, and enter, permissive host cells in order to gain access to the cellular machinery that is required for their replication. The very large number of mammalian viruses identified to date is reflected in the fact that almost every human and animal cell type is a target for infection by one, or commonly more than one, species of virus. As viruses have adapted to target certain cell types for their propagation, there is exquisite specificity in cellular tropism. This specificity is frequently, but not always, mediated by the first step in the viral replication cycle: attachment of viral surface proteins to receptors expressed on susceptible cells. Viral receptors may be protein, carbohydrate, and/or lipid. Many viruses can use more than one attachment receptor, and indeed may sequentially engage multiple receptors to infect a cell. Thus, it is useful to differentiate between attachment receptors, that simply allow viruses a foothold at the limiting membrane of a cell, and entry receptors that mediate delivery the viral genome into the cytoplasm. For some viruses the attachment factors that promote binding to permissive cells are very well defined, but the sequence of events that triggers viral entry is only now beginning to be understood. For other viruses, despite many efforts, the receptors remain elusive. In this chapter we will confine our review to viruses that infect mammals, with particular focus on human pathogens. We do not intend that this will be an exhaustive overview of viral attachment receptors; instead we will take a number of examples of well-characterized virus-receptor interactions, discuss supporting evidence, and highlight any controversies and uncertainties in the field. We will then conclude with a reflection on general principles of viral attachment, consider some exceptions to these principles, and make some suggestion for future research.

Viral Entry into Host Cells, edited by Stefan Pöhlmann and Graham Simmons.

INTRODUCTION

Despite striking differences in genome type, replication, morphology and tropism, all viruses carry structures, usually proteins, on the outer surface of the virion that mediate attachment to receptors expressed on the surface of target cells. There are two principal groups of viruses, with gross classification based on whether they are limited by a lipid membrane derived from a host cell membrane (enveloped viruses), or nonenveloped viruses, that have a protein exterior. Although most examples of viral surface structures are virally-encoded, others, such as adhesion molecules and proteoglycans, may be host cell-derived.

Attachment may be a relatively nonspecific process, by which surface viral structures associate reversibly with the cell limiting membrane via generalized biochemical properties such as charge, or may use more sophisticated bonding patterns to receptors that are specific for the viral family, genus, type, subtype or strain. A number of viruses use ubiquitously expressed receptors to attach to target cells, and replication is restricted by other factors, for example temperature sensitivity in the case of viruses that target the upper respiratory tract. Glycoproteins are major targets for viral attachment, and due to their regulated expression on different cell types, help provide the specificity required for viral tropism. Lipids are also used by a number of viruses for binding to target cells, particularly viruses that invade cells of the gastrointestinal tract. A great many viruses take advantage of the carbohydrate moieties present on both glycoproteins and glycolipids.

Because receptor binding is the essential first step in viral replication, and most viral attachment receptors are nonpolymorphic, there is strong selective pressure to maintain highly conserved receptor binding sequences. This is in the face of negative selective pressures such as those exerted by neutralizing antibodies. To overcome this, some viruses have adapted to shield the important receptor binding domains of their protein using a variety of strategies including protective glycan 'shields' and protein 'decoys'. Another strategy is to bury the receptor binding surface in a cleft or 'valley', occluding bulky immunoglobulin molecules. Furthermore, a number of viruses have evolved to use conformational changes, triggered by binding to a primary attachment receptor, to expose or create a previously hidden or 'cryptic' domain that then attaches to a second receptor, commonly the one that then mediates virus entry into the cell. For some viruses these interactions are very well defined, and contact sites have been mapped precisely to specific amino acid residues on the viral attachment protein, interacting with known domains on the cognate cellular receptor. Two very good examples of viruses where a great deal is known about the mechanism of virion attachment are the human immunodeficiency virus (HIV) and influenza A virus. Both of these viruses have benefited from intense study over a number of years and unparalleled levels of structural and functional information are available.

NONSPECIFIC VIRUS-RECEPTOR INTERACTIONS

Charge-Based Interactions

Often, viruses may find themselves in conditions that are unfavourable for cellular attachment, examples of which include the presence of competitive ligands in the extracellular milieu, conditions of flow such as are found in the vasculature, and repulsive forces of viral and cellular membranes in the case of enveloped viruses. Virus particles are often relatively labile, and inactivation of infectivity may ensue if entry into a cell is not achieved relatively rapidly. Thus, viruses need to limit random three-dimensional diffusion and expedite the receptor engagement process. To achieve this, many have chosen to anchor onto the cell glycocalyx, as this is the first physical structure to be encountered as the cell surface is approached. The glycocalyx is a ubiquitous carbohydrate 'umbrella', that contains negatively charged moieties in which the charge density is dependent on cell type and stage of differentiation and maturation. The charge within the glycocalyx is predominantly contributed by a group of glycoproteins called proteoglycans, that express sulfated glycan side chains termed glycosaminoglycans (GAGs). The number of glycan chains and the degree of sulfation varies according to the proteoglycan type, resulting in a wide spectrum of charge densities per molecule (Fig. 1). Moreover, the composition of the repeating dissacharide units allows for many different sulfated proteoglycans to exist, of which heparan sulfate and to a lesser extent, chondroitin sulfate can serve as viral attachment factors.

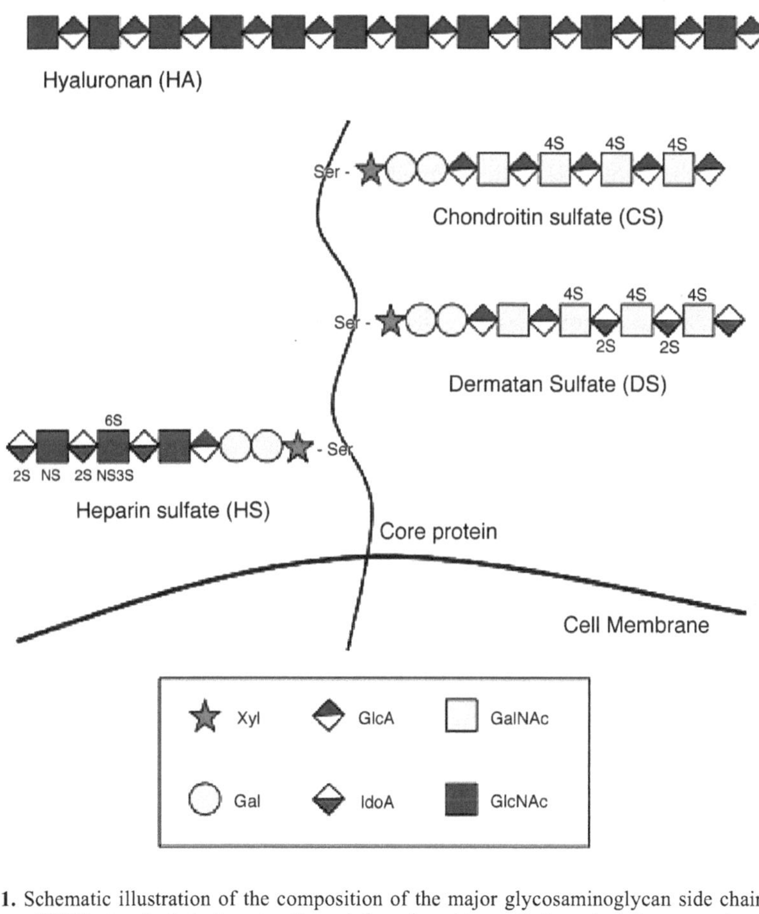

Figure 1. Schematic illustration of the composition of the major glycosaminoglycan side chains. Many viruses use HSPGs to attach to target cells and there is evidence that a select few may also associate with chondroitin sulfate containing proteoglycans.

A variety of both eukaryotic and prokaryotic pathogens, including viruses, bacteria, fungi and parasites have the ability to associate with proteoglycans. The requirement for proteoglycan binding appears to be a somewhat indistinct motif or patch of basic amino acids on the pathogen surface, but there is a degree of specificity linking the structure and charge of the pathogen motif and the type of proteoglycan used. GAGs are very promiscuous attachment receptors for viruses: many viruses have been identified that bind GAGs, including member of the *Retroviridae*, *Picornaviridae* and *Flaviviridae*[1-6] (Table 1), and the list is growing.

HSV

The first described and one of the best-characterized virus-proteoglycan interactions is probably that between herpes simplex virus type 1 (HSV-1) and a class of highly sulfated proteoglycans termed heparan sulfate proteoglycans (HSPGs). Herpes simplex virus attachment and entry into target cells requires the concerted action of multiple viral glycoprotein and cellular receptors, but the attachment of virions to permissive cells is initiated by binding of the viral structural proteins gB and gC to the disaccharide repeats of heparan sulphate.[7,8] In vitro, this association probably concentrates virions on the surface of target cells and facilitates subsequent interactions between HSV-1 glycoproteins and cellular receptors that are required for virus entry. Binding of gB or gC to HSPGs alone is insufficient to mediate virus entry, but in the absence of gC, infection decreases 10 fold,[9] and although HSPG binding is not an absolute requirement for virus entry the increased kinetics of infection after HSPG binding means they function as true attachment receptors. The recognition events that permit gB and gC to bind heparan sulfate are currently being dissected and it appears the minimal oligosaccharide recognized by HSV comprises as little as 10 monosaccharide units.[10] Interestingly, data indicate that gB and gC do not recognize the same receptor unit, moreover different gC proteins from different HSV types (HSV-1 versus HSV-2) also attach to distinct receptor domains,[11,12] however the importance of this in vivo is not established. Of the two HSV-1 glycoproteins that interact with HSPGs, gC may be more important in initial attachment, and while it is dispensable for growth in vitro, it is almost always expressed in primary patient isolates (reviewed in ref. 13). Within gC the basic and hydrophobic amino acids between 129 and 160, and residue 247, interact with heparan sulfate; mutation of these residues significantly reduces attachment of virions to cells.[14] This is consistent with the idea that positively charged viral domains promote binding to negatively charged heparan sulfate disaccharides. As mentioned earlier, there is some redundancy in the system and although the presence of gC enhances infectivity measurably, both gB and gC can bind to heparan sulfate. Mutation studies have shown if both proteins are removed, virus infectivity is severely impaired, however interpretation of these data is complicated because gB has a role in virus entry that is independent of the initial HSPG binding.[15] Although the relative contributions of gB and gC to virion binding are difficult to dissect, clearly the maintenance of both gB and gC in HSV-1 evolution indicates these proteins are indispensable for viral infectivity and pathogenesis.

As well as interacting with heparan sulfate containing GAGs, there is evidence that HSV can use chondroitin sulfate proteoglycans (CSPGs) as an auxiliary receptor in the absence of HSPGs[16,17] (Fig. 1). Studies designed to map the chondroitin sulfate

Table 1. Examples of attachment receptors for animal viruses

Family	Virus	Receptor	Reference
Adenoviridae	Adenovirus	CAR	53
		HSPGs	32
		sialic acid	128
Bunyaviridae	Hantavirus	β_3 integrins	129
Caliciviridae	Norovirus	HSPGs	130
Circoviridae	Circovirus	CSPGs	131
		HSPGs	
Filoviridae	Ebola virus	DC-SIGN	93
Flaviviridae	Dengue virus	DC-SIGN	88
		HSPGs	132
	HCV	DC-SIGN	90
		HSPGs	1
		SR-B1	97
Herpesviridae	CMV	HSPGs	133
		DC-SIGN	92
	HSV1 and 2	HSPGs	8
		CSPGs	18
	KSV (HHV8)	DC-SIGN	77
		HSPGs	76
		$\alpha_3\beta_1$, $\alpha_2\beta_1$	75
	HHV7	CD4	134
Orthomyxoviridae	Influenza A virus	α2,3 sialic acid (avian)	104
		α2,6 sialic acid (human)	96
		mannose receptor	
Paramyxoviridae	RSV	HSPGs	135
	Sendai virus	α2,3 sialic acid	136, 137
Picornaviridae	Coxsackie	CAR	53
		$\alpha_v\beta_3$	138
	Rhinovirus	ICAM-1 (CD54)	50
	Rhinovirus 87	sialic acid	139
	Rhinovirus 89	HSPGs	3
	FMDV	HSPGs	140
		$\alpha_v\beta_3$	141
	Echovirus	$\alpha_2\beta_1$, $\alpha_v\beta_3$	57 142
Poxviridae	Vaccinia virus	HSPG	143
		CSPG	144
Retroviridae	HIV	CD4	145
		HSPGs	6
		DC-SIGN	94
		Langerin	82
		mannose receptor	
	HTLV	DC-SIGN	146
		HSPGs	25
Reoviridae	Rotavirus	α2,6 or α2,3 sialic acid*	110
		$\alpha_2\beta_1$, α_4	61, 65
Rhabdoviridae	Rabies virus	NCAM	147

*Animal strains of rotavirus use terminal sialic acid for attachment to target cells but evidence to date suggests that human strains do not.

binding domains have shown that there is some redundancy between CSPG and HSPG attachment with the binding sites mostly overlapping, although subtle differences are apparent.[18] In addition to HSV, CSPGs can support the binding of other viruses to target cells (Table 1) and it is likely that many viruses that associate with HSPGs may also interact with chondroitin sulfate containing GAGs.

HIV

HIV-1 binds HSPGs via its surface envelope glycoprotein, gp120. The surfaces on gp120 mediating this interaction have been partially defined, and appear to consist of two structures: the V3 loop and the CD4-induced (CD4i) surface. These two regions form the chemokine receptor binding surface, and contain patches of positively charged amino acids that contribute to HSPG binding (Fig. 2). The ability of HIV to interact with HSPGs depends on several factors, and is linked to the tropism of the virus. The dominant determinant of HSPG binding is the charge on the V3 loop, and this appears to be linked to whether the virus uses CXCR4 or CCR5 as its entry coreceptor. CXCR4-using (X4) viruses tend

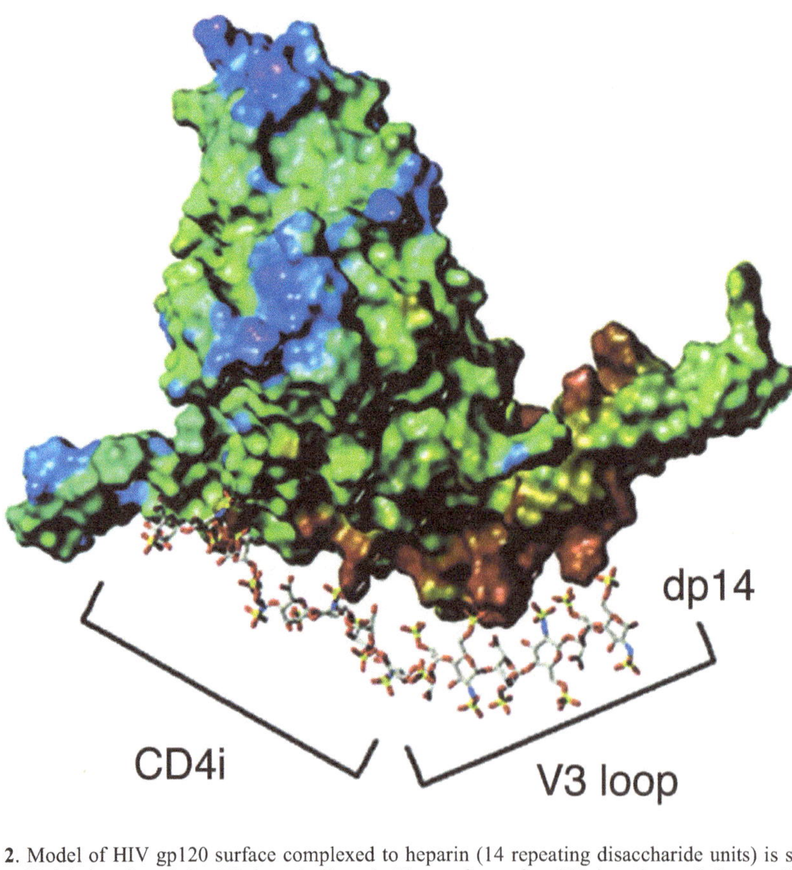

Figure 2. Model of HIV gp120 surface complexed to heparin (14 repeating disaccharide units) is shown with the CD4-induced and the V3 loop indicated. The surface of gp120 is color-coded according to electrostatic potential from blue (most negative) to red (most positive). Reprinted from Vives RR et al. J Biol Chem 2005; 280(22):21353-21357.[20]

to have a more basic V3 loop, and associate more tightly with HSPGs than CCR5-using (R5) viruses.[19] Moreover, the charge on the V3 loop is most likely increased during viral passage in HSPG-rich cell types such as human T-cell leukaemia virus (HTLV)-transformed T-cell lines. Thus, selection pressure on the virus to attach efficiently to HSPG-containing cell surfaces drives gp120 adaptation to a more positively charged surface. This selection, amongst others, results in viruses termed T-cell line adapted (TCLA). The charge on the V3 loop of TCLA viruses may reach +9, whereas that on an R5 primary isolate (PI) that has never been passaged in cell lines is typically +2 to +5. The affinity of monomeric X4 gp120 interaction with the prototype GAG heparin has been measured at 200nM in the absence of sCD4, and 17nM in the presence,[20] confirming the two site binding of gp120 to GAGs. Semi-quantitative analysis of X4 HIV virion binding to HSPG$^+$ cells (HeLa) demonstrates that, as expected, the overall avidity of the interaction is much higher than that for monomeric gp120, no doubt as a result of multivalent gp120-HSPG interactions.[6] Although there is a wealth of information regarding TCLA virus-HSPG association in vitro, little information is available regarding the importance of HSPGs in primary isolate HIV-1 attachment.[21] Moreover, very little is known concerning the use of HSPGs by HIV-1 on primary cells. Recently it has been observed that primary CD4$^+$ T cells express low levels of HSPGs under certain conditions,[2] although we have no insight into whether this makes these cells better targets for HIV-1 infection. There is evidence (our unpublished results) that CD4 is the dominant attachment receptor for TCLA virus on CD4$^+$ T cells, suggesting that other factors such as HSPGs may play a more minor (if any) role. However, it should be noted that primary isolates of HIV-1 tend to have a much lower affinity for CD4 than TCLA viruses[22,23] and so under these circumstances HSPGs may play a more significant role. Macrophages express the proteoglycan syndecan-II upon maturation, facilitating HIV-1 attachment to, and infection of, these cells.[4] It will be of interest to see whether the same is true of dendritic cells and other related cell types. Finally, nothing is known regarding the relationship between HIV-1 and HSPGs in vivo. One can speculate that HSPGs may promote HIV-1 infection in vivo by facilitating virus adsorption to target cell membranes, assuming that sufficient HSPG is expressed, that the viral V3 loop is sufficiently basic, and that the CD4i region is at least partially constitutively exposed. A recent study suggested that HSPGs might allow HIV-1 to be taken up into a protected intracellular environment and subsequently represented to permissive target cells.[5] However, it seems equally, if not more likely, that HSPGs ubiquitously expressed on epithelial and endothelial cell surfaces would trap HIV-1 onto (or into) a nonpermissive cellular environment that would lead to virus inactivation before viral 'rescue' by infection in trans of a permissive cell type.[22]

In a twist to the established HIV-1-GAG interaction story, it has been noted that HIV-1 can take up proteoglycans during budding from infected cells expressing these molecules at the plasma membrane.[24] The chemokine RANTES, when oligomeric, cross-linked the virions to target cell membranes and thereby enhanced viral infection of those cells. Thus, RANTES may have opposing effects on viral infection of CD4$^+$ cells: inhibition by coreceptor occupation but enhancement by increasing viral attachment.

HTLV-1

The surface Env subunit (gp46) of a related retrovirus, HTLV-1, binds HSPGs in an efficient manner, leading to enhanced HTLV pseudotype infection and HTLV-1 Env-mediated cell-cell fusion.[25] Moreover, even though human T cells express low levels of HSPGs after activation, this may be sufficient to increase infection by HTLV-1.

Although we believe the data in these two studies to be robust, there are caveats in their interpretation. First, it is unlikely that the HTLV-1 isolates used, and the Env derived from them, represents 'primary isolate'-derived material. More likely is that they have an adapted phenotype coming from production of the virus by cells expressing high levels of HSPGs and have thereby been selected for strong HSPG binding. A related observation that we (unpublished data) and others[26] have made, is that HTLV-1 transformed $CD4^+$ T-cell lines express high surface levels of HSPG, whereas nonHTLV-1-transformed T-cell lines express little or none. Thus, HTLV-1 infection (or transformation) may upregulate HSPGs on $CD4^+$ T cells, implying advantage for the virus. The second caveat is that HTLV-1 disseminates predominantly in vivo by direct cell-cell spread, probably via a 'virological synapse',[27] and not by release of cell-free virus or via cell-cell fusion. It remains to be seen, therefore, whether the presence of HSPGs on T cells is beneficial for HTLV-1 spread in vivo, or not.

Nonenveloped Viruses

The interaction between foot and mouth disease virus (FMDV) and GAGs has been characterized at the atomic level by crystallography of FMDV-heparan sulfate complexes.[28] The HSPG binding site of FMDV resides in a shallow depression located at the junction of the three major capsid proteins, and essentially no changes in protein conformation are required to accommodate the sugar. Basic and polar residues in VP1 (Lys193), VP2 (Thr134 and Arg135) and VP3 (Arg56 and Asn88) are crucial for heparan sulfate binding. Additional interactions between VP1, 2 and 3 and the HSPGs help to stabilize the complex and a total of nine amino acids make contact with the sulfate sugar. Studies comparing variants of FMDV have indicated that Arg56 in VP3 and Arg135 in VP2 may be important mediators of heparan sulfate binding between different strains of virus and changes that result in an Arg at position 56 are clearly associated with the acquisition of a high affinity HSPG binding phenotype.[29,30] Interestingly the adaptation to HSPG binding may be an in vitro phenomenon for FMDV and heparan sulfate binding is rapidly lost upon transfer into an animal model.[29]

There are of course other nonenveloped viruses, in addition to FMDV, that bind to GAGs to establish a foothold at the cell surface. Of particular interest are the adenoviruses that are being extensively modified, evaluated and tested as potential viral vectors to be used in gene therapy. Gene therapy naturally requires exquisitely refined cell targeting, so that the genetic material intended to correct the specific cellular defect is delivered to precisely the right tissue and cell. This requires that we fully understand the mechanisms of attachment of virus vectors in order to block native receptor binding and retarget virus to tissue specific sites. In addition to adenovirus binding to α_v integrins and CAR there is data indicating that Ad serotypes 2 and 5 can also interact with HS GAGs at the cell surface.[31,32] This finding has important implications for gene therapy because of the diversity of cell types that express these GAGs. Elucidating, and inhibiting, the molecular mechanisms of this interaction will be necessary to deliver adenoviral vectors to the correct tissue and prevent sequestering virus at inappropriate sites.

Host Cell Molecule Interactions

During budding from infected target cells, enveloped viruses wrap themselves in host cell membrane. If this takes place at the plasma membrane, as is the case for many viruses, then plasma membrane proteins, including adhesion molecules, may be incorporated into the viral envelope. This has been documented for several viruses, but the best characterized is probably HIV-1. Early studies demonstrated that mAbs to adhesion molecules inhibited HIV-1 Env-mediated cell-cell fusion.[33] Subsequently it was shown that $CD4^+$ T cells deficient in LFA-1 (leukocyte function antigen 1), obtained from Leukocyte Adhesion Deficiency Type-1 (LAD-1) patients, were deficient for HIV-induced syncytium formation.[34] Confirmatory evidence came from studies in which virions produced in cells over-expressing transgenic ICAM-1 were shown to be substantially (up to 10-fold) more infectious on $LFA\text{-}1^+$ target cells than virions produced in ICAM-1-low expressing cells.[35,36] This increase in viral infectivity probably results from the artificially high levels of ICAM-1 incorporated into the viral envelope. ICAM-1-enhanced infection probably also reflects the low CD4 affinity of primary isolate Env and the predicted small number of Env molecules on each HIV-1 virion.[37] A series of further studies by Tremblay's group has extended our understanding, at the molecular level, of the ICAM-1-LFA-1 interaction in the context of HIV infectivity.[38]

As well as classical adhesion molecules, HIV incorporates other cell surface proteins into the viral envelope, some of which influence viral infectivity. Studies have shown that incorporation of host-derived MHC class II (MHC-II) into the HIV envelope enhances infectivity and that virus produced in MHC-II+ve RAJI cells showed more rapid attachment and entry kinetics than virus produced in MHC-II-ve RAJI cells.[39] Moreover the presence of host derived CD80 (B7.1) and CD86 (B7.2) in HIV virions can enhance infectivity by facilitating attachment to CD28 and CTLA4 on the surface of target immune cells.[40] Many other enveloped viruses also incorporate host cell molecules during budding (reviewed in ref. 41) and this may be a general characteristic used by enveloped virus to enhance attachment.

SPECIFIC VIRUS-RECEPTOR INTERACTIONS

Immunoglobulin Superfamily Members

CD4

In vitro and in vivo, HIV infects predominantly $CD4^+$ T cells and cells of the monocyte/macrophage linage. Viral attachment to $CD4^+$ T cells and macrophages is mediated by the Env gp120 subunit, which along with the fusion protein gp41 result from proteolytic cleavage of the precursor gp160. gp120 binds CD4 by contact between 26 amino acids on gp120 and 22 residues of the D1 extracellular domain of the CD4 molecule.[42] Within CD4 these residues are localized between amino acids 25 and 64, but in gp120 are spread over about 6 discontinuous regions.[42] In gp120 it has been reported that Asp 368, Glu 370 and Trp 427 are particularly important and are conserved between both HIV-1 and HIV-2.[42,43] Interaction between monomeric gp120 and CD4 tends to be of high affinity (typically low nanomolar kD), but oligomerization into the intact Env spike has, in every case reported, reduced the affinity by up to two (or more) orders of magnitude.[22,44,45] The reason for this is not yet known, but the working hypothesis is that the CD4 binding surface on gp120 is partially obscured in the assembled trimer, potentially by the bulky V1V2 loop of the same gp120, or that of an adjacent gp120 protomer.[46]

After binding to CD4, gp120 undergoes a conformational change that exposes a binding site for the second receptor, a chemokine receptor (CKR), either CCR5 or CXCR4. The chemokine receptor binding site (CKRbs) of gp120 is spatially distinct from the CD4 binding site and is predominantly made up of amino acids in the 'bridging sheet' of gp120.[42,47] Although the chemokine receptors were initially termed 'secondary' receptors, mutations in the V1-V2 loop of gp120 can render HIV CD4-independent,[48] suggesting, along with other evidence, that the CKRs are most likely the primordial receptors and CD4 adaptation came later. The adaptation of HIV-1 to CD4-independence appears to be an in vitro phenomenon. By contrast, HIV-2 and SIV isolates have been found to be CD4-independent ex vivo, but they are rare and their fitness in vivo is unclear.

$CD4^+$ T cells express high levels of CD4, and as such this is likely to be the predominant HIV attachment receptor on this cell type. By contrast, the other $CD4^+$ permissive target cells, dendritic cells (DCs), Langerhans cells (LCs), monocytes/macrophages and their central nervous system (CNS) relatives microglia, express low levels of CD4. It has thus been proposed that HIV-1 attachment to these cells is likely to be via other types of structure. We have already mentioned HSPGs and adhesion molecules (above), but probably more important are the families of lectins that decorate the membranes of antigen presenting cells (Table 1), and these are discussed later.

ICAM-1

Some *Picornaviridae* family members, such as human rhinoviruses (HRV) and certain Coxsackie viruses, also use an Ig-like protein, in this case ICAM-1 (intercellular adhesion molecule 1, CD54), to attach to target cells. For HRV this interaction is very well defined. HRVs infect cells of the respiratory tract and are the major cause of the common cold. Of the approximately 100 known serotypes of human rhinovirus most utilize CD54 as the cellular attachment factor.[49,50] ICAM-1 was identified as the receptor for HRV using a mAb screen to inhibit infection followed by affinity chromatography. ICAM-1 is a ubiquitously expressed receptor that is found on many cell types, including the nasal epithelium to which HRV attaches, and the normal function of this protein is to mediate cell-cell interactions between ICAM-1 expressing cells and cells that express the cognate binding partner LFA-1 or Mac-1. ICAM-1 contains five Ig-like extracellular domains and the binding site for LFA-1 and HRV are within domains 1 and 2 (D1 and D2) but the pertinent amino acids required for HRV attachment are in the N-terminal D1 region.[51,52] Unlike LFA-1 binding, HRV attachment to ICAM-1 does not require divalent cation cofactors. Crystallography and cryo-electron microscopic modelling have identified the attachment sites in the BC, CD, DE and FG loops of D1 and charge complementarity is apparent between D1 and the HRV binding sites.[52] These charged interactions are very important for ICAM-1-HRV binding and are conserved amongst different rhinovirus groups. Human rhinoviruses are nonenveloped virions whose capsid consists of 60 protomers each of which is made up of a single unit of VP1, 2, 3 and 4. VP1, VP2 and VP3 comprise the capsid shell and the main feature of the rhinovirus capsid is the 'canyon'. The canyon, as the name suggests, is a depression or recess on the surface of the HRV virion that is constructed by residues from VP1, 2 and 3 and it is in this canyon that ICAM-1 binding interactions occur. The receptor binding site is necessarily highly conserved and is hidden at the bottom of the canyon, protected from neutralizing antibodies. Hypervariable residues line the canyon walls and it is thought that these are exposed to constant immune surveillance. Attachment of HRV to ICAM-1 is mediated by residues that lie deep within the recess on the canyon floor and because HRV attaches to the N terminal D1 region of ICAM-1, data suggests that the long extracellular Ig-like region must act as a probe to facilitate receptor-virus binding.

CAR (Adenovirus)

The coxsackie and adenovirus receptor (CAR), as the name suggests, facilitates cell surface attachment of Coxsackie viruses and group A, C, D, E and F adenoviruses. CAR was identified as the attachment receptor by a number of groups after being purified from extracts of cultured cells with intact Ad2 virions.[53-55] Not all adenoviruses require CAR to attach to host cells and adenovirus type 5 first recognizes MHC class I while serotypes 2 and 9 bypass CAR and MHC and bind directly integrins and enter host cells (Table 1). Morphologically, adenovirus capsids contain five prominent trimeric fibres that protrude from the vertices of the icosahedral capsid and are anchored into the virion via a penton base and are readily seen by electron microscopy. The distal end of the fibre contains the receptor binding knob region that interacts with the 46 kDa CAR. CAR and MHC class I alone are insufficient to permit viral infection and a second binding step involving integrin family members is required. CAR contains two Ig superfamily domains that are liganded at the N terminal V-like domains by the Ad-fibre knob[56] and crystallography has revealed that binding is multivalent and 3 CAR V domains associate with the surface exposed loop of a single Ad knob. Greater than 50% of the receptor-ligand binding is contributed by the AB loop of the knob region and the complex is stabilized by a series of hydrogen bonds. Although a great deal of structural information is known about adenovirus-CAR binding, the potential for designing therapeutic inhibitors is complicated by the fact that the normal physiological function of CAR is unknown and that expression is on the basolateral surface of epithelial cells and so drug accessibility to the site of virus-cell attachment may be problematic.

Integrins

Integrins are heterodimers of α and β subunits that are present on many cell types and whose function is to mediate interactions between cells and the extracellular matrix, and to serve as adhesion molecules for cell-cell interactions. Because they are so widely expressed on different tissues, integrins are often recognized as cellular attachment receptors by viruses. Some viruses use integrins exclusively to mediate binding and entry whereas others use them as coreceptors. Echovirus is a picornavirus that uses integrins as the primary receptor for cell binding and three different $\alpha\beta$ heterodimers can serve as attachment factors. Viruses bind to integrins by short virally encoded peptide sequences present in structural proteins that mimic the natural integrin ligands. $\alpha_2\beta_1$, also known as VLA-2, is the attachment receptor for echovirus serotypes 1 and 8 and was identified using mAbs to block infection of susceptible cells.[57] This was confirmed using cells expressing little $\alpha_2\beta_1$ that did not bind virus until transfected with the α_2 encoding cDNA. Binding of echovirus to $\alpha_2\beta_1$ is via a DGE tripeptide motif in VP1. Not all echoviruses use $\alpha_2\beta_1$ and serotype 22 associates with the α_v subunit presented as $\alpha_v\beta_3$ or $\alpha_v\beta_1$ and this is through a RGD domain also within VP1. The interaction of VP1 with α_v is relevant in vivo and RGD-containing peptides can inhibit virus replication in a mouse model.[58] The first structural integrin data for picornaviruses was obtained using echovirus[59] and the virus-binding site is located within the I domain of the α_2 subunit, in agreement with earlier recombinant protein work implicating residues 199-201 and 212-216 in binding.[60] The "I" (inserted) domain is present within a subclass of α subunits and contains the MIDAS (metal ion dependent adhesion site) that chelates divalent cations which act as cofactors for ligand binding. Interestingly, unlike collagen, echovirus binding does not require divalent cations and the residues that bind echovirus are on a different face to those that recognize collagen, indicating that the integrin associates with these two ligands by different mechanisms. Importantly for viral fitness, echovirus binds with a 10 fold higher affinity to integrins than do the natural ligands.[59] Separate binding domains, coupled with the higher affinity interactions, would help virus attach to target cells even in the presence of potentially competing ligands such as collagen and other components of the extracellular matrix. In addition, models predict that multiple $\alpha_2\beta_1$ heterodimers can bind to adjacent sites on the echovirus capsid without any apparent steric hindrance, leading to a higher affinity interaction that sequesters virus tightly to the cell surface.

Rotavirus, a nonenveloped virus that causes gastroenteritis in infant humans and animals, also binds to integrins as part of the infection process.[61] Rotavirus entry appears to be a multi-step process whereby most animal rotaviruses (and probably some human strains also) bind initially to a sialic acid containing carbohydrate and then interact with integrins to promote viral entry into the cytoplasm.[62,63] An exception to this is the porcine rotavirus strain CRW-8 that is integrin independent.[64] A number of integrins have now been identified that interact with rotavirus motifs present within the two outer capsid proteins, the spike protein VP4 and the major outer capsid protein VP7, and some of these promote virus binding while others are involved in post-attachment entry steps. Integrins that have been shown to mediate initial events in rotavirus attachment to target cells in vitro are $\alpha_2\beta_1$, $\alpha_4\beta_1$ and $\alpha_4\beta_7$.[61,65] Binding to these integrins is modulated by motifs within VP7 and the VP5* domain of VP4. $\alpha_2\beta_1$ binding by rotavirus occurs via a DGE sequence in VP5* and the I domain of α_2 is both necessary and sufficient for binding.[66,67] Up to 4 potential binding motifs are present within both VP4 and VP7 that can interact with $\alpha_4\beta_1$ and $\alpha_4\beta_7$ integrins.[64] The reason for the high degree of redundancy in the system is unclear, however it is clearly necessary for efficient infection, and blocking these interactions has a measurable effect on binding and entry. Whether rotavirus interacts sequentially or alternatively with all integrins present on the surface of target cells is not known. It is possible that the presence of multiple binding motifs within VP4 and VP7 is influenced by the expression levels of integrins in vivo during various stages of cellular differentiation and may also be an important determinant of species tropism.

Multispan Receptors

Tetraspanins

The role of multispan proteins as attachment receptors for animal viruses is relatively ill-defined when compared to other factors that promote virion binding. There is some evidence that multispan proteins can function as viral attachment proteins and clearly they are capable of acting as coreceptors for some viruses. Tetraspanins are four transmembrane receptors whose exact cellular function is unknown. Tetraspanins shuttle between endocytic membranes and the plasma membrane by virtue of their association with intracellular vesicles and at the cell surface tetraspanins form what is known as a "tetraspan web", where, because of their reported association with integrins, they are thought to be involved in cell-cell interactions. To date, the only tetraspanin implicated as a receptor for viral attachment is CD81.[68] Whether CD81 is strictly an attachment factor for hepatitis C virus (HCV), or also mediates viral entry, is controversial. It has been suggested that CD81 acts as a post-attachment entry receptor for HCV[69] and a number of other receptors including DC-SIGN, the LDL receptor, HSPGs and scavenger receptor class B type I (SR-B1) have also been implicated in attachment (reviewed in ref. 70). HCV experiments have been complicated by the absence, until recently, of a cell culture system to grow virus, or of pseudotyped virions. Because of this, studies have used truncated expressed forms of E2, the putative viral receptor binding protein, which, while very informative, is not necessarily reflective of the situation with intact virions. However, despite the limitations, there is solid data demonstrating that the E2 protein of HCV binds to an extracellular loop of CD81[68] and so it may function as the binding receptor. Moreover, reports have indicated that $CD81^+$ nonhepatic cell lines do not support HCV pseudovirus entry, adding some weight to the argument that CD81 promotes virus binding but that additional receptors regulate entry.[69,71,72] With the advances in techniques to study HCV it is anticipated that a great deal of new information about virus binding will become available over the next few years that could help clarify the area of tetraspanins in viral attachment.

Seven and 12 Transmembrane Receptors

As mentioned previously, seven transmembrane (TM) proteins (the G-protein coupled chemokine receptors CXCR4 and CCR5) act as coreceptors to allow infection of HIV-1 into $CD4^+$ target cells. Not all lentiviruses require CD4 and lab adapted isolates of HIV-2 as well as primary isolates of the simian immunodeficiency virus (SIV) are able to infect CD4 negative cells via CXCR4 and CCR5 respectively. This is not just an in vitro property of HIV-2 and it is reported that most primary patient isolates of HIV-2 use CXCR4 or CCR5 to infect target cells.[73] Thus, 7TM receptors are able to function as attachment factors for a small group of viruses.

Very recently a 12 TM protein has been identified as a putative viral receptor for Kaposi's sarcoma herpes virus (KSHV, also known as HHV8).[74] The receptor identified, xCT, is a human protein involved in L-cystine influx into cell and concomitant efflux of L-glutamine. xCT and was identified in a two-cell fusion assay whereby xCT expression in 3T3 cells rendered cells susceptible to fusion with cells expressing transfected KSHV glycoprotein. In addition to this, a viral entry assay demonstrated xCT- BHK-21 and K562 cells did not support viral entry whereas xCT transfected cells did. Thus, whether the effect on viral infectivity is due a direct effect on virus penetration or whether xCT functions along with GAGs, DC-SIGN and β_1 integrins[75-77] at the first stages of attachment is unclear but remains a possibility.

Lectins

DC-SIGN

Lectins are a family of molecules that bind carbohydrate moieties, many in a calcium (C-type-lectins) dependent manner (Fig. 3). DC-SIGN is a C-type lectin that was first cloned in 1992 from placental tissue, was demonstrated to exhibit an affinity for gp120 in the low nM range, but its function was unknown. In 2000, Geijtenbeek and coworkers identified DC-SIGN as a DC-expressed ligand for HIV-1 gp120. The interaction is based

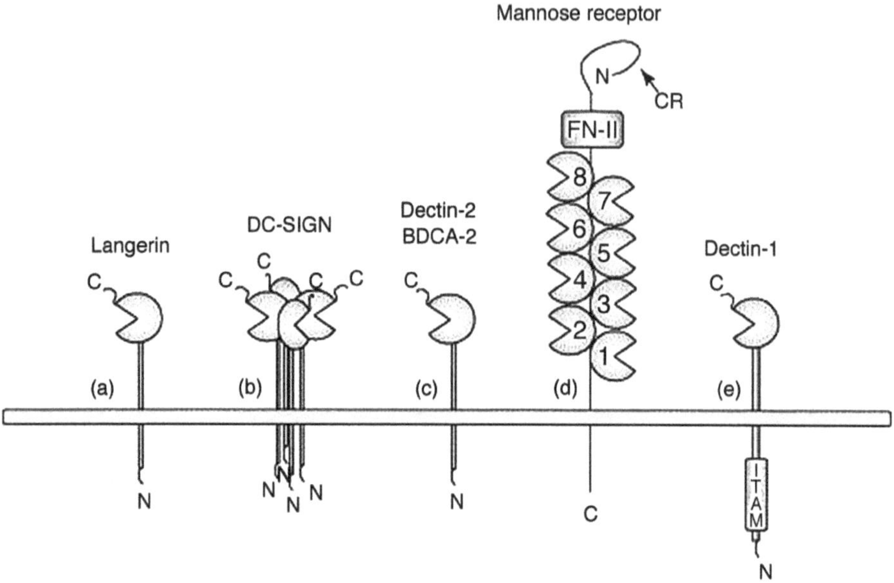

Figure 3. The C-type lectin-like receptors expressed by dendritic cells and macrophages. DC-SIGN, Langherin and the mannose receptor have all been shown to promote viral attachment to target cells. Reprinted from: McGreal EP et al. Curr Opin Immunol 2005; 17(1):18-24;[148] ©2005, with permission from Elsevier.

on recognition of terminal mannose moieties on gp120 and it has been confirmed that DC-SIGN binding to gp120 is carbohydrate dependent, with no appreciable effect of protein-protein interactions.[78] DC-SIGN binding to N-linked high mannose on gp120 is mediated by a characteristic carbohydrate recognition domain (CRD) and it is the Glu-Pro-Asn motif that confers the mannose specificity. Specifically, studies have shown that Glu347, Asn349, Glu351 and Asp355 are important for gp120 binding, while Gly346 is critical.[79] DC-SIGN cannot mediate HIV-1 entry, and is therefore a true attachment receptor. Nevertheless, if CD4 and a CKR are present on the same cell, DC-SIGN capture of virus can increase the efficiency of infection of that cell,[80] no doubt by enhancing viral attachment to the permissive cell surface followed by viral presentation to the entry receptors. Geijtenbeek and colleagues described an intriguing DC-SIGN-mediated phenomenon, that ligation of DC-SIGN by HIV-1 resulted in internalization of the virus into an intracellular compartment, within which HIV-1 remained infectious for several days. Subsequent studies appeared to confirm this, and hypothesized that HIV-1 uptake by DCs allowed viral storage in a 'safe' compartment until presentation to $CD4^+$ T cells in local secondary lymphoid tissue led to infection in trans. More recent studies have contested the concept that HIV-1 can remain infectious in a DC endosomal compartment for such an extended time, and instead demonstrate that these cells become infected at a low level.[81,82] Moreover, there is evidence that high affinity gp120-DC-SIGN binding is lost at lower pH[83] and while this provides a mechanism for virus escape from acidifying endosomes, it raises questions about the role of DC-SIGN as a mediator of HIV capture in the acidic environs of the vaginal mucosa. Whatever the mechanism of preservation of viral infectivity, the model for DC-mediated transfer of virus from the mucosae to the draining lymph nodes as a primary route of dissemination is compelling, and in vitro tissue culture models back this up.[84,85] Whether this trafficking occurs in a DC-SIGN-dependent manner alone or is mediated also by Langerhans cells that are prevalent at mucosal surfaces and can bind gp120 via their CLR, Langerin,[82,86] is unclear. It is therefore striking that antagonists of the DC-SIGN-gp120 interaction, such as mannan, have no inhibitory effect on HIV-1 infection following mucosal challenge in macaques:[87] clearly the in vivo relevance of this model remains to be confirmed.

DC-SIGN is a promiscuous viral attachment receptor, and also binds human cytomegalovirus (CMV), hepatitis C virus, Dengue virus and Ebola virus envelope glycoproteins (Table 1).[88-92] For CMV and Ebola virus, DC-SIGN appears to function in a manner similar to its role in HIV infection. In the case of Ebola virus, it seems DC-SIGN can mediate infection in cis and in trans[93] while CMV-associated DC-SIGN expressing cells are not infected directly but instead transfer virus, via DC-SIGN capture, to permissive target cells. Finally, for HCV both DC-SIGN and the liver/lymph node specific molecule L-SIGN (also termed DC-SIGNR) attach to the E2 glycoprotein of HCV and as well as permitting binding of transformed cell lines this interaction mediates HCV binding to primary cells.[90] Because of the restricted tissue expression of L-SIGN, in particular, this lectin may be an efficient way of targeting HCV to the liver.

Other Lectin Receptors

Antigen presenting cells carry multiple lectin receptors whose function is to sample the extracellular milieu. Amongst these, the mannose receptor is perhaps the best characterized, and like DC-SIGN it also acts an attachment factor for viruses via its specificity for terminal mannose, fucose and N-acetylglucosamine. The mannose receptor binds HIV gp120, and this may be an additional mechanism for infection of DCs and other APCs.[82] Mannose receptor binding of HIV gp120 can also explain the observation that blocking of gp120-DC-SIGN binding alone, or in combination with inhibitors of CD4 and the CKR, do not completely abolish binding of HIV to DCs.[82,94] Moreover, macrophages are a prevalent in vivo target for HIV infection and attachment of incoming virions to the mannose receptor may facilitate better viral replication kinetics in cis or infection in trans.[95] Mannose receptor binding has also been shown to promote attachment of influenza virus to macrophages[96] via high-mannose oligosaccharides on the viral hemagglutinin (HA) and a good correlation is seen between the degree of glysosylation of HA and the ability of flu virions to bind to macrophages. It seems very likely, that like DC-SIGN, the mannose receptor will be involved in the pathogenesis of other viruses that harbour mannosylated structural proteins.

Scavenger Receptors

Scavenger receptors are cell surface proteins that can bind and internalize modified lipoproteins such as acetylated LDL and have a broad ligand binding specificity. Initially described in cultured macrophages, scavenger receptors are found on many different cell types. Of the two classes of scavenger receptors identified (class A and B) class B type 1 (SR-B1) has been implicated in binding of HCV to target cells[97] and SR-B1 is expressed to high levels in the liver, placing it at the correct anatomical site to play a role in HCV infection.[98,99] Data from experiments performed with soluble proteins indicates that

HCV binding to SR-B1 is via the hypervariable region 1 of E2 but whether this region comprises all or part of the SR-B1 binding site in not known.[97] The exact role of SR-B1 in HCV attachment and entry is unclear and it has been shown that expression of SR-B1 and CD81 together are insufficient to mediate virus infection, indicating that additional molecules are likely to be involved.[72] One interesting possibility is that SR-B1 acts as a "nonclassical viral receptor" and facilitates infection of cells because of its ability to modulate the lipid composition of membranes.[70] In support of this theory the capacity of SR-B1 to act as an entry receptor for HCV is enhanced by SR-B1 mediated uptake of lipids from HDL.[72,100] Thus, whether SR-B1 acts as an attachment receptor or can also internalize bound virus is not known but the potentially novel role that SR-B1 may play in HCV infection makes it a fascinating topic of study.

Nonproteic Attachment Receptors

Sialic Acid

Many viruses recognize carbohydrate moieties for binding to cells (Table 1) and carbohydrates are well suited to this for a many reasons. Firstly, carbohydrate side chains can project relatively long distances from the cell surface and so may be one of the first structures encountered by incoming viruses. Secondly, carbohydrates are often charged and so attract charged residues in viral proteins that can form rapid and transient interactions. Finally, sugars such as sialic acids are ubiquitously expressed on cells and viruses that use carbohydrates tend to have a broad host range. Probably the best-studied virus that uses sialic acid to bind to target cells is influenza virus. Influenza virus infects cells of the upper respiratory tract via binding of the trimeric viral HA protein to sialyloligosaccharide receptors on the surface of target cells.[101-103] During viral assembly the HA undergoes proteolytic cleavage to produce HA1 and HA2 which remain associated via disulphide bonds. HA1 comprises the membrane distal globular head of the molecule and X-ray crystallographic studies have determined the receptor binding site of the HA resides in a pocket which is completely occupied by the sialic acid, supporting the idea that it is the only necessary component of the cell receptor for influenza.[102] Approximately 15 amino acids that are widely spaced on the linear molecule are required for direct binding of sialic acid, or help to stabilize the interaction.[102,104] Of these residue 226 is believed to be the most important in determining the sialic acid specificity, and a change from Leu to Glu at this position alters the viral preference from Neu5Acα2-6 to Neu5Acα2-3,[105] while a number of other amino acids, both proximal and distal are also important.[106-108] The simplicity of the carbohydrate receptor as well as structural studies on both the viral HA and the neuraminidase protein (required for releasing virions from the cell surface after viral exit) has opened the door for rational drug design and the inhibition of influenza infection and dissemination within the host.

Rotavirus expresses an outer capsid protein VP7 and a spike protein VP4 that work together to allow virus to bind to and infect mature, differentiated enterocytes on the small intestine. While animal strains of rotavirus bind in a sialic acid dependent manner to host cells, human rotaviruses (and some animal strains) are sialic-acid independent[63,109-111] and it is thought that another, as yet undefined, carbohydrate group may be the attachment receptor for human strains. Alternatively, the glycan receptor for human strains of rotavirus may be an internal (and thus neuraminidase-insensitive) sialic acid.[112,113] Terminal sialic acids, or related carbohydrates in the case of human rotaviruses, probably serve as the

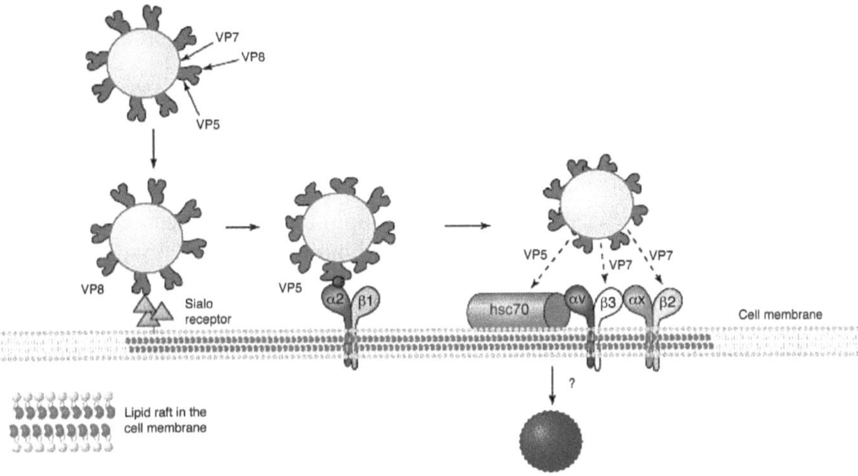

Figure 4. A model for the early stages of rotavirus attachment for a neuraminidase-sensitive strain of rotavirus. Initial binding is mediated by the VP8* domain of VP4 that binds to a sialic acid containing glycoprotein or glycolipid, perhaps with a contribution by the underlying galactose. A (possible) conformational change may then occur to facilitate an interaction between the DGE motif in rotavirus VP5* and the I domain of the $\alpha_2\beta_1$ integrin. After this a number of possible additional binding events may take place, involving other integrins and additional receptors, culminating in virus crossing the plasma membrane by a presently unknown mechanism. Reprinted from: Lopez S, Arias CF. Multistep entry of rotavirus into cells: Trends Microbiol 2004; 12(6):271-278; ©2004;[149] with permission from Elsevier.

attachment receptors for rotavirus and precede virus binding to integrins that then mediate virus entry.[61] This is the hypothesis behind the proposed two-step model for rotavirus attachment and entry[62,63] (and Fig. 4). The requirement for sialic acid in animal rotavirus binding was demonstrated in a series of studies involving enzymatic cleavage and competitive sialoconjugates and these studies demonstrated that rotaviruses are able to distinguish between different sialic acid-galactose linkages. For example simian viruses seem to prefer α2,3 linked sialic acid while bovine strains show a preference for α2,6 linkages[114-116] and both can distinguish between N-acetyl and N-glycolyl sialic acids.[113] The hemagglutinating and carbohydrate binding domain of VP4 has been mapped to VP8*[110,117] and crystallographic information about the structure of VP8* alone and liganded to sialic acid,[118] coupled with knowledge of the sugar specificity of rotavirus, has permitted early stage evaluation of synthetic receptor mimics that could be used to inhibit rotavirus infection and/or reduce disease severity.[119,115,120]

Other Carbohydrate Attachment Receptors

Although animal rotaviruses recognize terminal sialic acids, the hierarchy of preference for different sialic acid linkages hinted that the underlying glycan group may also be a component of the rotavirus attachment receptor. This has been confirmed by a number of groups who have shown that animal rotaviruses can bind to receptors where the terminal glycan is a galactose, or that galactose-binding inhibitors can effectively block viral attachment to cells in vitro[62,109,121] and it seems the minimal structural element required may be a terminal sialyl-galactose. Whether the carbohydrate receptor is present on a glycoprotein, glycolipid or ganglioside is unclear, and may vary between different species of rotavirus, but there is evidence that for porcine strains the putative in vivo receptor is a ganglioside.[122,123]

Other viruses can also use glycolipids to attach to target cell. For example, HIV binding to CD4-negative neuronal cells and intestinal cell lines (that do not express the preferred receptors) in vitro may take place via galactosyl ceramide (GalCer).[124,125] Antibodies that block gp120-GalCer binding or GalCer derivatives are able to inhibit infection of neural cells[124,126] and the galactosyl-lipid linkage has been demonstrated to be an important factor in gp120 binding. GalCer binds to HIV gp120 with high affinity and the binding site on HIV has been mapped to amino acids 206-275 in gp120.[127] It should be noted that with the exception of some strains of rotavirus and HIV infection of certain cell types, most viruses that use carbohydrate for attachment tend to exploit sialic acid, or acetylated sialic acid, to attach to host cells.

CONCLUSION

Viruses have subverted a wide variety of cell surface molecules as attachment receptors. In some cases these are also entry receptors, in other cases they are not sufficient to allow virus infection and entry receptors must also be engaged. Most of the work on viral attachment to cell surface receptors has been done in vitro, and we have generally little information on whether these receptors function equivalently, or even at all, in vivo. In many cases receptor expression in vitro explains the species and tissue tropism of viruses in vivo, but in other cases does not, implying the existence of other factors. Certainly in vitro work using primary tissues and nonadapted viruses is a step in the right direction, and indeed may be the only source of information where an animal model is not available.

Since attachment is the very first step in viral entry, targeting this step for prophylactic or therapeutic ends is attractive. Despite the potential, only a few drugs have been licensed to date that target this crucial first stages in the virus life cycle, and with further advances in our understanding of the molecular details that govern viral attachment, we hope to reveal more secrets that can help us to expand the area of viral intervention strategies.

REFERENCES

1. Barth H, Schafer C, Adah MI et al. Cellular binding of hepatitis C virus envelope glycoprotein E2 requires cell surface heparan sulfate. J Biol Chem 2003; 278(42):41003-41012.
2. Jones KS, Petrow-Sadowski C, Bertolette DC et al. Heparan sulfate proteoglycans mediate attachment and entry of human T-cell leukemia virus type 1 virions into CD4+ T cells. J Virol 2005; 79(20):12692-12702.
3. Vlasak M, Goesler I, Blaas D. Human rhinovirus type 89 variants use heparan sulfate proteoglycan for cell attachment. J Virol 2005; 79(10):5963-5970.
4. Saphire AC, Bobardt MD, Zhang Z et al. Syndecans serve as attachment receptors for human immunodeficiency virus type 1 on macrophages. J Virol 2001; 75(19):9187-9200.
5. Bobardt MD, Saphire AC, Hung HC et al. Syndecan captures, protects, and transmits HIV to T lymphocytes. Immunity 2003; 18(1):27-39.
6. Mondor I, Ugolini S, Sattentau QJ. Human immunodeficiency virus type 1 attachment to HeLa CD4 cells is CD4 independent and gp120 dependent and requires cell surface heparans. J Virol 1998; 72(5):3623-3634.
7. Spear PG, Shieh MT, Herold BC et al. Heparan sulfate glycosaminoglycans as primary cell surface receptors for herpes simplex virus. Adv Exp Med Biol 1992; 313:341-353.

8. WuDunn D, Spear PG. Initial interaction of herpes simplex virus with cells is binding to heparan sulfate. J Virol 1989; 63(1):52-58.
9. Herold BC, WuDunn D, Soltys N et al. Glycoprotein C of herpes simplex virus type 1 plays a principal role in the adsorption of virus to cells and in infectivity. J Virol 1991; 65(3):1090-1098.
10. Lycke E, Johansson M, Svennerholm B et al. Binding of herpes simplex virus to cellular heparan sulphate, an initial step in the adsorption process. J Gen Virol 1991; 72(Pt 5):1131-1137.
11. Trybala E, Liljeqvist JA, Svennerholm B et al. Herpes simplex virus types 1 and 2 differ in their interaction with heparan sulfate. J Virol 2000; 74(19):9106-9114.
12. Gerber SI, Belval BJ, Herold BC. Differences in the role of glycoprotein C of HSV-1 and HSV-2 in viral binding may contribute to serotype differences in cell tropism. Virology 1995; 214(1):29-39.
13. Spear PG. Herpes simplex virus: Receptors and ligands for cell entry. Cell Microbiol 2004; 6(5):401-410.
14. Mardberg K, Trybala E, Glorioso JC et al. Mutational analysis of the major heparan sulfate-binding domain of herpes simplex virus type 1 glycoprotein C. J Gen Virol 2001; 82(Pt 8):1941-1950.
15. Laquerre S, Argnani R, Anderson DB et al. Heparan sulfate proteoglycan binding by herpes simplex virus type 1 glycoproteins B and C, which differ in their contributions to virus attachment, penetration, and cell-to-cell spread. J Virol 1998; 72(7):6119-6130.
16. Banfield BW, Leduc Y, Esford L et al. Evidence for an interaction of herpes simplex virus with chondroitin sulfate proteoglycans during infection. Virology 1995; 208(2):531-539.
17. Bergefall K, Trybala E, Johansson M et al. Chondroitin sulfate characterized by the E-disaccharide unit is a potent inhibitor of herpes simplex virus infectivity and provides the virus binding sites on gro2C cells. J Biol Chem 2005; 280(37):32193-32199.
18. Mardberg K, Trybala E, Tufaro F et al. Herpes simplex virus type 1 glycoprotein C is necessary for efficient infection of chondroitin sulfate-expressing gro2C cells. J Gen Virol 2002; 83(Pt 2):291-300.
19. Moulard M, Lortat-Jacob H, Mondor I et al. Selective interactions of polyanions with basic surfaces on human immunodeficiency virus type 1 gp120. J Virol 2000; 74(4):1948-1960.
20. Vives RR, Imberty A, Sattentau QJ et al. Heparan sulfate targets the HIV-1 envelope glycoprotein gp120 coreceptor binding site. J Biol Chem 2005; 280(22):21353-21357.
21. Zhang YJ, Hatziioannou T, Zang T et al. Envelope-dependent, cyclophilin-independent effects of glycosaminoglycans on human immunodeficiency virus type 1 attachment and infection. J Virol 2002; 76(12):6332-6343.
22. Ugolini S, Mondor I, Sattentau QJ. HIV-1 attachment: Another look. Trends Microbiol 1999; 7(4):144-149.
23. Fouts TR, Trkola A, Fung MS et al. Interactions of polyclonal and monoclonal anti-glycoprotein 120 antibodies with oligomeric glycoprotein 120-glycoprotein 41 complexes of a primary HIV type 1 isolate: Relationship to neutralization. AIDS Res Hum Retroviruses 1998; 14(7):591-597.
24. Trkola A, Gordon C, Matthews J et al. The CC-chemokine RANTES increases the attachment of human immunodeficiency virus type 1 to target cells via glycosaminoglycans and also activates a signal transduction pathway that enhances viral infectivity. J Virol 1999; 73(8):6370-6379.
25. Pinon JD, Klasse PJ, Jassal SR et al. Human T-cell leukemia virus type 1 envelope glycoprotein gp46 interacts with cell surface heparan sulfate proteoglycans. J Virol 2003; 77(18):9922-9930.
26. Ohshiro Y, Murakami T, Matsuda K et al. Role of cell surface glycosaminoglycans of human T cells in human immunodeficiency virus type-1 (HIV-1) infection. Microbiol Immunol 1996; 40(11):827-835.
27. Igakura T, Stinchcombe JC, Goon PK et al. Spread of HTLV-I between lymphocytes by virus-induced polarization of the cytoskeleton. Science 2003; 299(5613):1713-1716.
28. Fry EE, Lea SM, Jackson T et al. The structure and function of a foot-and-mouth disease virus-oligosaccharide receptor complex. EMBO J 1999; 18(3):543-554.
29. Sa-Carvalho D, Rieder E, Baxt B et al. Tissue culture adaptation of foot-and-mouth disease virus selects viruses that bind to heparin and are attenuated in cattle. J Virol 1997; 71(7):5115-5123.
30. Fry EE, Newman JW, Curry S et al. Structure of Foot-and-mouth disease virus serotype A10 61 alone and complexed with oligosaccharide receptor: Receptor conservation in the face of antigenic variation. J Gen Virol 2005; 86(Pt 7):1909-1920.
31. Dechecchi MC, Melotti P, Bonizzato A et al. Heparan sulfate glycosaminoglycans are receptors sufficient to mediate the initial binding of adenovirus types 2 and 5. J Virol 2001; 75(18):8772-8780.
32. Dechecchi MC, Tamanini A, Bonizzato A et al. Heparan sulfate glycosaminoglycans are involved in adenovirus type 5 and 2-host cell interactions. Virology 2000; 268(2):382-390.
33. Hildreth JE, Orentas RJ. Involvement of a leukocyte adhesion receptor (LFA-1) in HIV-induced syncytium formation. Science 1989; 244(4908):1075-1078.
34. Pantaleo G, Butini L, Graziosi C et al. Human immunodeficiency virus (HIV) infection in CD4+ T lymphocytes genetically deficient in LFA-1: LFA-1 is required for HIV-mediated cell fusion but not for viral transmission. J Exp Med 1991; 173(2):511-514.

35. Fortin JF, Cantin R, Lamontagne G et al. Host-derived ICAM-1 glycoproteins incorporated on human immunodeficiency virus type 1 are biologically active and enhance viral infectivity. J Virol 1997; 71(5):3588-3596.
36. Rizzuto CD, Sodroski JG. Contribution of virion ICAM-1 to human immunodeficiency virus infectivity and sensitivity to neutralization. J Virol 1997; 71(6):4847-4851.
37. Zhu P, Chertova E, Bess Jr J et al. Electron tomography analysis of envelope glycoprotein trimers on HIV and simian immunodeficiency virus virions. Proc Natl Acad Sci USA 2003; 100(26):15812-15817.
38. Tremblay MJ, Fortin JF, Cantin R. The acquisition of host-encoded proteins by nascent HIV-1. Immunol Today 1998; 19(8):346-351.
39. Cantin R, Fortin JF, Lamontagne G et al. The acquisition of host-derived major histocompatibility complex class II glycoproteins by human immunodeficiency virus type 1 accelerates the process of virus entry and infection in human T-lymphoid cells. Blood 1997; 90(3):1091-1100.
40. Giguere JF, Bounou S, Paquette JS et al. Insertion of host-derived costimulatory molecules CD80 (B7.1) and CD86 (B7.2) into human immunodeficiency virus type 1 affects the virus life cycle. J Virol 2004; 78(12):6222-6232.
41. Cantin R, Methot S, Tremblay MJ. Plunder and stowaways: Incorporation of cellular proteins by enveloped viruses. J Virol 2005; 79(11):6577-6587.
42. Kwong PD, Wyatt R, Robinson J et al. Structure of an HIV gp120 envelope glycoprotein in complex with the CD4 receptor and a neutralizing human antibody. Nature 1998; 393(6686):648-659.
43. Olshevsky U, Helseth E, Furman C et al. Identification of individual human immunodeficiency virus type 1 gp120 amino acids important for CD4 receptor binding. J Virol 1990; 64(12):5701-5707.
44. Fouts TR, Binley JM, Trkola A et al. Neutralization of the human immunodeficiency virus type 1 primary isolate JR-FL by human monoclonal antibodies correlates with antibody binding to the oligomeric form of the envelope glycoprotein complex. J Virol 1997; 71(4):2779-2785.
45. Jeffs SA, Goriup S, Kebble B et al. Expression and characterisation of recombinant oligomeric envelope glycoproteins derived from primary isolates of HIV-1. Vaccine 2004; 22(8):1032-1046.
46. Kwong PD, Wyatt R, Sattentau QJ et al. Oligomeric modeling and electrostatic analysis of the gp120 envelope glycoprotein of human immunodeficiency virus. J Virol 2000; 74(4):1961-1972.
47. Berger EA. HIV entry and tropism: The chemokine receptor connection. Aids 1997; 11(Suppl A):S3-16.
48. Kolchinsky P, Mirzabekov T, Farzan M et al. Adaptation of a CCR5-using, primary human immunodeficiency virus type 1 isolate for CD4-independent replication. J Virol 1999; 73(10):8120-8126.
49. Staunton DE, Merluzzi VJ, Rothlein R et al. A cell adhesion molecule, ICAM-1, is the major surface receptor for rhinoviruses. Cell 1989; 56(5):849-853.
50. Greve JM, Davis G, Meyer AM et al. The major human rhinovirus receptor is ICAM-1. Cell 1989; 56(5):839-847.
51. Kolatkar PR, Bella J, Olson NH et al. Structural studies of two rhinovirus serotypes complexed with fragments of their cellular receptor. EMBO J 1999; 18(22):6249-6259.
52. Bella J, Kolatkar PR, Marlor CW et al. The structure of the two amino-terminal domains of human ICAM-1 suggests how it functions as a rhinovirus receptor and as an LFA-1 integrin ligand. Proc Natl Acad Sci USA 1998; 95(8):4140-4145.
53. Bergelson JM, Cunningham JA, Droguett G et al. Isolation of a common receptor for Coxsackie B viruses and adenoviruses 2 and 5. Science 1997; 275(5304):1320-1323.
54. Svensson U, Persson R, Everitt E. Virus-receptor interaction in the adenovirus system I. Identification of virion attachment proteins of the HeLa cell plasma membrane. J Virol 1981; 38(1):70-81.
55. Tomko RP, Xu R, Philipson L. HCAR and MCAR: The human and mouse cellular receptors for subgroup C adenoviruses and group B coxsackieviruses. Proc Natl Acad Sci USA 1997; 94(7):3352-3356.
56. Bewley MC, Springer K, Zhang YB et al. Structural analysis of the mechanism of adenovirus binding to its human cellular receptor, CAR. Science 1999; 286(5444):1579-1583.
57. Bergelson JM, Shepley MP, Chan BM et al. Identification of the integrin VLA-2 as a receptor for echovirus 1. Science 1992; 255(5052):1718-1720.
58. Zimmermann H, Eggers HJ, Nelsen-Salz B. Cell attachment and mouse virulence of echovirus 9 correlate with an RGD motif in the capsid protein VP1. Virology 1997; 233(1):149-156.
59. Xing L, Huhtala M, Pietiainen V et al. Structural and functional analysis of integrin alpha2I domain interaction with echovirus 1. J Biol Chem 2004; 279(12):11632-11638.
60. King SL, Kamata T, Cunningham JA et al. Echovirus 1 interaction with the human very late antigen-2 (integrin alpha2beta1) I domain. Identification of two independent virus contact sites distinct from the metal ion-dependent adhesion site. J Biol Chem 1997; 272(45):28518-28522.
61. Coulson BS, Londrigan SL, Lee DJ. Rotavirus contains integrin ligand sequences and a disintegrin-like domain that are implicated in virus entry into cells. Proc Natl Acad Sci USA 1997; 94(10):5389-5394.
62. Jolly CL, Beisner BM, Holmes IH. Rotavirus infection of MA104 cells is inhibited by Ricinus lectin and separately expressed single binding domains. Virology 2000; 275(1):89-97.

63. Mendez E, Lopez S, Cuadras MA et al. Entry of rotaviruses is a multistep process. Virology 1999; 263(2):450-459.
64. Graham KL, Fleming FE, Halasz P et al. Rotaviruses interact with alpha4beta7 and alpha4beta1 integrins by binding the same integrin domains as natural ligands. J Gen Virol 2005; 86(Pt 12):3397-3408.
65. Hewish MJ, Takada Y, Coulson BS. Integrins alpha2beta1 and alpha4beta1 can mediate SA11 rotavirus attachment and entry into cells. J Virol 2000; 74(1):228-236.
66. Londrigan SL, Graham KL, Takada Y et al. Monkey rotavirus binding to alpha2beta1 integrin requires the alpha2 I domain and is facilitated by the homologous beta1 subunit. J Virol 2003; 77(17):9486-9501.
67. Graham KL, Halasz P, Tan Y et al. Integrin-using rotaviruses bind alpha2beta1 integrin alpha2 I domain via VP4 DGE sequence and recognize alphaXbeta2 and alphaVbeta3 by using VP7 during cell entry. J Virol 2003; 77(18):9969-9978.
68. Pileri P, Uematsu Y, Campagnoli S et al. Binding of hepatitis C virus to CD81. Science 1998; 282(5390):938-941.
69. Cormier EG, Tsamis F, Kajumo F et al. CD81 is an entry coreceptor for hepatitis C virus. Proc Natl Acad Sci USA 2004; 101(19):7270-7274.
70. Cocquerel L, Voisset C, Dubuisson J. Hepatitis C virus entry: Potential receptors and their biological functions. J Gen Virol 2006; 87(Pt 5):1075-1084.
71. Hsu M, Zhang J, Flint M et al. Hepatitis C virus glycoproteins mediate pH-dependent cell entry of pseudotyped retroviral particles. Proc Natl Acad Sci USA 2003; 100(12):7271-7276.
72. Bartosch B, Vitelli A, Granier C et al. Cell entry of hepatitis C virus requires a set of coreceptors that include the CD81 tetraspanin and the SR-B1 scavenger receptor. J Biol Chem 2003; 278(43):41624-41630.
73. Reeves JD, Hibbitts S, Simmons G et al. Primary human immunodeficiency virus type 2 (HIV-2) isolates infect CD4-negative cells via CCR5 and CXCR4: Comparison with HIV-1 and simian immunodeficiency virus and relevance to cell tropism in vivo. J Virol 1999; 73(9):7795-7804.
74. Kaleeba JA, Berger EA. Kaposi's sarcoma-associated herpesvirus fusion-entry receptor: Cystine transporter xCT. Science 2006; 311(5769):1921-1924.
75. Akula SM, Pramod NP, Wang FZ et al. Integrin alpha3beta1 (CD 49c/29) is a cellular receptor for Kaposi's sarcoma-associated herpesvirus (KSHV/HHV-8) entry into the target cells. Cell 2002; 108(3):407-419.
76. Akula SM, Wang FZ, Vieira J et al. Human herpesvirus 8 interaction with target cells involves heparan sulfate. Virology 2001; 282(2):245-255.
77. Rappocciolo G, Jenkins FJ, Hensler HR et al. DC-SIGN is a receptor for human herpesvirus 8 on dendritic cells and macrophages. J Immunol 2006; 176(3):1741-1749.
78. Hong PW, Flummerfelt KB, de Parseval A et al. Human immunodeficiency virus envelope (gp120) binding to DC-SIGN and primary dendritic cells is carbohydrate dependent but does not involve 2G12 or cyanovirin binding sites: Implications for structural analyses of gp120-DC-SIGN binding. J Virol 2002; 76(24):12855-12865.
79. Su SV, Hong P, Baik S et al. DC-SIGN binds to HIV-1 glycoprotein 120 in a distinct but overlapping fashion compared with ICAM-2 and ICAM-3. J Biol Chem 2004; 279(18):19122-19132.
80. Lee B, Leslie G, Soilleux E et al. cis Expression of DC-SIGN allows for more efficient entry of human and simian immunodeficiency viruses via CD4 and a coreceptor. J Virol 2001; 75(24):12028-12038.
81. Nobile C, Petit C, Moris A et al. Covert human immunodeficiency virus replication in dendritic cells and in DC-SIGN-expressing cells promotes long-term transmission to lymphocytes. J Virol 2005; 79(9):5386-5399.
82. Turville SG, Cameron PU, Handley A et al. Diversity of receptors binding HIV on dendritic cell subsets. Nat Immunol 2002; 3(10):975-983.
83. Snyder GA, Ford J, Torabi-Parizi P et al. Characterization of DC-SIGN/R interaction with human immunodeficiency virus type 1 gp120 and ICAM molecules favors the receptor's role as an antigen-capturing rather than an adhesion receptor. J Virol 2005; 79(8):4589-4598.
84. Stahl-Hennig C, Steinman RM, Tenner-Racz K et al. Rapid infection of oral mucosal-associated lymphoid tissue with simian immunodeficiency virus. Science 1999; 285(5431):1261-1265.
85. Hu J, Gardner MB, Miller CJ. Simian immunodeficiency virus rapidly penetrates the cervicovaginal mucosa after intravaginal inoculation and infects intraepithelial dendritic cells. J Virol 2000; 74(13):6087-6095.
86. Valladeau J, Ravel O, Dezutter-Dambuyant C et al. Langerin, a novel C-type lectin specific to Langerhans cells, is an endocytic receptor that induces the formation of Birbeck granules. Immunity 2000; 12(1):71-81.
87. Veazey RS, Shattock RJ, Pope M et al. Prevention of virus transmission to macaque monkeys by a vaginally applied monoclonal antibody to HIV-1 gp120. Nat Med 2003; 9(3):343-346.
88. Tassaneetrithep B, Burgess TH, Granelli-Piperno A et al. DC-SIGN (CD209) mediates dengue virus infection of human dendritic cells. J Exp Med 2003; 197(7):823-829.
89. Lozach PY, Lortat-Jacob H, de Lacroix de Lavalette A et al. DC-SIGN and L-SIGN are high affinity binding receptors for hepatitis C virus glycoprotein E2. J Biol Chem 2003; 278(22):20358-20366.
90. Pohlmann S, Zhang J, Baribaud F et al. Hepatitis C virus glycoproteins interact with DC-SIGN and DC-SIGNR. J Virol 2003; 77(7):4070-4080.

91. Gardner JP, Durso RJ, Arrigale RR et al. L-SIGN (CD 209L) is a liver-specific capture receptor for hepatitis C virus. Proc Natl Acad Sci USA 2003; 100(8):4498-4503.
92. Halary F, Amara A, Lortat-Jacob H et al. Human cytomegalovirus binding to DC-SIGN is required for dendritic cell infection and target cell trans-infection. Immunity 2002; 17(5):653-664.
93. Alvarez CP, Lasala F, Carrillo J et al. C-type lectins DC-SIGN and L-SIGN mediate cellular entry by Ebola virus in cis and in trans. J Virol 2002; 76(13):6841-6844.
94. Geijtenbeek TB, Kwon DS, Torensma R et al. DC-SIGN, a dendritic cell-specific HIV-1-binding protein that enhances trans-infection of T cells. Cell 2000; 100(5):587-597.
95. Nguyen DG, Hildreth JE. Involvement of macrophage mannose receptor in the binding and transmission of HIV by macrophages. Eur J Immunol 2003; 33(2):483-493.
96. Reading PC, Miller JL, Anders EM. Involvement of the mannose receptor in infection of macrophages by influenza virus. J Virol 2000; 74(11):5190-5197.
97. Scarselli E, Ansuini H, Cerino R et al. The human scavenger receptor class B type I is a novel candidate receptor for the hepatitis C virus. EMBO J 2002; 21(19):5017-5025.
98. Acton S, Rigotti A, Landschulz KT et al. Identification of scavenger receptor SR-BI as a high density lipoprotein receptor. Science 1996; 271(5248):518-520.
99. Landschulz KT, Pathak RK, Rigotti A et al. Regulation of scavenger receptor, class B, type I, a high density lipoprotein receptor, in liver and steroidogenic tissues of the rat. J Clin Invest 1996; 98(4):984-995.
100. Voisset C, Callens N, Blanchard E et al. High density lipoproteins facilitate hepatitis C virus entry through the scavenger receptor class B type I. J Biol Chem 2005; 280(9):7793-7799.
101. Wilson IA, Skehel JJ, Wiley DC. Structure of the haemagglutinin membrane glycoprotein of influenza virus at 3 A resolution. Nature 1981; 289(5796):366-373.
102. Weis W, Brown JH, Cusack S et al. Structure of the influenza virus haemagglutinin complexed with its receptor, sialic acid. Nature 1988; 333(6172):426-431.
103. Paulson JC, Sadler JE, Hill RL. Restoration of specific myxovirus receptors to asialoerythrocytes by incorporation of sialic acid with pure sialyltransferases. J Biol Chem 1979; 254(6):2120-2124.
104. Wiley DC, Skehel JJ. The structure and function of the hemagglutinin membrane glycoprotein of influenza virus. Annu Rev Biochem 1987; 56:365-394.
105. Rogers GN, Paulson JC. Receptor determinants of human and animal influenza virus isolates: Differences in receptor specificity of the H3 hemagglutinin based on species of origin. Virology 1983; 127(2):361-373.
106. Gambaryan AS, Robertson JS, Matrosovich MN. Effects of egg-adaptation on the receptor-binding properties of human influenza A and B viruses. Virology 1999; 258(2):232-239.
107. Connor RJ, Kawaoka Y, Webster RG et al. Receptor specificity in human, avian, and equine H2 and H3 influenza virus isolates. Virology 1994; 205(1):17-23.
108. Daniels RS, Douglas AR, Skehel JJ et al. Antigenic analyses of influenza virus haemagglutinins with different receptor-binding specificities. Virology 1984; 138(1):174-177.
109. Superti F, Donelli G. Gangliosides as binding sites in SA-11 rotavirus infection of LLC-MK2 cells. J Gen Virol 1991; 72(Pt 10):2467-2474.
110. Fukudome K, Yoshie O, Konno T. Comparison of human, simian, and bovine rotaviruses for requirement of sialic acid in hemagglutination and cell adsorption. Virology 1989; 172(1):196-205.
111. Ciarlet M, Estes MK. Human and most animal rotavirus strains do not require the presence of sialic acid on the cell surface for efficient infectivity. J Gen Virol 1999; 80(Pt 4):943-948.
112. Guo CT, Nakagomi O, Mochizuki M et al. Ganglioside GM(1a) on the cell surface is involved in the infection by human rotavirus KUN and MO strains. J Biochem (Tokyo) 1999; 126(4):683-688.
113. Delorme C, Brussow H, Sidoti J et al. Glycosphingolipid binding specificities of rotavirus: Identification of a sialic acid-binding epitope. J Virol 2001; 75(5):2276-2287.
114. Willoughby RE. Rotaviruses preferentially bind O-linked sialylglycoconjugates and sialomucins. Glycobiology 1993; 3(5):437-445.
115. Kiefel MJ, Beisner B, Bennett S et al. Synthesis and biological evaluation of N-acetylneuraminic acid-based rotavirus inhibitors. J Med Chem 1996; 39(6):1314-1320.
116. Willoughby RE, Yolken RH. SA11 rotavirus is specifically inhibited by an acetylated sialic acid. J Infect Dis 1990; 161(1):116-119.
117. Fuentes-Panana EM, Lopez S, Gorziglia M et al. Mapping the hemagglutination domain of rotaviruses. J Virol 1995; 69(4):2629-2632.
118. Dormitzer PR, Sun ZY, Wagner G et al. The rhesus rotavirus VP4 sialic acid binding domain has a galectin fold with a novel carbohydrate binding site. EMBO J 2002; 21(5):885-897.
119. Fazli A, Bradley SJ, Kiefel MJ et al. Synthesis and biological evaluation of sialylmimetics as rotavirus inhibitors. J Med Chem 2001; 44(20):3292-3301.
120. Liakatos A, Kiefel MJ, Fleming F et al. The synthesis and biological evaluation of lactose-based sialylmimetics as inhibitors of rotaviral infection. Bioorg Med Chem 2006; 14(3):739-757.

121. Willoughby RE, Yolken RH, Schnaar RL. Rotaviruses specifically bind to the neutral glycosphingolipid asialo-GM1. J Virol 1990; 64(10):4830-4835.
122. Rolsma MD, Gelberg HB, Kuhlenschmidt MS. Assay for evaluation of rotavirus-cell interactions: Identification of an enterocyte ganglioside fraction that mediates group A porcine rotavirus recognition. J Virol 1994; 68(1):258-268.
123. Rolsma MD, Kuhlenschmidt TB, Gelberg HB et al. Structure and function of a ganglioside receptor for porcine rotavirus. J Virol 1998; 72(11):9079-9091.
124. Harouse JM, Collman RG, Gonzalez-Scarano F. Human immunodeficiency virus type 1 infection of SK-N-MC cells: Domains of gp120 involved in entry into a CD4-negative, galactosyl ceramide/3' sulfo-galactosyl ceramide-positive cell line. J Virol 1995; 69(12):7383-7390.
125. Yahi N, Baghdiguian S, Moreau H et al. Galactosyl ceramide (or a closely related molecule) is the receptor for human immunodeficiency virus type 1 on human colon epithelial HT29 cells. J Virol 1992; 66(8):4848-4854.
126. Bhat S, Spitalnik SL, Gonzalez-Scarano F et al. Galactosyl ceramide or a derivative is an essential component of the neural receptor for human immunodeficiency virus type 1 envelope glycoprotein gp120. Proc Natl Acad Sci USA 1991; 88(16):7131-7134.
127. Bhat S, Mettus RV, Reddy EP et al. The galactosyl ceramide/sulfatide receptor binding region of HIV-1 gp120 maps to amino acids 206-275. AIDS Res Hum Retroviruses 1993; 9(2):175-181.
128. Arnberg N, Edlund K, Kidd AH et al. Adenovirus type 37 uses sialic acid as a cellular receptor. J Virol 2000; 74(1):42-48.
129. Gavrilovskaya IN, Shepley M, Shaw R et al. beta3 Integrins mediate the cellular entry of hantaviruses that cause respiratory failure. Proc Natl Acad Sci USA 1998; 95(12):7074-7079.
130. Tamura M, Natori K, Kobayashi M et al. Genogroup II noroviruses efficiently bind to heparan sulfate proteoglycan associated with the cellular membrane. J Virol 2004; 78(8):3817-3826.
131. Misinzo G, Delputte PL, Meerts P et al. Porcine circovirus 2 uses heparan sulfate and chondroitin sulfate B glycosaminoglycans as receptors for its attachment to host cells. J Virol 2006; 80(7):3487-3494.
132. Chen Y, Maguire T, Hileman RE et al. Dengue virus infectivity depends on envelope protein binding to target cell heparan sulfate. Nat Med 1997; 3(8):866-871.
133. Neyts J, Snoeck R, Schols D et al. Sulfated polymers inhibit the interaction of human cytomegalovirus with cell surface heparan sulfate. Virology 1992; 189(1):48-58.
134. Black JB, Pellett PE. Human herpesvirus 7. Rev Med Virol 1999; 9(4):245-262.
135. Krusat T, Streckert HJ. Heparin-dependent attachment of respiratory syncytial virus (RSV) to host cells. Arch Virol 1997; 142(6):1247-1254.
136. Holmgren J, Svennerholm L, Elwing H et al. Sendai virus receptor: Proposed recognition structure based on binding to plastic-adsorbed gangliosides. Proc Natl Acad Sci USA 1980; 77(4):1947-1950.
137. Markwell MA, Svennerholm L, Paulson JC. Specific gangliosides function as host cell receptors for Sendai virus. Proc Natl Acad Sci USA 1981; 78(9):5406-5410.
138. Roivainen M, Piirainen L, Hovi T et al. Entry of coxsackievirus A9 into host cells: Specific interactions with alpha v beta 3 integrin, the vitronectin receptor. Virology 1994; 203(2):357-365.
139. Uncapher CR, DeWitt CM, Colonno RJ. The major and minor group receptor families contain all but one human rhinovirus serotype. Virology 1991; 180(2):814-817.
140. Jackson T, Ellard FM, Ghazaleh RA et al. Efficient infection of cells in culture by type O foot-and-mouth disease virus requires binding to cell surface heparan sulfate. J Virol 1996; 70(8):5282-5287.
141. Berinstein A, Roivainen M, Hovi T et al. Antibodies to the vitronectin receptor (integrin alpha V beta 3) inhibit binding and infection of foot-and-mouth disease virus to cultured cells. J Virol 1995; 69(4):2664-2666.
142. Nelsen-Salz B, Eggers HJ, Zimmermann H. Integrin alpha(v)beta3 (vitronectin receptor) is a candidate receptor for the virulent echovirus 9 strain Barty. J Gen Virol 1999; 80(Pt 9):2311-2313.
143. Chung CS, Hsiao JC, Chang YS et al. A27L protein mediates vaccinia virus interaction with cell surface heparan sulfate. J Virol 1998; 72(2):1577-1585.
144. Hsiao JC, Chung CS, Chang W. Vaccinia virus envelope D8L protein binds to cell surface chondroitin sulfate and mediates the adsorption of intracellular mature virions to cells. J Virol 1999; 73(10):8750-8761.
145. Maddon PJ, Dalgleish AG, McDougal JS et al. The T4 gene encodes the AIDS virus receptor and is expressed in the immune system and the brain. Cell 1986; 47(3):333-348.
146. Ceccaldi PE, Delebecque F, Prevost MC et al. DC-SIGN facilitates fusion of dendritic cells with human T-cell leukemia virus Type 1-infected cells. J Virol 2006; 80(10):4771-4780.
147. Thoulouze MI, Lafage M, Schachner M et al. The neural cell adhesion molecule is a receptor for rabies virus. J Virol 1998; 72(9):7181-7190.
148. McGreal EP, Miller JL, Gordon S. Ligand recognition by antigen-presenting cell C-type lectin receptors. Curr Opin Immunol 2005; 17(1):18-24.
149. Lopez S, Arias CF. Multistep entry of rotavirus into cells: A Versaillesque dance. Trends Microbiol 2004; 12(6):271-278.

CHAPTER 2

PICORNAVIRUS ENTRY

Jeffrey M. Bergelson*,[1] and Carolyn B. Coyne[2]

[1]Department of Pediatrics, University of Pennsylvania Perelman School of Medicine, Division of Infectious Diseases, Children's Hospital of Philadelphia, Philadelphia, Pennsylvania, USA; [2]Department Microbiology and Molecular Genetics, University of Pittsburgh School of Medicine, Pittsburgh, Pennsylvania, USA
**Corresponding Author: Jeffrey M. Bergelson—Email: bergelson@email.chop.edu*

Abstract: The essential event in picornavirus entry is the delivery of the RNA genome to the cytoplasm of a target cell, where replication occurs. In the past several years progress has been made in understanding the structural changes in the virion important for uncoating and RNA release. In addition, for several viruses the endocytic mechanisms responsible for internalization have been identified, as have the cellular sites at which uncoating occurs. It has become clear that entry is not a passive process, and that viruses initiate specific signals required for entry. And we have begun to recognize that for a given virus, there may be multiple routes of entry, depending on the particular target cell and the receptors available on that cell.

INTRODUCTION

The picornaviruses are a family of small, nonenveloped viruses with a single strand positive-sense RNA genome (Table 1). A number of picornaviruses are significant pathogens in humans and animals: polioviruses and other members of the enterovirus group cause febrile illnesses and infections of the central nervous system; rhinoviruses are a major cause of the common cold; and foot and mouth disease virus causes economically important disease in cattle. Picornaviruses replicate within the cytoplasm, and transfected viral RNA is sufficient to initiate infection; therefore, the critical event in the entry process is delivery of RNA into the cell. The RNA is encased in a protein capsid, which must be sufficiently tough to protect the RNA from harsh environmental conditions (for example, enteroviruses enter the host by crossing the intestinal mucosa, and must survive exposure to gastric acid and intestinal proteases). Nonetheless, at an appropriate moment in the virus life cycle, the capsid must permit the release of RNA into the cell. The major

Viral Entry into Host Cells, edited by Stefan Pöhlmann and Graham Simmons.

Table 1. Selected picornaviruses

Genus	Example	Host
Enterovirus	Poliovirus	Humans
	Echovirus	"
	Coxsackievirus A	"
	Coxsackievirus B	"
	Enterovirus	"
Rhinovirus	Rhinovirus	"
Parechovirus	Parechovirus	"
Hepatovirus	Hepatitis A virus	"
Aphthovirus	Foot and mouth disease virus	Ruminants
Cardiovirus	Theiler's encephalomyelitis virus	Mice
	Encephalomyocarditis virus	Mice

questions in picornavirus entry concern how the virus capsid is disassembled to release the RNA (a process referred to as "uncoating"), and how the RNA is translocated across the cellular membranes that separate the cytoplasm from the external environment. Many aspects of the entry process remain poorly understood.

PICORNAVIRUS STRUCTURE

The typical picornavirus capsid is an icosahedral structure constructed of 12 pentamers, each pentamer composed of 5 copies of each of the four viral structural proteins, VP1-4.[1,2] VP1-3 (each approximately 30 kD in mass) form the external surface of the capsid. VP4, a small (approximately 7 kD) protein with an N-terminal myristic acid modification, is located on the inner surface, possibly in contact with the viral genome. The surfaces of many rhinoviruses and enteroviruses are marked by deep depressions (or a continuous canyon) surrounding the five-fold axis of symmetry; these serve as the site of attachment for cellular receptors (Fig. 1A). In contrast, foot and mouth disease virus does not have a canyon, and the receptor attaches to a flexible loop exposed on the virus surface.[3]

In the structures of both rhinoviruses and enteroviruses, a hydrophobic pocket is evident beneath the canyon floor, accessible to the outside environment through a small pore (Fig. 1B). In most instances this pocket is filled by a small molecule (termed "pocked factor"), evident as electron density in the crystal structure, and believed to be a fatty acid. A number of antiviral drugs [such as disoxaril (WIN51711) and pleconaril] bind within the pocket, presumably displacing the natural pocket factor;[4-7] these drugs stabilize the virion and prevent infection, in part by blocking the uncoating process.[5,7] It has thus been suggested that occupation of the pocket acts to regulate the stability of the virion. The N-termini of VP1-3 form a stabilizing network on the inner surface of the poliovirus capsid,[9] which must be disrupted during the uncoating process. At physiologic temperatures the virion structure is dynamic, and undergoes reversible conformational changes (described as "breathing") during which VP4 and the N-terminus of VP1 become exposed and accessible to neutralizing antibodies and proteolytic enzymes.[8,9]

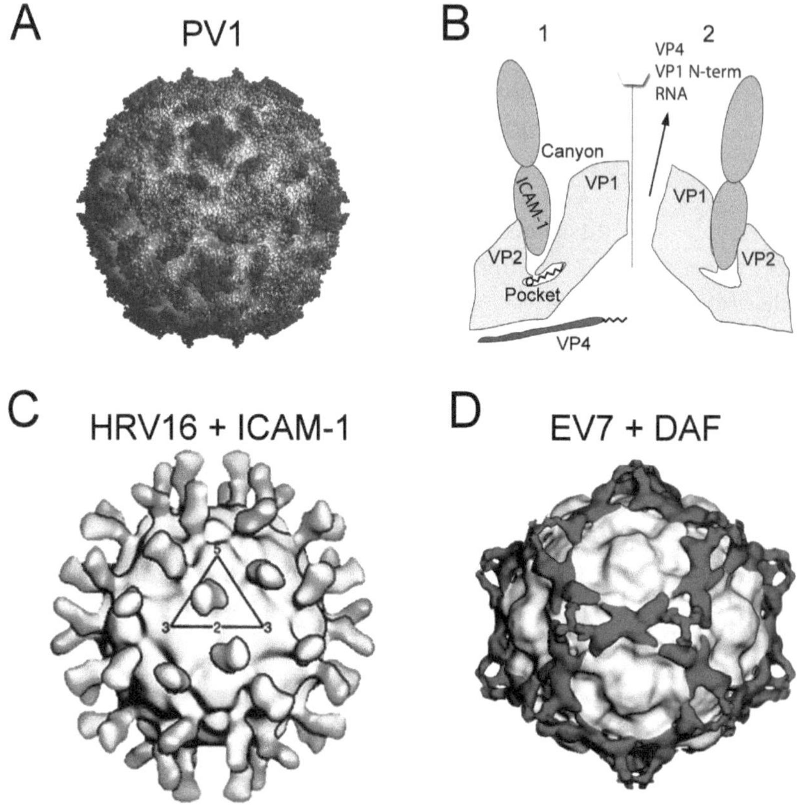

Figure 1. The picornavirus canyon and receptor interactions. A) Structure of poliovirus 1 determined by X-ray crystallography. Prominences on the viral surface are dark, depressions are light. The canyon surrounds the star-shaped prominence at the 5-fold axis. (Rasmol image courtesy of Dr. J.-Y. Sgro, UW-Madison, from data in PDB-ID: 2PLV, ref. 111. See http://virology.wisc.edu/virusworld). B) Model of receptor interaction with the canyon. ICAM-1 binds first to the south wall of the canyon (1). Subsequent interaction with the north wall of the canyon leads to conformational change in VP1, with displacement of pocket factor; formation of a channel at the 5-fold axis may permit exit of VP4, the VP1 N-terminus, and RNA. Adapted by permission from Macmillan Publishers Ltd: EMBO J ©1999.[17] C) Structure of human rhinovirus 16 complexed with ICAM-1 (dark); the ICAM-1 N-terminus is inserted in the canyon. Reprinted from reference 14, with permission from Macmillan Publishers Ltd: Nat Struct Biol ©2008. D) Structure of echovirus 7 complexed with DAF; DAF lies across the virus surface, near the 2-fold axis of symmetry, but does not interact with the canyon. Reprinted from reference 18, with permission from the National Academy of Sciences, USA, ©2002.

These transient movements, which are inhibited by the stabilizing antiviral drugs,[9] may contribute to the irreversible changes that occur during uncoating (see below).

ATTACHMENT TO A RECEPTOR

The first event in the entry process is attachment of virus to a receptor on the target cell. Receptors trap virus at the cell surface, increasing the likelihood that infection will occur. In some cases, contact with the receptor may destabilize the virion and trigger the uncoating process; in others, endocytic uptake of the virus-receptor complex may deliver the virion to an intracellular compartment where uncoating can occur. In addition, receptors may activate intracellular signaling pathways that are required for virus internalization and intracellular trafficking. (In the discussion that follows we will not attempt to make distinctions between "true receptors", "coreceptors," and "attachment factors"; if a virus binds to a molecule on the cell surface, and if that molecule contributes in an important way to infection by the virus, we consider it a receptor).

Specific receptors have been identified for a number of picornaviruses (Table 2). The receptor for the major group of rhinoviruses (ICAM-1), the poliovirus receptor (PVR), and the coxsackievirus and adenovirus receptor (CAR) are members of the immunoglobulin superfamily, and function in cellular adhesion. The N-terminal domains of these receptors bind to a site within the viral canyon[10-14] (Fig. 1B,C). The close contact between the viruses and these receptors leads to a conformational change in the virion; attachment of the receptor may distort the base of the canyon, resulting in expulsion of the pocket factor, and destabilization of the capsid. Kinetic studies of rhinovirus-ICAM-1 and poliovirus-PVR interaction each reveal two classes of binding sites on the viral surface, differing both in association rate and affinity;[15,16] one possible explanation is that the initial binding of receptor in the canyon is followed by a deeper, higher-affinity interaction that leads to expulsion of pocket factor and subsequent uncoating events.[17]

Table 2. Picornavirus receptors

Virus	Receptor
Poliovirus	PVR (CD155)[89]
Rhinovirus (major group)	ICAM-1[90-92]
Rhinovirus (minor group)	LDL receptor[93]
Coxsackie B viruses	CAR[73,74]
	Decay accelerating factor (DAF, CD55)[76,77]
	Heparan sulfate[94]
Echovirus 1	Integrin $\alpha2\beta1$[67]
Echovirus 6, 7	Decay accelerating factor (CD55)[95,96]
	Heparan sulfate[97]
	? others
Coxsackie A viruses	ICAM[98]
	DAF[99]
	Integrin $\alpha v\beta3$[100]
	MHC-I -associated GRP78[101]
Enterovirus 71	SCARB2[102]
	PSGL-1[103]
	Heparan sulfate[104]
Foot and mouth disease virus	Integrins $\alpha v\beta1$, $\alpha v\beta3$, $\alpha v\beta6$, $\alpha v\beta8$[105-108]
	Heparan sulfate[109]
Encephalomyocarditis virus	Mouse VCAM[110]

A number of enteroviruses and rhinoviruses bind to decay accelerating factor (DAF), a complement regulatory protein. DAF does not insert in the canyon, but instead drapes across the virus surface in a way that differs for different viruses [echovirus (EV) 7 (Fig 1D),[18] EV12,[19] and CVB3[110]]. Unlike ICAM, CAR, and PVR, DAF does not appear to induce conformational changes in the capsid,[20,21] and for virtually all DAF-binding viruses, infection is thought to require interaction with a second receptor that induces uncoating. Other picornaviruses have been found to infect by interacting with members of the integrin family of homodimeric adhesion receptors, members of the LDL receptor family, and with heparin sulfate on the cell surface; it is likely that these receptors do not induce uncoating, but instead serve to target virus to an appropriate endosomal compartment, where other factors trigger the uncoating process.

THE A PARTICLE, A LIKELY UNCOATING INTERMEDIATE

After attachment to receptors on the cell surface, many enteroviruses and rhinoviruses undergo a conformational change to form altered (A) particles,[22-25] reflected by a change in sucrose gradient sedimentation rate, from approximately 160S to 135S [(Fig. 2), reviewed in ref. 26]. Appearance of A particles is followed by the appearance of 80S particles, empty capsids from which the RNA has been released.

Several lines of evidence suggest that the poliovirus A particle is an intermediate of the uncoating process: infection is blocked by drugs that stabilize the virion and prevent A

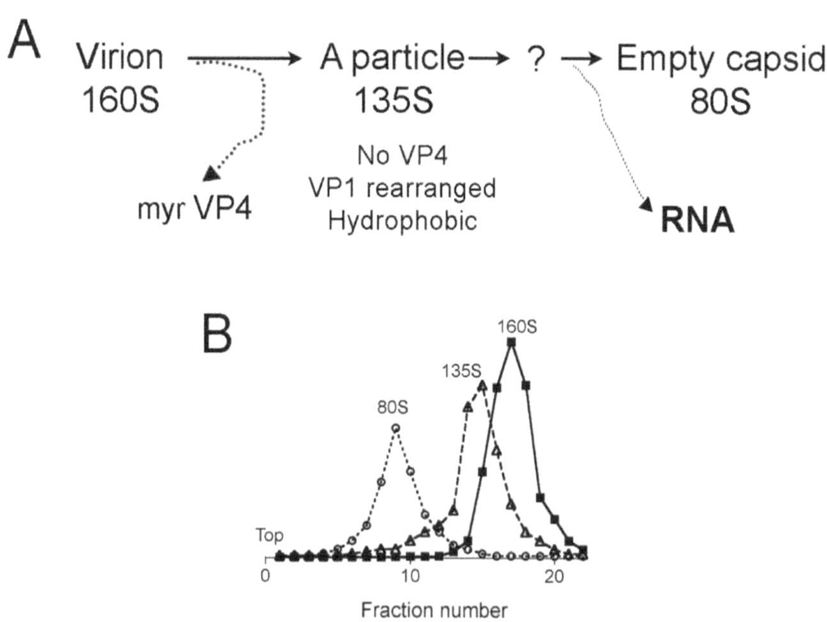

Figure 2. Poliovirus uncoating intermediates. A) 160S virion is converted to A particles and empty capsids, releasing RNA. B) Native virions (160S), A-particles (135S), and 80S particles are separated on sucrose gradients.

particle formation;[27] A particles are infectious when delivered to susceptible cells (although they infect very inefficient because they do not bind to PVR);[28] and kinetic experiments indicate that during infection, RNA release from A particles is faster than from native virions (by several minutes—approximately the time required for A particles to form).[29] Infection by certain cold-adapted poliovirus mutants occurs without evident accumulation of A particles, which has led some ixnvestigators to suggest that they are not true intermediates of the uncoating process;[30] however, subsequent work suggests that at low temperature, A particles do not accumulate because they are formed more slowly than they are eliminated by subsequent events in uncoating.[29] It is likely that the uncoating process involves additional intermediate structures that are not stable enough to be detected.

Transition to the A particle involves release of myristylated VP4, which is inserted into cell membranes, and exposure of the hydrophobic N-terminus of VP1, which becomes available to tether the virion to cellular membranes. Interaction of VP4 and VP1 with membranes leads to the formation of pores, which may serve as portals for RNA transfer into the cytosol.[31-33] Release of VP4 from the interior of the capsid is probably the irreversible step that distinguishes capsid "breathing" from uncoating.

The structure of the poliovirus 135S particle has been determined by cryo-electron microscopy.[34,35] The A particle is 4% larger than the native virion (accounting in part for its slower sedimentation in sucrose gradients). The expansion of the virion creates several gaps between capsid subunits, which may permit the extrusion of VP4 and movement of the VP1 N-terminus from the interior to the virion surface. (The movement of capsid subunits has been referred to as a "molecular tectonic" model of conformational change[34]). However, the gaps do not seem sufficiently large to allow the RNA to exit, suggesting that RNA may not be released from the 135S particle itself, but from a form of the virion that has not been identified.

A-particles, as well as empty capsids, have been observed during infection by a number of human rhinoviruses,[23,36] but no structures have been reported for a rhinovirus A particle. In vitro, exposure of different rhinovirus serotypes to ICAM-1 in vitro has several possible outcomes, which are likely to depend on the strength of the interactions between capsid protomers, as well as the particular structural distortions caused by receptor engagement. Some serotypes are relatively stable, forming a virus-receptor complex without proceeding to uncoat; others are triggered to release RNA without the appearance of a stable A particle intermediate. Interaction of HRV3 with ICAM-1 in vitro leads to expansion of the capsid (by 4%, as observed for the poliovirus A particle), with separation of the capsid protomers near the receptor attachment site;[37] in this case, the receptor remains attached to the intermediate form of the capsid, which retains VP4 and has not yet undergone full conversion to an A particle. Interaction of poliovirus with PVR probably results in the transient formation of a similar expanded structure before the exposure of VP1 and release of VP4 lead to an irreversible change.

THE 80S EMPTY CAPSID

The 80S forms of HRV2[38] (which binds to the LDL receptor) and HRV14[39] (which binds ICAM-1) are remarkable for the opening of a channel at the 5-fold axis, which may permit the release of RNA. The N-termini of VP1, buried in the native capsid, are extruded at the 5-fold axis of the 80S particle, and VP4 molecules—released from their interior location—appear to remain associated with the capsid near the 5-fold axis. Thus VP4 and

VP1 would be well situated to fix the virus to the cell membrane, possibly forming a pore through which RNA, exiting the channel formed at the 5-fold axis, might enter the cell.

However, in the structure of the poliovirus 135S particle, the N-terminus of VP3 forms a plug at the five-fold axis that would interfere with RNA release,[34] and there is no other evident portal through which RNA could pass. If the 135S and 80S particles really are formed upstream and downstream of RNA release, the structural evidence suggests that poliovirus RNA must be released from a transient intermediate form that has not been identified so far.

CELL BIOLOGY OF VIRUS ENTRY: ENDOCYTIC MECHANISMS

Clathrin-Mediated Endocytosis

Cells take up macromolecules by a variety of endocytic mechanisms [reviewed in ref. 40] and viruses probably exploit all of these to enter cells (Fig. 3) [reviewed in ref. 41]. Clathrin-mediated endocytosis (often referred to as receptor-mediated endocytosis) is a process by which membrane proteins and their associated ligands are internalized in vesicles surrounded by a basket-like coat of clathrin. The assembly of clathrin coats depends on a number of adaptor, accessory, and regulatory proteins, and the release of clathrin-coated

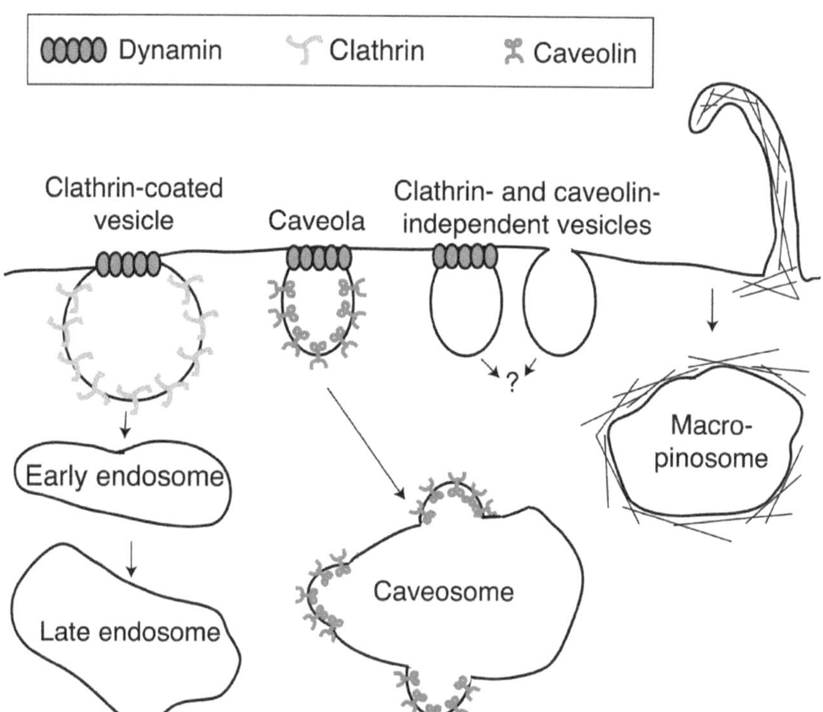

Figure 3. Mechanisms of virus entry. The best-described endocytic mechanisms involve clathrin-coated vesicles and caveolae. Modified from references 40 and 41.

vesicles into the cell requires the activity of dynamin, a large GTPase. Clathrin-coated vesicles deliver their cargoes to early endosomes, from which they may be recycled back to the plasma membrane, or proceed further to more acidic late endosomes and lysosomes. Clathrin-mediated endocytosis is the best-understood endocytic mechanism. Many classic studies of virus entry have focused on viruses (such as Semliki Forest virus, influenza virus, and adenovirus) that are internalized in clathrin-coated vesicles, and then move to endosomes, where endosomal acidification triggers the events that permit membrane penetration.

Caveolar Endocytosis

Clathrin-independent endocytic mechanisms are less well understood. The best-studied of these involves caveolae—vesicles with cholesterol-rich membranes, coated by the integral membrane protein caveolin. Caveolae are seen in electron micrographs as flask-shaped invaginations of the plasma membrane; they are largely static, but in response to ligand-induced signals they detach from the membrane (in a process requiring dynamin) and enter the cell. SV40 enters cells in caveolae, and moves to a distinctive caveolin-rich intracellular vesicle called the caveosome. Although it is known that there is vesicular traffic involving caveolae, caveosomes, and early endosomes,[42] the functions of the caveosome, and its role in endocytosis, is largely undefined.

Clathrin- and Caveolin-Independent Pathways

It is clear that viruses also enter cells by mechanisms that are independent of both clathrin and caveolin. Some of these are sensitive to depletion of cellular cholesterol, others require dynamin function; however, in the absence of specific marker proteins, these pathways are difficult to define at present. Flotillin has recently been identified as a protein important for at least one clathrin- and caveolin-independent endocytic process.[43] One distinctive pathway, macropinocytosis, is a receptor-independent process in which actin rearrangements lead to ruffling of the plasma membrane, and fusion of ruffles leads to internalization of extracellular fluid in actin-coated vesicles.

Methods for Studying Picornavirus Entry

Our current understanding of picornavirus entry depends on a consensus between morphologic studies, in which the movement of virions is followed by electron microscopy or fluorescence microscopy, and functional studies in which inhibitors of specific cellular processes are used to block entry and infection. Each of these approaches is subject to important limitations. Picornavirus particle/PFU ratios are very high (100-1,000), so it is always difficult to know that viral particles observed by microscopy are the ones that give rise to infection. And inhibitors, despite their supposed specificity, often have unanticipated effects.

Commonly used pharmacologic inhibitors include chlorpromazine (which inhibits clathrin recycling and formation of clathrin-coated vesicles), bafilomycin (which inhibits endosomal acidification), filipin (which disrupts caveolae by sequestering cholesterol), and amiloride (which inhibits macropinocytosis by an uncertain mechanism). Dominant-negative mutants and siRNAs directed against components of endocytic processes (e.g., clathrin and clathrin accessory proteins, caveolin, dynamin) are thought to be highly specific, but the most reliable information comes from studies that combine a variety of approaches.

POLIOVIRUS: DOES "ENTRY" INVOLVE INTERNALIZATION OF THE VIRION?

In vitro, poliovirus is converted to A particles when it is exposed to recombinant receptor at 37°C.[44] With high concentrations of receptor, virus is converted to 80S empty capsids.[13] Conversion occurs even at neutral pH. (Coxsackievirus B3[21] and some rhinoviruses[45] also undergo conformational changes when exposed to soluble receptors in vitro.) It is conceivable that uncoating could be completed at the cell surface; as poliovirus contacts its receptor, conformational changes may occur, and VP1 and VP4 may create pores in the plasma membrane through which RNA can enter the cell. Whether poliovirus does enter the cell, or whether it injects its RNA at the cell surface, has been a subject of continuing controversy.

Experiments in the 1980s suggested that poliovirus entered cells in clathrin-coated vesicles, and that uncoating occurred within acidified endosomes. Poliovirus was observed within clathrin-coated vesicles in electron micrographs,[46] and agents that interfere with endosomal acidification (such as chloroquine and monensin) were reported to prevent (or at least delay) RNA release from the capsid.[47,48] However, as discussed above, demonstrating a virus particle in a particular vesicle does not confirm that the vesicle is important for infection, and drugs may have nonspecific effects. Experiments with bafilomycin, a powerful inhibitor of the endosomal proton pump, subsequently demonstrated that poliovirus replication does not require endosomal acidification.[49] Furthermore, inhibition of dynamin, a GTPase required for internalization of clathrin-coated vesicles, does not block poliovirus replication in HeLa cells.[50] It thus appears that classical endocytosis and endosomal acidification are not essential.

Two recent studies support this view, and suggest that poliovirus may follow different entry routes in different cells. In the first, the entry of individual polio virions (with distinct fluorescent labels incorporated in RNA and capsid proteins) was examined in live HeLa cells.[51] Virions were seen to enter the cell in small vesicles and release their RNA molecules while very close to the cell surface. Entry did not require clathrin, caveolin, or flotillin, and the identity of the entry vesicles remains uncertain. However, an inhibitor of tyrosine kinases blocked entry, suggesting a possible role for kinase signaling.

In a second recent study,[52] we have found evidence for a different entry pathway in a different cell type: in microvascular endothelial cells derived from human brain, poliovirus entry requires both dynamin and caveolin (but not clathrin), and virus is detected within the cells in vesicles containing caveolin. Conversion to A particles appears to occur after contact with PVR on the cell surface, but the release of RNA from A particles does not appear to occur until virus has entered the cell. It is unclear what intracellular stimulus is required for the A particle to proceed further in the uncoating process.

FOOT AND MOUTH DISEASE VIRUS (FMDV) UNCOATS IN AN ACIDIC ENDOSOME

FMDV is transmitted by the respiratory route, and unlike poliovirus, has not evolved to resist gastric acidity. The capsid is labile even in mildly acidic conditions, dissociating into pentamers at pH 6.5.[53] However, contact with receptors on the cell surface does not cause FMDV to undergo conformational changes or release VP4.[54] Once internalized, virions are rapidly degraded to 12S pentameric capsid subunits with no intermediate 135S or 80S structures detected.

After attachment to the receptor, integrin $\alpha v\beta 6$, FMDV enters the cell in clathrin-coated vesicles, then traffics to early endosomes.[55,56] FMDV can infect cells lacking specific receptors by an antibody-dependent, Fc receptor-mediated pathway, suggesting that specific receptors are not required to trigger uncoating.[57] Because FMDV is acid-labile, delivery to endosomes is sufficient for uncoating to occur. In contrast, for acid-stable viruses like poliovirus and coxsackievirus, contact with a specific receptor is needed to trigger uncoating.

RHINOVIRUSES USE BOTH RECEPTOR-MEDIATED AND ENDOSOMAL TRIGGERS FOR UNCOATING

Human rhinoviruses (HRV) are also susceptible to acid, although some are more stable than others. Twelve of the approximately 100 rhinovirus serotypes (the minor group, which includes HRV2 and HRV14) bind to the members of the low density lipoprotein receptor (LDLR) family. LDLR does not bind within the canyon, but interacts with a star-shaped prominence on the viral surface. Once bound to the receptor, virus enters the cell (probably in clathrin-coated vesicles,[58] although clathrin-independent mechanisms have also been suggested[59]), and traffics to endosomes, where acidification leads to capsid disassembly and release of RNA. The receptor serves primarily to deliver the virus to the acidic endosomal compartment, and does not itself induce essential conformational changes. Under acidic conditions virus is released from the LDL receptor, so it is likely that uncoating within the endosome occurs when the virus and receptor are no longer in contact.[60]

Nearly 90 rhinovirus serotypes (the major group) bind to the N-terminal domain of ICAM-1, which interacts with the viral canyon. Attachment to ICAM is followed by internalization of virions to the endosomal compartment, and exposure to low pH. HRV3 and HRV14—both of which lack the pocket factor proposed to stabilize the capsid—are induced to form A particles when they interact with ICAM, even at neutral pH.[13,45] HRV16, which does have a pocket factor, remains stable when bound to the receptor at neutral pH. However, under mildly acidic conditions (pH 5.5-6.0, similar to the environment within an endosome), contact with receptor induces uncoating of HRV16.[61] Thus, both the receptor and the endosomal environment contribute to the uncoating of some major group HRV. Experiments with agents that inhibit endosomal acidification suggest that HRV16 and other viruses that are more stable to receptor-induced changes are more dependent on endosomal acidification for uncoating and infection.[61]

Some major group RV (HRV54[62] and variants of HRV89[63]) can infect ICAM-deficient cells by binding to an alternate receptor, heparan sulfate proteoglycan. The relevant mutations in HRV89 are localized at the interfaces between capsid protomers as well as at the binding site for heparin.[64] Because ICAM itself destabilizes the capsid, ICAM-dependent infection by these viruses is not blocked by bafilomycin, a proton pump inhibitor that prevents endosomal acidification; in contrast, heparan sulfate does not destabilize the capsid, and heparan sulfate-dependent infection is blocked by bafilomycin. Thus, it appears that the strength of protomer-protomer interactions helps determine the strength of the signal required for uncoating: the most stable virions require both receptor-mediated disruption and low pH, but less stable virions uncoat in response to milder stimuli.

If uncoating occurs in an endosomal vesicle, RNA must traverse the endosomal membrane to reach the cytosol. In the case of HRV2, RNA is released within late endosomes; uncoating virions induce size-selective pores that are likely to permit RNA to escape.[65] In contrast, HRV14 causes a larger-scale disruption of the endosomal membrane releasing 135S and 80S particles into the cytosol;[66] whether the 135S particles undergo further uncoating remains unclear.

ECHOVIRUS 1 NEEDS CAVEOLAE FOR ENTRY

Echovirus 1 (EV1) binds to an integrin receptor ($\alpha 2\beta 1$);[67,68] cryo-electron microscopy reveals that the $\alpha 2$ I-domain inserts into the viral canyon.[69] In vitro, interaction with soluble I domain does not trigger A particle formation,[69] but A particles are formed during infection,[70] and it is unclear whether other parts of the integrin, or an additional factor, are required. Clustering of $\alpha 2\beta 1$ by antibodies leads to activation of protein kinase C (PKCα), and internalization of the receptor in vesicles containing caveolin (thought to be caveolae and caveosomes).[71] EV1 also causes clustering of integrins, and it is internalized with the receptor in caveolar vesicles.[72] Entry and infection are blocked by inhibitors of PKCα, suggesting that a receptor-induced PKC signal is required; entry also requires dynamin, as well as the activity of one or more tyrosine kinases.[70]

Although virus rapidly enters the cell, A particles are not detectable for 1-2 hrs, suggesting that uncoating does not begin until virus enters the cell and has been transported to caveosomes. Viral proteins and viral RNA remain concentrated in caveosomes for several hours, and do not appear to traffic elsewhere before replication begins. It is possible that EV1 uncoating occurs in caveosomes, but it is not clear what initiates the uncoating process within these neutral-pH vesicles.

COXSACKIEVIRUSES: MULTIPLE RECEPTORS SERVE DISTINCT FUNCTIONS

CAR and DAF Serve Distinct Functions in Virus Entry

All group B coxsackieviruses (CVB) bind to the coxsackievirus and adenovirus receptor (CAR),[73-75] but a subset of CVB also bind to decay accelerating factor (DAF).[76,77] CAR, an immunoglobulin family member like PVR and ICAM-1, binds to the CVB canyon.[14] Contact with CAR in vitro is sufficient to trigger A particle formation,[21] and expression of CAR permits transfected rodent cells to become infected.[73,74] A subset of CVB also bind to a second receptor, decay accelerating factor (DAF).[76,77] Unlike CAR, DAF does not initiate conversion of virus to A particles,[21] and although virus binds to human DAF on transfected rodent cells, no infection occurs. There has been a continuing question about why CVB and other enteroviruses and rhinoviruses bind to DAF, when interaction with DAF appears insufficient for infection.

A CVB4 isolate has been reported to enter cells by a lipid-raft-dependent mechanism, independent of clathrin-mediated processes.[78] In contrast, CVB3 entry into HeLa cells has been reported to occur by a clathrin-mediated pathway, and to depend on endosomal acidification.[79] In neither of these studies was it determined whether the virus isolate bound DAF. Other investigators have reported that a DAF-binding isolate of EV11 enters cells by a mechanism different from that used by an EV11 isolate that does not bind DAF.[80] It will be interesting to determine whether DAF-binding and DAF-independent isolates enter HeLa cells by different routes, but important insights into DAF function come from studies of polarized epithelial cells, a cell type with—perhaps—greater relevance to viral pathogenesis.

CVB and other enteroviruses are transmitted by the fecal-oral route, and must cross the intestinal mucosa to initiate infection. The intestines are lined by polarized epithelial cells, with distinct apical and basolateral surfaces; although virus in the intestinal lumen is free to interact with apical cell membranes, intercellular tight junctions prevent virus access to basolateral membranes, and interfere with penetration of virus into deeper cell layers. CAR is a component of the tight junction,[81] and it is absent from the apical surface and inaccessible to virus (Fig. 4A). As a result, application of CAR-dependent viruses to the apical surface results in little, if any, infection.[81] In contrast, DAF is highly

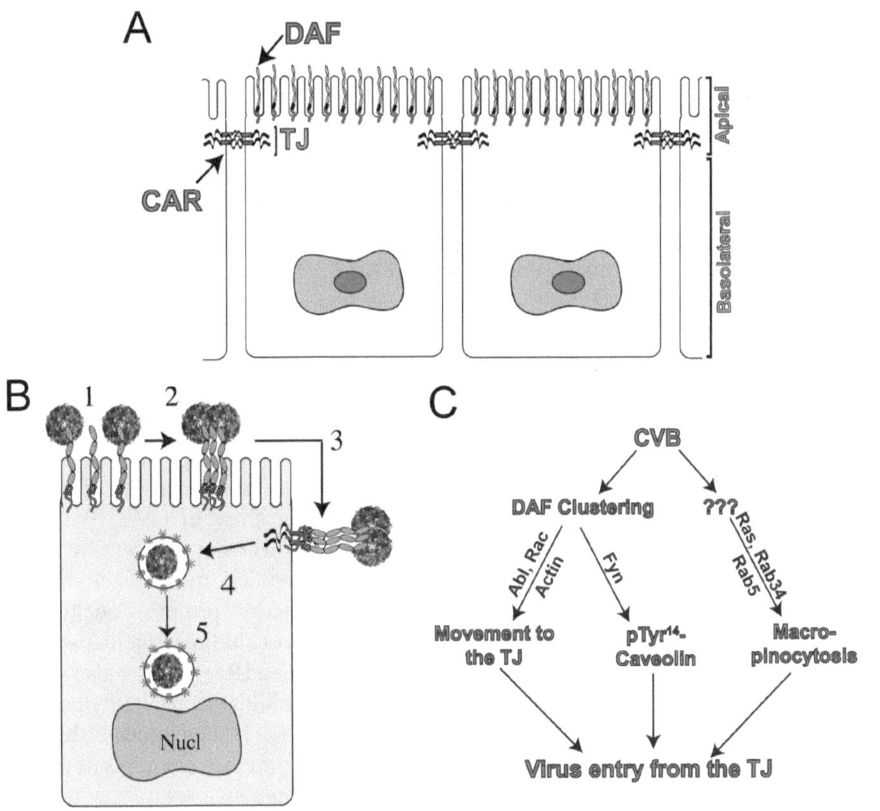

Figure 4. Coxsackievirus interaction with polarized epithelial cells. A) Tight junctions (TJ) separate the apical membrane from the basolateral membrane, and prevent passage of solutes across the epithelial layer. CAR is a component of the tight junction; DAF is expressed on the apical surface. B) CVB3 binds to DAF on the apical surface of a polarized epithelial cell (1); DAF clustering (2) initiates signals that permit virus movement to the tight junction (3). At the junction, virus interacts with CAR and is converted to A-particles, then enters the cell in a caveolin-containing vesicle (4) and moves to the perinuclear region (5). C) CVB3-induced signals required for entry. Clustering of DAF leads to activation of Abl and Rac, inducing actin rearrangements that permit virus movement the TJ. DAF clustering activates Fyn, leading to phosphorylation of caveolin. CVB3 also induces macropinocytosis, which requires activation of Ras, as well as Rab5 and Rab34. It is not yet known whether the signals required for macropinocytosis are induced by DAF clustering or by another mechanism.

expressed on the apical membrane, and DAF-binding CVB3 and CVB5 isolates infect polarized epithelium.[82] We suspect that DAF provides a means for CVB, and possibly for other enteroviruses, to interact with intestinal epithelium as virus is transmitted from person to person.

The CVB entry route in polarized epithelial cells is complex[83] (Fig 4B). Virus binds to DAF on the apical cell surface then moves to the tight junction (TJ); there it interacts with CAR and is converted to A particles. (CAR is essential; when CAR is depleted with siRNA, virions move to the TJ, but fail to undergo conversion to A particles, and fail to enter the cell.) Virus then enters the cell, appearing first in caveolin-containing vesicles in the cell periphery, then moving to a perinuclear location. It appears that uncoating does not occur until virus has moved deep into the cell. We have found (unpublished observations) that in the presence of nocodazole, an inhibitor of microtubule formation, virus does not move to the perinuclear compartment, 135S particles are not converted to 80S empty capsids, and infection is prevented; however, nocodazole has no effect on virus replication when viral RNA is transfected directly into the cell. Conversion to A particles and release of RNA thus occur at different sites within (or on the surface of) the cell.

VIRUS-INDUCED SIGNALS ARE REQUIRED FOR ENTRY

DAF Mediated (and Other) Signals Required for CVB Entry

DAF does more than permit CVB to attach to polarized epithelium. It also initiates multiple intracellular signals required for entry.[83] By clustering DAF, virus activates c-Abl, a tyrosine kinase responsible for initiating cytoskeletal rearrangements that permit virus movement to the TJ. At the same time, DAF mediates activation of Fyn, a Src family kinase that induces phosphorylation of caveolin, permitting virus entry from the TJ (Fig. 4C).

Although internalization of caveolae is thought to require the activity of dynamin GTPase, dynamin is not required for caveolin-dependent entry of CVB. We do not know whether virus enters the cell in typical caveolae, or whether it enters by another caveolin-dependent mechanism. CVB entry is accompanied by the activation of macropinocytosis, and the internalization of a tight junction protein—occludin—in macropinosomes.[84] We have found that CVB entry requires occludin, as well as a number of regulatory molecules (including PI3 kinase, Rab34, Rab5, and Ras) that are also required for macropinocytosis of occludin; furthermore, drugs that inhibit macropinocytosis, such as amiloride and rottlerin, prevent CVB entry from the TJ. It thus appears that CVB entry in polarized cells involves a distinctive process that combines elements of caveolar endocytosis with features generally associated with macropinocytosis.

PVR SIGNALS IN POLIOVIRUS ENTRY

As mentioned above, we have recently found that poliovirus enters human brain microvascular endothelial cells (HBMEC) by a caveolin-dependent mechanism that is clearly different from the entry route in HeLa cells. HBMEC are polarized cells that replicate features of the blood-brain barrier, and they may provide insights into the mechanisms by which poliovirus spreads from the bloodstream to the central nervous system. Strikingly, the function of PVR in infection of HBMEC is not limited to its role in virus attachment and uncoating. PVR ligation by virus initiates signals within the cell that are required for entry from the cell surface. The PVR cytoplasmic domain is phosphorylated after contact with virus, and phosphorylated PVR recruits and activates a protein tyrosine phosphatase, SHP-2, that is essential for virus entry and infection. PVR ligation also triggers activation of Rho GTPase, leading to dramatic rearrangements of the actin cytoskeleton that are important for entry as well.

Receptor-induced signals are thus required for caveolin-dependent entry by at least three picornaviruses—EV1, PV, and CVB3. Because caveolar endocytosis is triggered by specific ligands, such signals may be particularly important for virus entry by caveolin-dependent mechanisms. Nonetheless, We believe it likely that a virus may evolve to use a specific receptor not only because the receptor is expressed on particular target cells, but also because its intrinsic signaling capacity serves to prime the cell for entry and infection.

CONCLUSION AND FUTURE PERSPECTIVES

The past several years have provided a number of advances in our understanding of picornavirus entry. Progress has been made in understanding the structural changes in the virion important for uncoating. For a number of viruses, the endocytic mechanisms responsible for internalization have been identified, as have the sites at which uncoating occurs. It has become clear that entry is not a passive process, and that viruses initiate specific signals required for entry. And we have begun to recognize that for a given virus, there may be multiple routes of entry, depending on the particular target cell and the receptors available on that cell.

Important questions remain. To understand how RNA exits the picornavirus capsid and traverses the membrane will require that structural biologists capture the moment of RNA release; recent structural studies of poliovirus bound to a membrane-receptor complex[85,86] are an important step in that direction. Much of our recent understanding of the cell biology of entry has come from the use of fluorescence microscopy, as well as the use of siRNA and dominant-negative mutants to dissect the cellular components required for entry. Understanding the dynamic process by which viruses enter the cell, disassemble, and traffic to appropriate intracellular compartments is likely to benefit greatly from imaging techniques that make it possible to track individual virions and viral genomes in a living cell.[87] The application of broad siRNA screens[88] to the entry process is likely to enhance our appreciation of the complex interactions between virus entry mechanisms and cellular signaling pathways.

REFERENCES

1. Hogle JM, Chow M, Filman DJ. Three-dimensional structure of poliovirus at 2.9 Å resolution. Science 1985; 229:1358-1365.
2. Rossmann MG, Arnold E, Erickson JW et al. Structure of a human common cold virus and functional relationship to other picornaviruses. Nature 1985; 317:145-153.
3. Acharya R, Fry E, Stuart D et al. The three-dimensional structure of foot-and-mouth disease virus at 2.9 Å resolution. Nature 1989; 337:709-716.
4. Chapman MS, Minor I, Rossmann MG et al. Human rhinovirus 14 complexed with antiviral compound R 61837. J Mol Biol 1991; 217:455-463.

5. Fox MP, Otto MJ, McKinlay MA. Prevention of rhinovirus and poliovirus uncoating by WIN 51711, a new antiviral drug. Antimicrob Agents Chemother 1986; 30:110-116.
6. Grant RA, Hiremath CN, Filman DJ et al. Structures of poliovirus complexes with anti-viral drugs: implications for viral stability and drug design. Curr Biol 1994; 4:784-797.
7. Rogers JM, Diana GD, McKinlay MA. Pleconaril. A broad spectrum antipicornaviral agent. Adv Exp Med Biol 1999; 458:69-76.
8. Li Q, Yafal AG, Lee YM et al. Poliovirus neutralization by antibodies to internal epitopes of VP4 and VP1 results from reversible exposure of these sequences at physiological temperature. J Virol 1994; 68:3965-3970.
9. Lewis JK, Bothner B, Smith TJ et al. Antiviral agent blocks breathing of the common cold virus. Proc Natl Acad Sci USA 1998; 95:6774-6778.
10. Olson NH, Kolatkar PR, Oliveria MA et al. Structure of a human rhinovirus complexed with its receptor molecule. Proc Natl Acad Sci USA 1993; 90:507-511.
11. Belnap DM, McDermott Jr BM, Filman DJ et al. Three-dimensional structure of poliovirus receptor bound to poliovirus. Proc Natl Acad Sci USA 2000; 97:73-78.
12. He Y, Bowman VD, Mueller S et al. Interaction of the poliovirus receptor with poliovirus. Proc Natl Acad Sci USA 2000; 97:79-84.
13. Xing L, Tjarnlund K, Lindqvist B et al. Distinct cellular receptor interactions in poliovirus and rhinoviruses. EMBO J 2000; 19:1207-1216.
14. He Y, Chipman PR, Howitt J et al. Interaction of coxsackievirus B3 with the full length coxsackievirus-adenovirus receptor. Nat Struct Biol 2001; 8:874-878.
15. Casasnovas JM, Springer TA. Kinetics and thermodynamics of virus binding to receptor: studies with rhinovirus, intercellular adhesion molecule-1 (ICAM-1), and surface plasmon resonance. J Biol Chem 1995; 270:13216-13224.
16. McDermott Jr BM, Rux AH, Eisenberg RJ et al. Two distinct binding affinities of poliovirus for its cellular receptor. J Biol Chem 2000; 275:23089-23096.
17. Kolatkar PR, Bella J, Olson NH et al. Structural studies of two rhinovirus serotypes complexed with fragments of their cellular receptor. EMBO J 1999; 18:6249-6259.
18. He Y, Lin F, Chipman PR et al. Structure of decay-accelerating factor bound to echovirus 7: a virus-receptor complex. Proc Natl Acad Sci USA 2002; 99:10325-10329.
19. Pettigrew DM, Williams DT, Kerrigan D et al. Structural and functional insights into the interaction of echoviruses and decay-accelerating factor. J Biol Chem 2006; 281:5169-5177.
20. Powell RM, Ward T, Evans DJ et al. Interaction between echovirus 7 and its receptor, decay-accelerating factor (CD55): evidence for a secondary cellular factor in A-particle formation. J Virology 1997; 71:9306-9312.
21. Milstone A, Petrella J, Sanchez M et al. Interaction with coxsackievirus and adenovirus receptor (CAR), but not with decay accelerating factor (DAF), induces A particle formation in a DAF-binding coxsackievirus B3 isolate. J Virol 2005; 79:655-660.
22. Crowell RL, Philipson L. Specific alterations of coxsackievirus B3 eluted from HeLa cells. J Virology 1971; 8:509-515.
23. Lonberg-Holm K, Korant BD. Early interaction of rhinoviruses with host cells. J Virol 1972; 9:29-40.
24. De Sena J, Mandel B. Studies on the in vitro uncoating of poliovirus. II. Characteristics of the membrane-modified particle. Virology 1977; 78:554-566.
25. Rosenwirth B, Eggers HJ. Early processes of echovirus 12-infection: elution, penetration, and uncoating under the influence of rhodanine. Virology 1979; 97:241-255.
26. Fricks CE, Hogle JM. Cell-induced conformational change in poliovirus: externalization of the amino terminus of VP1 is responsible for liposome binding. J Virology 1990; 64:1934-1945.
27. McSharry JJ, Caliguiri LA, Eggars HJ. Inhibition of uncoating of poliovirus by arildone, a new antiviral drug. Virology 1979; 97:307-315.
28. Curry S, Chow M, Hogle JM. The poliovirus 135S particle is infectious. Journal of Virology 1996; 70:7125-7131.
29. Huang Y, Hogle JM, Chow M. Is the 135S poliovirus particle an intermediate during cell entry? J Virology 2000; 74:8757-8761.
30. Dove AW, Racaniello VR. Cold-adapted poliovirus mutants bypass a postentry replication block. J Virology 1997; 71:4728-4735.
31. Tosteson MT, Chow M. Characterization of the ion channels formed by poliovirus in planar lipid membranes. J Virol 1997; 71:507-511.
32. Tosteson MT, Wang H, Naumov A et al. Poliovirus binding to its receptor in lipid bilayers results in particle-specific, temperaturesensitive channels. J Gen Virol 2004; 85:1581-1589.
33. Danthi P, Tosteson M, Li QH et al. Genome delivery and ion channel properties are altered in VP4 mutants of poliovirus. J Virol 2003; 77:5266-5274.

34. Belnap DM, Filman DJ, Trus BL et al. Molecular tectonic model of virus structural transitions: the putative cell entry states of poliovirus. J Virol 2000; 74:1342-1354.
35. Bubeck D, Filman DJ, Cheng N et al. The structure of the poliovirus 135S cell entry intermediate at 10-angstrom resolution reveals the location of an externalized polypeptide that binds to membranes. J Virol 2005; 79:7745-7755.
36. Lonberg-Holm K, Noble-Harvey J. Comparison of in vitro and cell-mediated alteration of a human Rhinovirus and its inhibition by sodium dodecyl sulfate. J Virol 1973; 12:819-826.
37. Xing L, Casasnovas JM, Cheng RH. Structural analysis of human rhinovirus complexed with ICAM-1 reveals the dynamics of receptor-mediated virus uncoating. J Virol 2003; 77:6101-6107.
38. Hewat EA, Neumann E, Blaas D. The concerted conformational changes during human rhinovirus 2 uncoating. Mol Cell 2002; 10:317-326.
39. Hewat EA, Blaas D. Cryoelectron microscopy analysis of the structural changes associated with human rhinovirus type 14 uncoating. J Virol 2004; 78:2935-2942.
40. Conner SD, Schmid SL. Regulated portals of entry into the cell. Nature 2003; 422:37-44.
41. Marsh M, Helenius A. Virus entry: open sesame. Cell 2006; 124:729-740.
42. Pelkmans L, Burli T, Zerial M et al. Caveolin-stabilized membrane domains as multifunctional transport and sorting devices in endocytic membrane traffic. Cell 2004; 118:767-780.
43. Glebov OO, Bright NA, Nichols BJ. Flotillin-1 defines a clathrin-independent endocytic pathway in mammalian cells. Nat Cell Biol 2006; 8:46-54.
44. Yafal A, Kaplan G, Racaniello V et al. Characterization of poliovirus conformational alteration mediated by soluble cell receptors. Virology 1993; 197:501-505.
45. Hoover-Litty H, Greve JM. Formation of rhinovirus-soluble ICAM-1 complexes and conformational changes in the virion. J Virology 1993; 67:390-397.
46. Zeichhardt H, Wetz K, Willingmann P et al. Entry of poliovirus type 1 and Mouse Elberfeld (ME) virus into HEp-2. Cells: Receptor-mediated endocytosis and endosomal or lysosomal uncoating. J Virology 1985; 66:483-492.
47. Madshus IH, Olsnes S, Sandvig K. Mechanism of entry into the cytosol of poliovirus type 1: requirement for low pH. J Cell Biol 1984; 98:1194-1200.
48. Madshus IH, Olsnes S, Sandvig K. Requirements for poliovirus entry into cells at low pH. EMBO J 1984; 3:1945-1950.
49. Perez L, Carrasco L. Entry of poliovirus into cells does not require a low-pH step. J Virology 1993; 67:4543-4548.
50. DeTulleo L, Kirchhausen T. The clathrin endocytic pathway in viral infection. EMBO J 1998; 17:4585-4593.
51. Brandenburg B, Lee LY, Lakadamyali M et al. Imaging poliovirus entry in live cells. PLoS Biology 2007; 5:183-196.
52. Coyne CB, Kim KS, Bergelson JM. Poliovirus entry into human brain microvascular cells requires receptor-induced activation of SHP-2. EMBO Journal 2007; 26:submitted.
53. Brown F, Cartwright B. Dissociation of foot-and-mouth disease virus into its nucleic acid and protein components. Nature 1961; 192:1163-1164.
54. Baxt B, Bachrach H. Early interactions of foot-and-mouth disease virus with cultured cells. Virology 1980; 104:42-55.
55. O'Donnell V, LaRocco M, Duque H et al. Analysis of foot-and-mouth disease virus internalization events in cultured cells. J Virol 2005; 79:8506-8518.
56. Berryman S, Clark S, Monaghan P et al. Early events in integrin alphavbeta6-mediated cell entry of foot-and-mouth disease virus. J Virol 2005; 79:8519-8534.
57. Mason P, Baxt B, Brown F et al. Antibody-complexed foot-and-mouth disease virus, but not poliovirus, can infect normally insusceptible cells via the Fc receptor. Virology 1993; 192:568-577.
58. Snyers L, Zwickl H, Blaas D. Human rhinovirus type 2 is internalized by clathrin-mediated endocytosis. J Virol 2003; 77:5360-5369.
59. Bayer N, Schober D, Huttinger M et al. Inhibition of clathrin-dependent endocytosis has multiple effects on human rhinovirus serotype 2 cell entry. J Biol Chem 2001; 276:3952-3962.
60. Brabec M, Baravalle G, Blaas D et al. Conformational changes, plasma membrane penetration, and infection by human rhinovirus type 2: role of receptors and low pH. J Virol 2003; 77:5370-5377.
61. Nurani G, Lindqvist B, Casasnovas JM. Receptor priming of major group human rhinoviruses for uncoating and entry at mild low-pH environments. J Virol 2003; 77:11985-11991.
62. Khan AG, Pichler J, Rosemann A et al. Human rhinovirus type 54 infection via heparan sulfate is less efficient and strictly dependent on low endosomal pH. J Virol 2007; 81:4625-4632.
63. Vlasak M, Goesler I, Blaas D. Human rhinovirus type 89 variants use heparan sulfate proteoglycan for cell attachment. J Virol 2005; 79:5963-5970.
64. Reischl A, Reithmayer M, Winsauer G et al. Viral evolution toward change in receptor usage: adaptation of a major group human rhinovirus to grow in ICAM-1-negative cells. J Virol 2001; 75:9312-9319.

65. Brabec M, Schober D, Wagner E et al. Opening of size-selective pores in endosomes during human rhinovirus serotype 2 in vivo uncoating monitored by single-organelle flow analysis. J Virol 2005; 79:1008-1016.
66. Schober D, Kronenberger P, Prchla E et al. Major and minor receptor group human rhinoviruses penetrate from endosomes by different mechanisms. J Virol 1998; 72:1354-1364.
67. Bergelson JM, Shepley MP, Chan BMC et al. Identification of the integrin VLA-2 as a receptor for echovirus 1. Science 1992; 255:1718-1720.
68. King SL, Kamata T, Cunningham JA et al. Echovirus 1 interaction with the human very late antigen-2 (integrin $\alpha2\beta1$) I domain: identification of two independent virus contact sites distinct from the metal ion-dependent adhesion site. J Biol Chem 1997; 272:28518-28522.
69. Xing L, Huhtala M, Pietiainen V et al. Structural and functional analysis of integrin alpha2I domain interaction with echovirus 1. J Biol Chem 2004; 279:11632-11638.
70. Pietiainen V, Marjomaki V, Upla P et al. Echovirus 1 endocytosis into caveosomes requires lipid rafts, dynamin II, and signaling events. Mol Biol Cell 2004; 15:4911-4925.
71. Upla P, Marjomaki V, Kankaanpaa P et al. Clustering induces a lateral redistribution of alpha 2 beta 1 integrin from membrane rafts to caveolae and subsequent protein kinase C-dependent internalization. Mol Biol Cell 2004; 15:625-636.
72. Marjomaki V, Pietiainen V, Matilainen H et al. Internalization of echovirus 1 in caveolae. J Virol 2002; 76:1856-1865.
73. Bergelson JM, Cunningham JA, Droguett G et al. Isolation of a common receptor for coxsackie B viruses and adenoviruses 2 and 5. Science 1997; 275:1320-1323.
74. Tomko RP, Xu R, Philipson L. HCAR and MCAR: The human and mouse cellular receptors for subgroup C adenoviruses and group B coxsackieviruses. Proc. Natl. Acad. Sci 1997; 94:3352-3356.
75. Martino TA, Petric M, Weingartl H et al. The coxsackie-adenovirus receptor (CAR) is used by reference strains and clinical isolates representing all six serotypes of coxsackievirus group B, and by swine vesicular disease virus. Virology 2000; 271:99-108.
76. Bergelson JM, Mohanty JG, Crowell RL et al. Coxsackievirus B3 adapted to growth in RD cells binds to decay-accelerating factor (CD55). J Virology 1995; 69:1903-1906.
77. Shafren DR, Bates RC, Agrez MV et al. Coxsackieviruses B1, B3, and B5 use decay accelerating factor as a receptor for cell attachment. J Virology 1995; 69:3873-3877.
78. Triantafilou K, Triantafilou M. Lipid-raft-dependent Coxsackievirus B4 internalization and rapid targeting to the Golgi. Virology 2004; 326:6-19.
79. Chung SK, Kim JY, Kim IB et al. Internalization and trafficking mechanisms of coxsackievirus B3 in HeLa cells. Virology 2005; 333:31-40.
80. Stuart AD, Eustace HE, McKee TA et al. A novel cell entry pathway for a DAF-using human enterovirus is dependent on lipid rafts. J Virol 2002; 76:9307-9322.
81. Cohen CJ, Shieh JT-C, Pickles RJ et al. The coxsackievirus and adenovirus receptor is a transmembrane component of the tight junction. Proc. Natl. Acad. Sci. USA 2001; 98:15191-15196.
82. Shieh JTC, Bergelson JM. Interaction with decay-accelerating factor facilitates coxsackievirus B infection of polarized epithelial cells. J Virology 2002; 76:9474-9480.
83. Coyne CB, Bergelson JM. Virus-induced Abl and Fyn kinase signals permit coxsackievirus entry through epithelial tight junctions. Cell 2006; 124:119-131.
84. Coyne CB, Shen L, Turner JR et al. Coxsackievirus entry from epithelial tight junctions requires occludin and the small GTPases Rab34 and Rab5. Cell Host and Microbe 2007; 2:181-192.
85. Bubeck D, Filman DJ, Hogle JM. Cryo-electron microscopy reconstruction of a poliovirus-receptor-membrane complex. Nat Struct Mol Biol 2005; 12:615-618.
86. Tuthill TJ, Bubeck D, Rowlands DJ et al. Characterization of early steps in the poliovirus infection process: receptor-decorated liposomes induce conversion of the virus to membrane-anchored entry-intermediate particles. J Virol 2006; 80:172-180.
87. Brandenburg B, Zhuang X. Virus trafficking—learning from single-virus tracking. Nat Rev Microbiol 2007; 5:197-208.
88. Pelkmans L, Fava E, Grabner H et al. Genome-wide analysis of human kinases in clathrin- and caveolae/raft-mediated endocytosis. Nature 2005; 436:78-86.
89. Mendelsohn CL, Wimmer E, Racaniello VR. Cellular receptor for poliovirus: Molecular cloning, nucleotide sequence, and expression of a new member of the immunoglobulin superfamily. Cell 1989; 56:855-865.
90. Greve JM, Davis G, Meyer AM et al. The major human rhinovirus receptor is ICAM-1. Cell 1989; 56:839-847.
91. Staunton DE, Merluzzi VJ, Rothlein R et al. A cell adhesion molecule, ICAM-1, is the major surface receptor for rhinoviruses. Cell 1989; 56:849-853.
92. Tomassini JE, Graham D, DeWitt CM et al. cDNA cloning reveals that the major group rhinovirus receptor on HeLa cells is intercellular adhesion molecule 1. Proc Natl Acad Sci U S A 1989; 86:4907-4911.

93. Hofer F, Gruenberger M, Kowalski H et al. Members of the low density lipoprotein receptor family mediate cell entry of a minor-group common cold virus. Proc. Natl. Acad. Sci 1994; 91: 1839-1842.
94. Zautner AE, Korner U, Henke A et al. Heparan sulfates and coxsackievirus-adenovirus receptor: each one mediates coxsackievirus B3 PD infection. J Virol 2003; 77:10071-10077.
95. Bergelson JM, Chan M, Solomon K et al. Decay-accelerating factor, a glycosylphosphatidylinositol-anchored complement regulatory protein, is a receptor for several echoviruses. Proc. Natl. Acad. Sci. USA 1994; 91:6245-6248.
96. Ward T, Pipkin PA, Clarkson NA et al. Decay accelerating factor (CD55) identified as a receptor for echovirus 7 using CELICS, a rapid immuno-focal cloning method. EMBO J 1994; 13: 5070-5074.
97. Goodfellow IG, Sioofy AB, Powell RM et al. Echoviruses bind heparan sulfate at the cell surface. J Virol 2001; 75:4918-4921.
98. Shafren DR, Dorahy DJ, Greive SJ et al. Mouse cells expressing human intercellular adhesion molecule-1 are susceptible to infection by coxsackievirus A21. J Virol 1997; 71:785-789.
99. Shafren DR, Dorahy DJ, Ingham RA et al. Coxsackievirus A21 binds to decay-accelerating factor but requires intercellular adhesion molecule 1 for cell entry. J Virol 1997; 71:4736-4743.
100. Roivainen M, Piirainen L, Hovi T et al. Entry of coxsackievirus A9 into host cells: specific interactions with alpha v beta 3 integrin, the vitronectin receptor. Virology 1994; 203:357-365.
101. Triantafilou K, Fradelizi D, Wilson K et al. GRP78, a coreceptor for coxsackievirus A9, interacts with major histocompatability complex class I molecules which mediate virus internalization. J Virol 2002; 76:633-643.
102. Yamayoshi S, Yamashita Y, Li J et al. Scavenger receptor B2 is a cellular receptor for enterovirus 71. Nat Med 2009; 15:798-801.
103. Nishimura Y, Shimojima M, Tano Y et al. 2009. Human P-selectin glycoprotein ligand-1 is a functional receptor for enterovirus 71. Nat Med 2009; 15:794-797.
104. Tan CW, Poh CL, Sam IC, Chan YF. Enterovirus 71 uses cell surface heparan sulfate glycosaminoglycan as an attachment receptor. J Virol 2013; 87:611-620.
105. Jackson T, Mould AP, Sheppard D et al. Integrin alphavbeta1 is a receptor for foot-and-mouth disease virus. J Virol 2002; 76:935-941.
106. Berinstein A, Roivainen M, Hovi T et al. Antibodies to the vitronectin receptor (integrin alpha V beta 3) inhibit binding and infection of foot-and-mouth disease virus to cultured cells. J Virol 1995; 69:2664-2666.
107. Jackson T, Clark S, Berryman S et al. Integrin alphavbeta8 functions as a receptor for foot-and-mouth disease virus: role of the beta-chain cytodomain in integrin-mediated infection. J Virol 2004; 78:4533-4540.
108. Jackson T, Sheppard D, Denyer M et al. The epithelial integrin alphavbeta6 is a receptor for foot-and-mouth disease virus. J Virol 2000; 74:4949-4956.
109. Jackson T, Ellard FM, Ghazaleh RA et al. Efficient infection of cells in culture by type O foot-and-mouth disease virus requires binding to cell surface heparan sulfate. J Virol 1996; 70: 5282-5287.
110. Huber SA. VCAM-1 is a receptor for encephalomyocarditis virus on murine vascular endothelial cells. J Virol 1994; 68:3453-3458.
111. Filman DJ, Syed R, Chow M et al. Structural factors that control conformational transitions and serotype specificity in type 3 poliovirus. EMBO J 1989; 8:1567-1579.
112. Yoder JD, Cifuente JO, Pan J et al. The crystal structure of a coxsackievirus B3-RD variant and a refined 9-angstrom cryo-electron microscopy reconstruction of the virus complexed with decay-accelerating factor (DAF) provide a new footprint of DAF on the virus surface. J Virol 2012; 86:12571-12581.

CHAPTER 3

REOVIRUS RECEPTORS, CELL ENTRY, AND PROAPOPTOTIC SIGNALING

Pranav Danthi,[†,1] Geoffrey H. Holm,[†,2] Thilo Stehle[3,4] and Terence S. Dermody*,[3,5,6]

[1]Department of Biology, Indiana University, Bloomington, Indiana, USA; [2]Department of Biology, Colgate University, Hamilton, New York, USA; [3]Department of Pediatrics, Vanderbilt University School of Medicine, Nashville, Tennessee, USA; [4]Interfakultäres Institut für Biochemie, Eberhard-Karls Universität, Tübingen, Germany; [5]Lamb Center for Pediatric Research, Vanderbilt University School of Medicine, Nashville, Tennessee, USA; [6]Department of Pathology, Microbiology and Immunology, Vanderbilt University School of Medicine, Nashville, Tennessee, USA

[†]These authors contributed equally to this manuscript.

**Corresponding Author: Terence S. Dermody—Email: terry.dermody@vanderbilt.edu*

Abstract: Mammalian orthoreoviruses (reoviruses) are members of the *Reoviridae*. Reoviruses contain 10 double-stranded (ds) RNA gene segments enclosed in two concentric protein shells, called outer capsid and core. These viruses serve as a versatile experimental system for studies of viral replication events at the virus-cell interface, including engagement of cell-surface receptors, internalization and disassembly, and activation of the innate immune response, including NF-κB-dependent cellular signaling pathways. Reoviruses also provide a model system for studies of virus-induced apoptosis and organ-specific disease in vivo.

Reoviruses attach to host cells via the filamentous attachment protein, σ1. The σ1 protein of all reovirus serotypes engages junctional adhesion molecule-A (JAM-A), an integral component of intercellular tight junctions. The σ1 protein also binds to cell-surface carbohydrate, with the type of carbohydrate bound varying by serotype. Following attachment to JAM-A and carbohydrate, reovirus internalization is mediated by β1 integrins, most likely via clathrin-dependent endocytosis. In the endocytic compartment, reovirus outer-capsid protein σ3 is removed by acid-dependent cysteine proteases in most cell types. Removal of σ3 results in the exposure of a hydrophobic conformer of the viral membrane-penetration protein, μ1, which pierces the endosomal membrane and delivers transcriptionally active reovirus core particles into the cytoplasm.

Reoviruses induce apoptosis in both cultured cells and infected mice. Perturbation of reovirus disassembly using inhibitors of endosomal acidification or protease

Viral Entry into Host Cells, edited by Stefan Pöhlmann and Graham Simmons.

activity abrogates apoptosis. The μ1-encoding M2 gene is genetically linked to strain-specific differences in apoptosis-inducing capacity, suggesting a function for μ1 in induction of death signaling. Reovirus disassembly leads to activation of transcription factor NF-κB, which modulates apoptotic signaling in numerous types of cells. Inhibition of NF-κB nuclear translocation using either pharmacologic agents or expression of transdominant forms of IκB blocks reovirus-induced apoptosis, suggesting an essential role for NF-κB activation in the death response. Multiple effector pathways downstream of NF-κB-directed gene expression execute reovirus-induced cell death. This chapter will focus on the mechanisms by which reovirus attachment and disassembly activate NF-κB and stimulate the cellular proapoptotic machinery.

INTRODUCTION

For several virus families, significant progress has been made in understanding the molecular events associated with viral entry into host cells. Viral entry steps include stable attachment of the virus to the cell surface, penetration of the virus into the interior of the cell, disassembly of the viral capsid, and activation of the viral genetic program. These steps are essential for the virus to traverse the extracellular environment to the cellular compartment in which viral transcription and replication occur. Viral entry mechanisms also have relevance to viral pathogenesis as these events often determine target cell selection within the host, which dictates the site of virus-induced disease. Moreover, entry steps can induce intracellular signaling cascades that influence whether cells enter into an antiviral state or undergo apoptosis. In this chapter, we consider mechanisms of reovirus cell entry and describe our current understanding about how these entry events initiate proapoptotic signaling.

PATHOGENESIS OF REOVIRUS INFECTION

Members of the *Reoviridae* family are nonenveloped viruses containing genomes of 9-12 segments of double-stranded (ds) RNA[1] (Fig. 1). This family includes mammalian orthoreoviruses (reoviruses), orbiviruses, and rotaviruses. For reoviruses, the viral proteins are designated with a Greek letter corresponding to the size of the encoding genome segment: sigma (σ) for proteins encoded by small genome segments, mu (μ) by medium segments, and lambda (λ) by large segments. Each of the genome segments encodes a single protein with the exception of the S1 gene, which encodes the viral attachment protein, σ1, and a small nonstructural protein, σ1s. Like other members of the *Reoviridae*, reovirus particles are formed from concentric protein shells. Two such shells exist for reoviruses, called outer capsid and core.[1]

Reoviruses can infect many mammalian species, including humans, although they are rarely associated with disease.[1,2] Three reovirus serotypes have been recognized based on neutralization and hemagglutination profiles. Each is represented by a prototype strain, type 1 Lang (T1L), type 2 Jones (T2J), and type 3 Dearing (T3D), which differ primarily in σ1 sequence.[3,4] The pathogenesis of reovirus infections has been most extensively studied using newborn mice, in which serotype-specific patterns of disease have been identified.[1,2] The best characterized of these models is reovirus pathogenesis in the murine central nervous system (CNS).

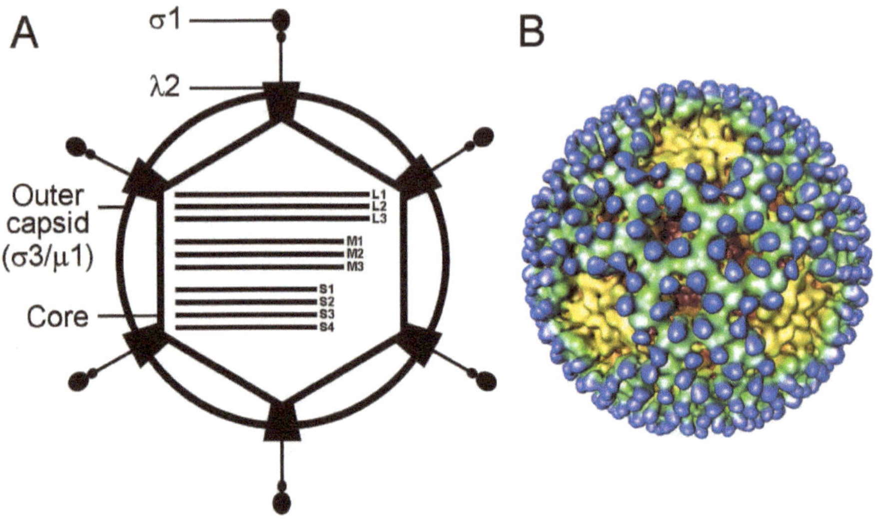

Figure 1. The reovirus virion. A) Schematic of a reovirus virion. Reovirus virions are composed of two concentric protein shells, outer capsid and core. The core contains the viral genome, which consists of 10 segments of double-stranded RNA. B) Cryo-EM image reconstruction of a reovirus virion at 23 Å resolution. Note the finger-like projections of σ3 (blue) that sit atop a layer of μ1 (green). The λ2 protein (yellow) forms a pentamer at each of the virion fivefold symmetry axes. Figure modified from: Nason E et al, J Virol 75:6625-6634; ©2001 with permission from the American Society for Microbiology.[74]

Following oral or intramuscular inoculation of newborn mice, strains of serotype 1 and serotype 3 reoviruses invade the CNS. However, these strains disseminate in the host by different routes and have distinct pathologic consequences. Serotype 1 reovirus spreads to the CNS hematogenously and infects ependymal cells,[5,6] resulting in hydrocephalus.[7] In contrast, serotype 3 reovirus spreads to the CNS by neural routes and infects neurons,[5,6,8] causing lethal encephalitis.[7,9] Studies using T1L x T3D reassortant viruses have shown that the pathways of viral spread[5] and tropism for neural tissues[6,10] segregate with the viral S1 gene. Since the S1 gene encodes attachment protein σ1,[11,12] these studies suggest that σ1 dictates the CNS cell types that serve as targets for reovirus infection, presumably by its capacity to bind to receptors expressed by specific CNS cells.

ATTACHMENT RECEPTORS: CELL-SURFACE SIALIC ACID AND JUNCTIONAL ADHESION MOLECULE-A

The σ1 protein is a filamentous, trimeric molecule about 480 Å in length with distinct head-and-tail morphology[13,14] (Fig. 2). Independent domains of the protein mediate binding to different types of cell-surface receptors. Sequences in the N-terminal σ1 tail bind to carbohydrate, which is known to be sialic acid in either α2,3 or α2,6 linkages for serotype 3 reoviruses.[15-19] The C-terminal σ1 head binds to junctional adhesion molecule-A (JAM-A, previously called JAM or JAM1),[20] a member of

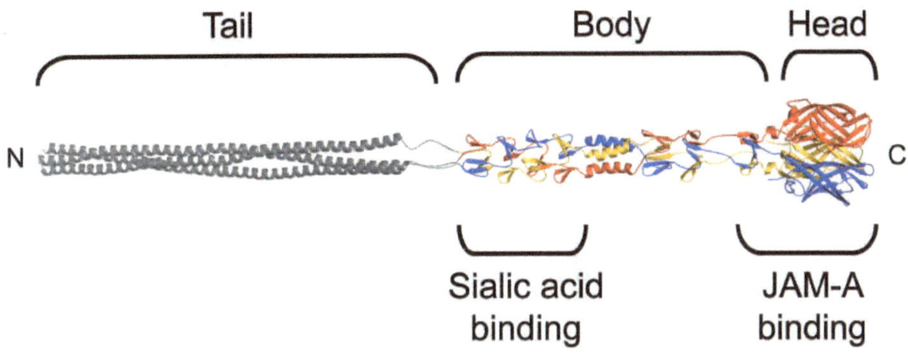

Figure 2. Attachment protein σ1. Full-length model of σ1 generated by adding a trimeric α-helical coiled-coil to the N-terminus of the crystallized σ1 fragment.[26] The three monomers of the crystallized fragment are shown in red, yellow, and blue; the model is shown in grey. Regions of the molecule that interact with sialic acid and JAM-A are indicated.

the immunoglobulin (Ig) superfamily that regulates formation of intercellular tight junctions.[21-23] The σ1 tail partially inserts into the virion, while the head projects away from the virion surface.[13,24]

The crystal structure of the C-terminal half of T3D σ1 (residues 170-455) reveals a homotrimer with an unusual structural fold[25,26] (Fig. 2). N-terminal residues in the crystallized fragment (170-309) form the body domain, which consists of seven β-spiral repeats interrupted by a short stretch of α-helix. β-spiral repeats are also observed in the adenovirus fiber[27] and avian reovirus σC.[28] C-terminal residues form the compact head domain (310-455), which consists of an 8-stranded β-barrel. Sequence analysis, coupled with the crystallographic data, has facilitated the development of a model of full-length σ1[25] (Fig. 2). The σ1 tail is predicted to contain ~20 heptad repeats of an N-terminal α-helical coiled-coil.[3,4]

Both murine (m) and human (h) homologs of JAM-A function as reovirus receptors.[20] The crystal structure of the extracellular region of hJAM-A consists of two concatenated immunoglobulin domains (D1, membrane distal and D2, membrane proximal)[29] (Fig. 3). Two monomers form a symmetrical dimer that is stabilized by extensive ionic and hydrophobic contacts between the D1 domains. Like the structures of reovirus σ1 and adenovirus fiber, the structures of JAM-A and the coxsackievirus and adenovirus receptor (CAR) are strikingly similar.[30] These observations suggest that reovirus and adenovirus use similar mechanisms of attachment. In concordance with this prediction, the σ1 head binds to the membrane-distal D1 domain of monomeric JAM-A,[31,32] analogous to the mechanism by which the adenovirus fiber binds to CAR.[33,34]

The presence of discrete receptor-binding domains in σ1 suggests that reoviruses employ a multiple-step binding process similar to that used by some herpesviruses[35,36] and retroviruses.[37,38] Binding studies using isogenic point-mutant viruses T3SA+ and T3SA-, which vary only in the capacity to engage sialic acid,[39] support this hypothesis. Kinetic analyses using inhibitors of sialic acid and JAM-A binding demonstrate that sialic acid is engaged first in the adsorption process, as the inhibitory effect of sialic acid analogs on infection by T3SA+ occurs at early but not late timepoints.[39] However, a σ1-specific monoclonal antibody (mAb) that blocks virus binding to JAM-A inhibits

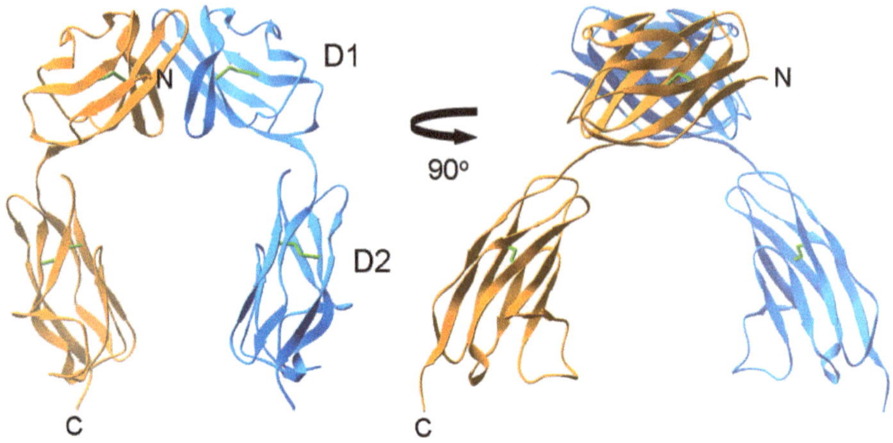

Figure 3. Crystal structure of the JAM-A extracellular region. Ribbon drawings of the hJAM-A dimer, with one monomer shown in yellow and the other in blue. Disulfide bonds are shown in green. The D1 domain is distal to the cell membrane. The two views differ by rotation of 90° along a vertical axis. Figure modified from: Prota AE et al, Proc Natl Acad Sci USA 100:5366-5371; ©2003 with permission from the National Academy of Sciences, USA.[29]

viral infectivity at both early and late times during adsorption.[39] Thus, reovirus binding to sialic acid enhances virus attachment through rapid adhesion of the virus to the cell surface where access to JAM-A is thermodynamically favored.

INTERNALIZATION RECEPTORS: β1 INTEGRINS

Although engagement of JAM-A is required for high-affinity attachment of reovirus to cells, binding to this receptor does not appear to stimulate viral internalization. Expression of a JAM-A mutant lacking a cytoplasmic tail in nonpermissive cells confers full susceptibility to reovirus infection, suggesting that cell-surface molecules other than JAM-A mediate viral internalization following attachment.[40] Outer-capsid protein λ2, which serves as the structural base for σ1,[13,24] contains integrin-binding sequences,[41] raising the possibility that integrins mediate reovirus endocytosis. Integrins are heterodimeric cell-surface molecules that consist of α and β subunits.[42] Integrins function to mediate cellular adhesion to the extracellular matrix, regulate cellular trafficking, and transduce both outside-in and inside-out signaling events.[43] Consistent with a role for integrins in reovirus internalization, a β1 integrin-specific antibody, but not antibodies specific for other integrin subunits, inhibits reovirus infection.[40] In comparison to isogenic cells expressing β1 integrin, uptake of reovirus into β1-deficient mouse embryonic stem cells is substantially diminished[40] (Fig. 4). This defect in uptake is associated with a parallel reduction in infectivity. Additionally, mutations in the NPXY motifs in the cytoplasmic tail of β1 integrin result in mislocalization of virions to lysosomes.[44] These data provide strong evidence that β1 integrins facilitate reovirus internalization and suggest that viral entry occurs by interactions of reovirus virions with independent attachment and entry receptors on the cell surface.

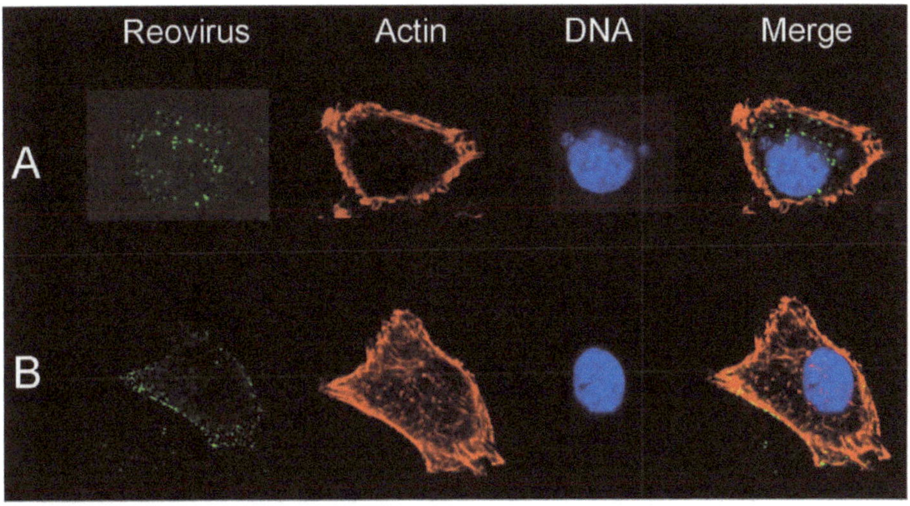

Figure 4. β1 integrin enhances reovirus entry into cells. (A) GD25β1A (β1 +/+) and (B) GD25 (β1 -/-) cells were chilled, adsorbed with T1L virions, and incubated at 4°C for 1 h. Nonadherent virus was removed, warm medium was added, and cells were incubated at 37°C for 30 min. Cells were fixed, stained for reovirus (green), actin (red), and DNA (blue), and imaged using confocal immunofluorescence microscopy. Representative digital fluorescence images of the same field are shown in each row. Figure modified from: Maginnis MS et al, J Virol 80:2760-2770; ©2006 with permission from the American Society for Microbiology.[40]

STEPWISE DISASSEMBLY IN THE ENDOCYTIC COMPARTMENT

Following attachment to cell-surface receptors, reovirus virions are delivered into the endocytic pathway (Fig. 5). Although conclusive evidence for the mechanism of internalization is lacking, current data support a role for clathrin-dependent endocytosis in reovirus cell entry. Thin-section EM images show virions in structures that resemble clathrin-coated pits on the cell surface and in clathrin-coated vesicles in the cytoplasm,[45-48] suggesting clathrin-dependent uptake. Reovirus virions and clathrin colocalize during internalization,[49] providing further evidence that reovirus entry is mediated by a mechanism involving clathrin.

Vesicles containing internalized reovirus virions are transported via microtubules[50] and accumulate in late endosomes.[45-48,50,51] In the endocytic compartment, reovirus virions undergo stepwise disassembly forming sequential disassembly intermediates, the first of which is the infectious subvirion particle (ISVP) (Fig. 5). ISVPs are characterized by the loss of outer-capsid protein σ3, a conformational change in attachment protein σ1, and cleavage of outer-capsid protein μ1 to form particle-associated fragments, δ and ϕ. Following further processing, ISVP-like particles (called ISVP*s) penetrate endosomal membranes, leading to release of transcriptionally active core particles, which lack μ1 and σ1, into the cytoplasm. Thus, the disassembly process consists of a highly coordinated series of events that are dependent on host cell functions that act upon discrete components of the viral outer capsid.

Insight into mechanisms of reovirus disassembly was first provided by using pharmacologic inhibitors of endosomal acidification and protease function. Treatment of murine L929 (L) cells[45,52,53] with the weak base ammonium chloride (AC), which raises the pH of endosomes and lysosomes,[54,55] blocks replication of reovirus when infection

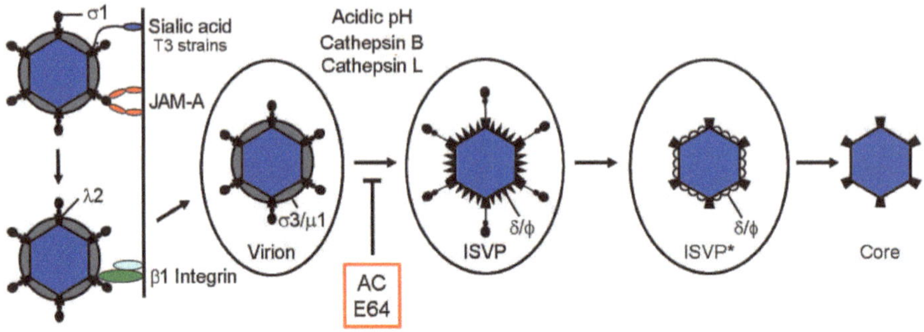

Figure 5. Stepwise disassembly of reovirus. Following attachment to cell-surface carbohydrate (sialic acid for serotype 3 [T3] strains) and JAM-A, reovirus uses β1 integrin to enter cells by receptor-mediated endocytosis. In the endocytic compartment, the viral outer capsid undergoes acid-dependent proteolysis. The first disassembly intermediate is the ISVP, which is characterized by loss of σ3 and cleavage of μ1 into particle-associated fragments δ and ϕ. The ISVP then undergoes further conformational changes to form the ISVP*. The ISVP* is characterized by loss of attachment protein σ1 and conformational rearrangements of the μ1 cleavage fragments to expose hydrophobic residues. The μ1 fragments mediate viral penetration of the endosomal membrane, releasing the transcriptionally active core into the cytoplasm. Treatment of cells with either AC or E64 blocks virion-to-ISVP conversion.

is initiated with virions. However, ISVPs generated in vitro by treatment of virions with the intestinal serine proteases chymotrypsin or trypsin can infect AC-treated cells.[45] This finding indicates that the block to reovirus replication mediated by AC occurs following internalization but prior to disassembly. Similarly, treatment of L cells with E64, an inhibitor of cysteine proteases,[56] arrests infection by reovirus virions but not by ISVPs,[57-60] suggesting that one or more endocytic cysteine proteases effects reovirus disassembly in host cells.

A CRITICAL ROLE FOR CATHEPSINS IN REOVIRUS ENDOSOMAL DISASSEMBLY

The major cysteine proteases in the endocytic compartment of fibroblasts such as L cells are cathepsins B, H, and L, with cathepsin L being the most abundant in several cell types.[56,61-64] Infection of either L cells treated with the cathepsin L inhibitor A-Phe-Tyr(*t*-Bu)-diazomethyl ketone or cathepsin L-deficient mouse embryo fibroblasts results in inefficient proteolytic disassembly and decreased viral yields. In contrast, L cells treated with the cathepsin B inhibitor CA-074Me and cathepsin B-deficient mouse embryo fibroblasts support reovirus disassembly and growth. However, removal of both cathepsin B and cathepsin L activity completely abrogates disassembly and growth of reovirus. Concordantly, cathepsin L mediates reovirus disassembly more efficiently than cathepsin B in vitro.[65] These results demonstrate that either cathepsin L or cathepsin B is required for reovirus entry into murine fibroblasts and indicate that cathepsin L is the primary mediator of reovirus disassembly in these cells. However, proteases other than cathepsin B and cathepsin L are capable of ISVP formation in other types of cells. In P388D cells, a macrophage-like cell line, cathepsin S, an acid-independent cysteine protease,[66] mediates uncoating of some strains of reovirus.[67] Titers of reovirus in mice lacking cathepsin B, L, or S are decreased at early times post-infection, indicating the importance of these proteases in reovirus replication in vivo.[68] However, other proteases in the enteric tract or airway also facilitate reovirus infection.[69]

OUTER-CAPSID PROTEIN σ3, THE MAJOR TARGET FOR ENDOSOMAL PROTEASE ACTIVITY

The first step in the disassembly of reovirus virions is the proteolytic removal of outer-capsid protein σ3.[45,57] The σ3 protein acts as a cap to protect μ1,[24] which is the viral protein that mediates membrane penetration.[70-72] Reovirus σ3 is a bi-lobed protein with its N-terminus in a virion-proximal smaller lobe bound to μ1 and its C-terminus in a virion-distal larger lobe[73,74] (Fig. 6A). Treatment of reovirus virions in vitro with either cathepsin B or cathepsin L leads to an initial cleavage of σ3 most likely near the C-terminus of the protein.[65] During proteolysis by cathepsin L, subsequent cleavages occur between residues 243-244 and 250-251,[65] which are physically located near the σ3 C-terminus[73] (Fig. 6A,B). Because of this proximity, the small end fragment released following initial cathepsin L cleavage likely exposes the other two sites, rendering them sensitive to subsequent cleavage events. The C-terminus therefore appears to act as a "safety latch" that controls access to internal,

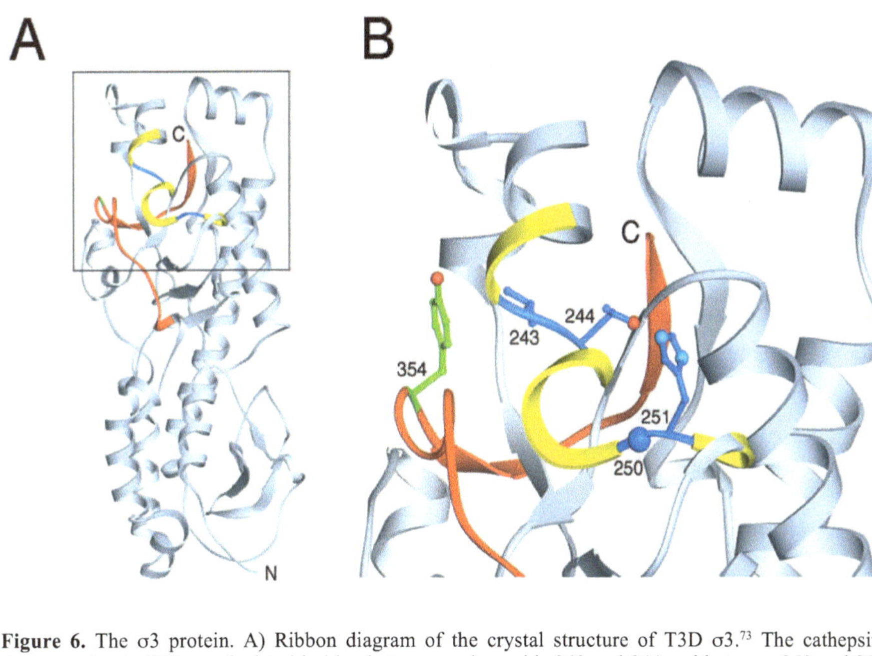

Figure 6. The σ3 protein. A) Ribbon diagram of the crystal structure of T3D σ3.[73] The cathepsin L cleavage sites in T1L are depicted in blue between amino acids 243 and 244 and between 250 and 251.[65] Surrounding residues, from amino acids 241 to 253, are shown in yellow. The C-terminal residues of σ3, from amino acids 340 to 365, are colored red. Tyrosine 354, which is altered in several PI,[75] D-EA,[59] and ACA-D viruses,[76] is colored green. The virion-distal end of σ3 is at the top of the figure, and the virion-proximal end and N-terminus are at the bottom. B) An enlarged view of the boxed region of σ3 indicated in panel B is shown using the same color scheme. Amino acids 243, 244, 250, 251, and 354 are depicted in ball-and-stick representation. Figure and legend modified from: Ebert DH et al, J Biol Chem 277:24609-24617; ©2002 with permission from the American Society for Biochemistry and Molecular Biology.[65]

proteolytically sensitive sites in σ3. Because reovirus disassembly in some cell types is an acid-dependent process, the safety latch might be primed for movement at acidic pH. Numerous reovirus mutants, including those selected during persistent infection (PI viruses)[75] and those selected for resistance to either AC (ACA-D viruses)[76] or E64 (D-EA viruses)[59] have mutations adjacent to the C-terminus of the protein (Fig. 6B). Residues at positions 198 and 354 are particularly important for regulating protease susceptibility.[77] Changes at these positions may mediate structural alterations in the safety latch that provide enhanced access to the cleavage sites located more internally in the protein.

MEMBRANE-PENETRATION PROTEIN μ1

The μ1 protein is genetically and biochemically linked to penetration of endosomal membranes by reovirus disassembly intermediates. Most of the μ1 protein on mature virions is autocatalytically cleaved near the N-terminus to generate two fragments, μ1N and μ1C[78,79] (Fig. 7). This cleavage is not required for virion assembly[80] and may occur physiologically during the transition from the ISVP to the ISVP*.[81] In ISVPs, μ1C is

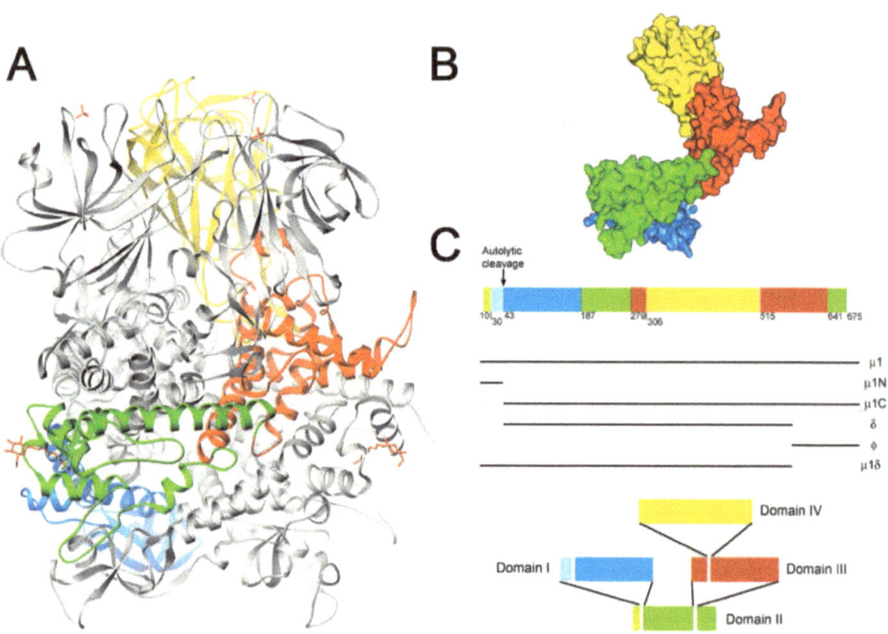

Figure 7. The μ1 protein. A) Ribbon diagram of the crystal structure of the T1L μ1 trimer without bound σ3.[88] One μ1 subunit is colored by domain (domain I, light and dark blue [μ1N, μ1C]; domain II, light and dark green [μ1N, μ1C]; domain III, red; domain IV, yellow); the other two μ1 subunits are shown in gray. β-octyl glucosides and sulfate ions present in the structure are shown in red and yellow. B) Surface-shaded representation of an isolated μ1 subunit. Colors and orientation are as in (A). C) Domain segmentation of the amino acid sequence as determined from the three-dimensional structure. Domain color code as in (A) and (B). The central domain II contains domains I and III as "inserts," and domain III similarly contains domain IV. Figure and legend modified from: Liemann S et al, Cell 108:283-295; ©2002 with permission from Elsevier.[88]

further cleaved by either endocytic[45,65] or intestinal[82] proteases to form fragments δ and ϕ, which remain particle-associated.[83] However, the role of this cleavage in viral penetration is not understood, as core particles recoated with mutant forms of μ1 incapable of δ/ϕ cleavage can establish productive infection.[84] In addition, μ1 is not cleaved at the δ/ϕ junction in ISVPs generated in the presence of alkyl sulfate detergents (dpSVPs), yet dpSVPs are infectious.[58]

Transformation from the ISVP to the ISVP* in vitro is triggered by differential cationic concentration or interactions with membranes.[85,87] In contrast to ISVPs, ISVP*s lack σ1 and have an altered conformer of μ1 in which internal hydrophobic residues are exposed. ISVP*s are capable of membrane penetration and transcription initiation.[85,87] The conformational change in μ1 may be the driving force for both the loss of σ1 and the initiation of transcription.[88] Mechanisms underlying these events are unknown, but it is possible that μ1 rearrangement induces a conformational change in λ2, the pentameric turret that anchors σ1, causing σ1 release.

Cleavage of intact μ1 to form μ1N and μ1C is required for the generation of the ISVP*.[80,81] Particles recoated with mutant forms of μ1 incapable of μ1N/μ1C cleavage can facilitate each of the entry steps, including μ1 conformational changes and transcription initiation, but are deficient in membrane penetration.[80] In addition to σ1, the N-terminal μ1 fragment μ1N is released from the ISVP* and forms membrane pores that recruit virus particles.[85,86] The ϕ fragment may act as a μ1N chaperone. However, the full nature of the membrane-penetration complex is not completely understood.

The conformational changes in μ1 that accompany viral disassembly are thought to expose internal hydrophobic residues and release μ1N from the particle as a consequence of massive rearrangement in the μ1 structure.[80,85,87,88] The cleavage of μ1 to form μ1N and μ1C is necessary for productive infection.[80] However, it is not clear whether membrane penetration is accomplished by soluble or particle-associated μ1N, perhaps acting in concert with other regions of the molecule. For example, an anion-binding site in domain IV lies on the outermost, solvent-exposed surface of the ISVP[88] (Fig. 7). This site may bind to phospholipid head groups bringing the virus particle into proximity with the endosomal membrane. This association also might trigger rearrangements in μ1 revealing the myristoylated μ1N and the internal hydrophobic residues.

VIRAL DETERMINANTS OF APOPTOSIS INDUCTION BY REOVIRUS

Reovirus elicits the morphological and biochemical features of apoptosis in both cultured cells[89-91] and in the murine CNS[92,93] and heart.[93-95] Insights into mechanisms by which reovirus induces apoptosis first emerged from studies of strain-specific differences in the efficiency of apoptosis induction. Reovirus strain T3D induces apoptosis to a much greater extent than T1L in a variety of cell types.[89,90,96] Experiments using T1L x T3D reassortant viruses implicated the S1 and M2 genes as the key determinants of differences in the capacity of reovirus strains to induce apoptosis in L cells[89] and Madin-Darby canine kidney (MDCK) cells.[90] Linear regression analysis of these data pointed to a primary role for the S1 gene in apoptosis induction with a minor contribution from the M2 gene. These findings were corroborated by studies analyzing genetic determinants of differences in apoptosis efficiency displayed by T1L and another serotype 3 strain, type 3 Abney (T3A).[97] Since no other viral gene segments are significantly associated with differences in apoptosis induction by serotype 1 and serotype 3 reovirus strains, these studies collectively pinpoint important functions for the S1 and M2 genes in the induction of this cellular response.

The S1 gene encodes viral attachment protein σ1 and nonstructural protein σ1s from distinct but overlapping reading frames.[98-100] The role of both proteins in apoptosis induction has been investigated. Reovirus strains T3C84 and T3C84-MA, which differ in the expression of σ1s,[101] do not differ in apoptosis-inducing capacity following infection of L cells or MDCK cells,[101] indicating that the σ1s protein is not required for apoptosis induction following infection of cultured cells. In contrast to those in vitro results, σ1s appears to influence the kinetics and severity of apoptosis induction in both the heart and CNS of infected mice.[102] However, since the viral strains compared in that study were not isogenic, it is not possible to exclude the involvement of other viral determinants in the observed differences in viral pathology. Studies using σ1s-null viruses recovered by reverse genetics identified an essential role for σ1s in establishing viremia and promoting viral dissemination to sites of secondary replication.[103,104] However, the precise role of σ1s in apoptosis induction in vivo remains unknown.

The σ1 protein binds to two cell-surface receptors, the proteinaceous receptor JAM-A[20] and a carbohydrate receptor. Strains of all three reovirus serotypes bind to JAM-A,[20,29,31,105] but only serotype 3 strains bind to sialic acid.[16,18,106] Sialic-acid-binding strain T3SA+ induces significantly higher levels of apoptosis than isogenic nonsialic-acid-binding strain T3SA-.[96] Consistent with these findings, removal of cell-surface sialic acid with neuraminidase, or blockade of virus binding to cell-surface sialic acid using a soluble competitor, sialyllactose, abolishes the capacity of T3SA+ to induce apoptosis.[96] These data indicate that the capacity to bind to sialic acid enhances the efficiency of apoptosis induced by reovirus infection. However, sialic acid binding is not sufficient to induce apoptosis. Blockade of σ1 binding to JAM-A by using either σ1- or JAM-A-specific mAbs also diminishes the apoptosis-inducing capacity of sialic-acid-binding reoviruses.[20,89] Thus, attachment of σ1 to both sialic acid and JAM-A is required for efficient induction of apoptosis.

Despite the role of the reovirus receptors in influencing apoptosis efficiency, receptor binding alone is not sufficient for apoptosis induction. Blockade of viral disassembly using either AC or E64 diminishes reovirus-induced apoptosis[107] (Fig. 8A). On the other hand, inhibition of de novo viral RNA synthesis using ribavirin does not affect apoptosis induced by reovirus[107] (Fig. 8B). Thus, in addition to sialic acid- and JAM-A-mediated attachment of reovirus to cells, replication steps that occur during or after viral disassembly but before the cytoplasmically delivered core becomes transcriptionally active also contribute to reovirus-induced apoptosis. Since the M2-encoded μ1 protein functions in virus-induced endosomal membrane penetration following disassembly but prior to synthesis of viral RNA,[80,83,88] the deleterious effects of reovirus disassembly inhibitors on apoptosis induction suggest a functional link between the M2 gene and differences in the efficiency of apoptosis exhibited by different reovirus strains in previous genetic studies.[89,90,97]

Further evidence for a role of the M2-encoded μ1 protein in apoptosis induction was gathered in a study in which productive reovirus infection was initiated in JAM-A-negative, Fc receptor-expressing Chinese hamster ovary cells (CHO-B1) using reovirus preincubated with capsid-specific mAbs.[108] Fc-mediated infection of CHO-B1 cells was found to induce apoptosis in an antibody dose-dependent manner[108] (Fig. 9). Furthermore, antibody-directed binding of reovirus to Fc receptors expressed on CHO-B1 cells was not sufficient for reovirus-induced apoptosis. Analogous to apoptosis initiated following uptake via cognate receptors, apoptosis induced following the Fc-receptor dependent pathway is sensitive to inhibitors of viral disassembly but is not perturbed

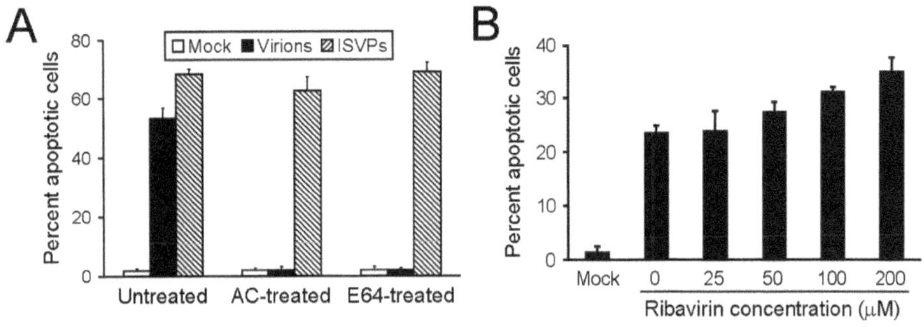

Figure 8. Reovirus-induced apoptosis in cells treated with inhibitors of viral replication. A) HeLa cells were either mock infected or infected with T3SA+ virions or ISVPs at an MOI of 100 PFU/cell and incubated in the absence or presence of 10 mM AC or 200 μM E64. B) HeLa cells were either mock infected or infected with T3SA+ at an MOI of 1,000 PFU/cell prior to incubation in the absence or presence of ribavirin at the concentrations shown. Mock-infected cells were incubated in the presence of 200 μM ribavirin. After incubation for 48 h (A and B), cells were stained with acridine orange. The results are expressed as the mean percentage of cells undergoing apoptosis for three independent experiments. Error bars indicate SD. Figure modified from: Connolly JL, Dermody TS, J Virol 76:1632-1641; ©2002 with permission from the American Society for Microbiology.[107]

by inhibitors of viral replication.[108] These data suggest that regardless of the receptor used to initiate infection, reovirus-induced apoptosis requires events that occur during or after viral disassembly but prior to viral RNA synthesis. Uptake via Fc receptors also allows nonsialic acid-binding reovirus strains to efficiently induce apoptosis,[108]

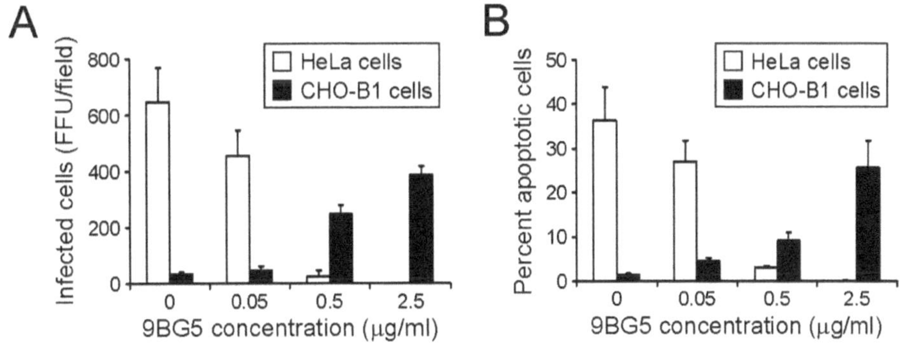

Figure 9. JAM-A-independent uptake of reovirus via Fc receptors leads to apoptosis. A) Reovirus T3D virions were incubated overnight with the indicated concentration of mAb 9BG5 and adsorbed to either HeLa cells or CHO-B1 cells at an MOI of 100 PFU/cell. After incubation at 37°C for 18 h, cells were fixed using methanol. Infected cells were visualized by immunostaining with polyclonal rabbit anti-reovirus serum. Reovirus-infected cells were quantified by counting fluorescent cells. The results are presented as mean fluorescent focus units (FFU)/field. B) HeLa cells or CHO-B1 cells were adsorbed with 100 PFU/cell of either virus or virus-antibody complex, harvested 48 h after infection, and stained with acridine orange. The results are expressed as the mean percentage of cells undergoing apoptosis for three independent experiments. Error bars indicate SD. Figure modified from: Danthi P et al, J Virol 80:1261-1270; ©2006 with permission from the American Society for Microbiology.[108]

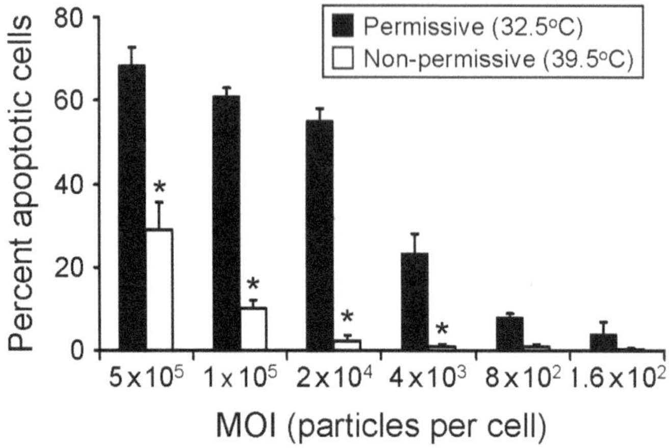

Figure 10. Apoptosis induced by temperature-sensitive μ1 mutant virus tsA279.64. HeLa cells were adsorbed with tsA279.64 grown at permissive or nonpermissive temperatures at the MOIs shown. After incubation at 37°C for 48 h, cells were harvested and stained with acridine orange. The results are expressed as the mean percentage of cells undergoing apoptosis for three independent experiments. Error bars indicate SD. *, $P < 0.05$ by Student's t-test relative to virions grown at permissive temperature at an equivalent MOI. Figure modified from: Danthi P et al, J Virol 80:1261-1270; ©2006 with permission from the American Society for Microbiology.[108]

suggesting that σ1-mediated differences in the efficiency of apoptosis induction are eliminated following Fc-receptor mediated uptake. Collectively, these findings suggest that signaling pathways initiated as a result of ligation of σ1 to sialic acid and JAM-A are dispensable for reovirus-induced apoptosis.

Differences in the capacity of T1L x T3D reassortant viruses to induce apoptosis following Fc-mediated uptake segregate strictly with the μ1-encoding M2 gene,[108] providing further support for a role of the μ1 protein in apoptosis induction. In addition, temperature-sensitive reovirus mutant tsA279.64, which bears a lesion in the M2 gene,[109] is defective in apoptosis induction[108] (Fig. 10). Particles of this virus strain when grown at nonpermissive temperature contain a misfolded, membrane penetration-defective μ1 protein.[109] In comparison to particles assembled at permissive temperature, those assembled at nonpermissive temperature are less efficient inducers of apoptosis. Moreover, reovirus μ1 mutants recovered by reverse genetics induce apoptosis less efficiently than does wild-type virus.[110,111] These findings suggest that reovirus membrane-penetration protein μ1 induces proapoptotic signaling events during or after endosomal membrane penetration.

SIGNALING PATHWAYS ACTIVATED BY REOVIRUS INFECTION

NF-κB

A critical component of the intracellular signal transduction apparatus activated following reovirus infection is nuclear factor-κB (NF-κB), a family of structurally related transcription factors that play important roles in cell growth and survival. Reovirus activates NF-κB in cell culture beginning at 2-4 h post-infection and achieves maximal levels of

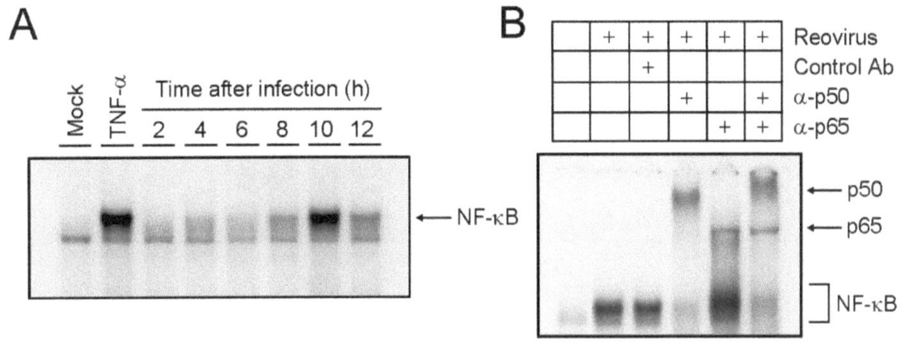

Figure 11. Reovirus activates NF-κB. A) NF-κB DNA-binding activity following reovirus infection. HeLa cells were either mock-infected or infected with T3D at an MOI of 100 PFU/cell and incubated at 37°C for the times shown. Uninfected cells also were treated with 20 ng/ml TNF-α for 1 h. Nuclear extracts were prepared and incubated with a [^{32}P]-labeled oligonucleotide consisting of the NF-κB consensus binding sequence. Incubation mixtures were resolved by acrylamide gel electrophoresis, dried, and exposed to film. NF-κB-containing complexes are indicated. B) NF-κB complexes activated by reovirus contain p50 and p65/RelA subunits. Nuclear extracts were prepared 10 h after viral adsorption. Extracts were incubated with no antibody, a control antibody, p50-specific antiserum (α-p50), p65-specific antiserum (α-p65), or both p50- and p65-specific antisera. NF-κB complexes unaffected by antibody and complexes demonstrating retarded mobility with antibodies to p50 or p65 are indicated. Figure modified from: Connolly JL et al, J Virol 74:2981-2989; ©2000 with permission from the American Society for Microbiology.[91]

activation at 8-10 h post-infection[91] (Fig. 11A). Electrophoretic mobility shift assays using antisera specific for p50 and p65 identified both of these subunits in the NF-κB complexes activated during reovirus infection (Fig. 11B). Concordantly, cells devoid of either p50 or p65 do not activate NF-κB following reovirus infection,[91] supporting the involvement of these NF-κB subunits in the complexes activated by reovirus. A second phase of NF-κB regulation occurs in some cell types following viral RNA synthesis and involves downregulation of NF-κB signaling through a mechanism dependent on the inhibition of IκBα degradation.[112] This sophisticated manipulation of a central cell fate-determining transcription factor emphasizes the importance of NF-κB activation status in the reovirus replication cycle.

Mechanisms leading to NF-κB activation following reovirus infection are not completely understood. The efficiency with which reovirus activates NF-κB is influenced by viral attachment to both sialic acid[96] and JAM-A.[39] Viral disassembly also is required for NF-κB activation, but subsequent events in viral replication are dispensable.[107] This finding suggests that replication steps following formation of ISVPs but before commencement of RNA synthesis are responsible for activating NF-κB. Intriguingly, NF-κB activation following reovirus infection occurs over a much longer time course than that elicited by other NF-κB agonists such as TNF-α,[113] suggesting that the viral agonist is constitutively active, similar to Epstein-Barr virus latent membrane protein 1[114] or human T-cell leukemia virus Tax.[115]

NF-κB activation can either potentiate[116-118] or inhibit apoptosis[119-121] depending on the nature of the NF-κB agonist. Three independent lines of evidence suggest that activation of NF-κB by reovirus is proapoptotic in cultured cells.[91] First, treatment of HeLa cells with proteasome inhibitors to block NF-κB activation following reovirus infection

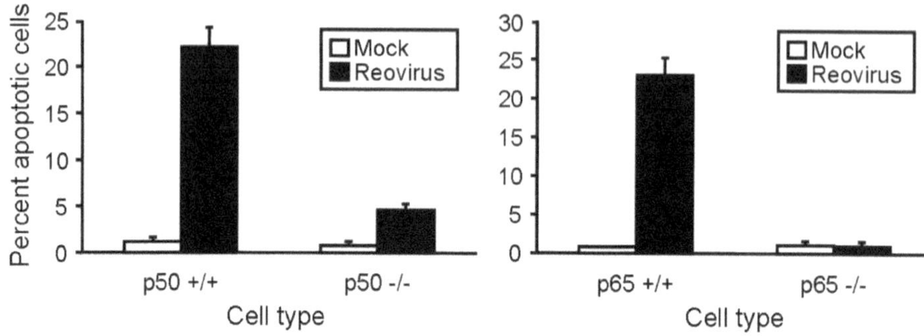

Figure 12. Apoptosis induced by reovirus is diminished in cells lacking NF-κB subunits p50 and p65. Cells were either mock-infected or infected with T3D at an MOI of 100 PFU/cell. After incubation at 37°C for 48 h, cells were stained with acridine orange. The results are expressed as the mean percentage of cells undergoing apoptosis for three independent experiments. Error bars indicate SD. Figure modified from: Connolly JL et al, J Virol 74:2981-2989; ©2000 with permission from the American Society for Microbiology.[91]

substantially diminishes reovirus-induced apoptosis. Second, transient expression of a dominant-negative form of IκB that constitutively represses NF-κB activation significantly reduces levels of apoptosis induced by reovirus. Third, apoptosis by reovirus is substantially diminished in mouse embryonic fibroblasts (MEFs) lacking either of the NF-κB subunits p50 or p65 (Fig. 12). Together, these data indicate that NF-κB plays an essential role in the mechanism by which reovirus induces apoptosis of host cells.

MAPKs

Mitogen-activated protein kinases (MAPKs) are important signal transducers that respond to a wide variety of stimuli. Several MAPKs transduce signals initiated by reovirus infection. Reovirus activates c-Jun NH_2-terminal kinase (JNK) by 10-12 h post-infection, and this activation is sustained for at least 20-30 h.[122] Strain-specific differences in the capacity of reovirus to activate JNK and its downstream effector c-Jun correlate with the capacity of those strains to induce apoptosis, suggesting that JNK activation is required for apoptotic signaling. Additionally, cells lacking MEK kinase 1 (MEKK1), an upstream activator of JNK, or engineered to express a kinase-inactive form of MEKK1 do not phosphorylate JNK or undergo apoptosis in response to reovirus infection.[123] Pharmacologic inhibitors of JNK inhibit reovirus-induced apoptosis[123] but do not block reovirus growth,[123,124] indicating that JNK activity is not required for reovirus replication. Interestingly, although JNK phosphorylates and activates c-Jun in response to reovirus infection, c-Jun activation is not required for apoptosis. These data indicate that JNK contributes to apoptosis induction via a mechanism independent of its activation of c-Jun,[123] possibly through its effect on mitochondrial signaling pathways.

The p38 MAPK is activated by reovirus between 4-8 h post-infection; [125] however, p38 becomes downregulated at late times (24-48 h) post-infection.[122] Pharmacologic inhibitors of p38 MAPK reduce reovirus growth in cells that express an activated Ras pathway, indicating that this pathway is important for replication of the virus.[124] Inhibition of p38 also blocks reovirus-induced secretion of the proinflammatory cytokines IL-1β and TNF-α.[125] However, pharmacologic inhibitors of p38 have no effect on reovirus-induced apoptosis,[122] indicating that this pathway is distinct from NF-κB-mediated death signaling during reovirus infection.[91] Finally, although reovirus infection activates the MAPK extracellular signal-regulated kinase (ERK) at early (10-30 min) and late (24 h) times post-infection, pharmacological inhibitors of ERK do not inhibit reovirus-induced apoptosis.[122] The role of ERK activation in reovirus replication is unknown.

Ras

Reovirus replication is potentiated in transformed cells containing an activated Ras pathway.[126,127] Expression of constitutively active signal transducers in the Ras pathway, including the vErbB oncogene and v-Ras-h, render nonpermissive cells permissive for reovirus replication.[128,129] This activity is mediated by the small G protein Ral and its guanine nucleotide exchange factors (RalGEFs) and p38 MAPK.[124] Activated Ras also can inhibit the antiviral protein kinase R (PKR),[130] which blocks protein synthesis and inhibits reovirus growth.[127] These observations suggest that the increased yield of reovirus following infection of transformed cells is due in part to suppression of PKR. The capacity of reovirus to preferentially infect and kill transformed cells has led to several studies evaluating its potential as an oncolytic agent.[129,131-133]

IRF-3

Innate immune responses are stimulated following reovirus infection, leading to the activation of transcription factor IRF-3 and the subsequent induction of a type I interferon (IFN) response. Reovirus activates IRF-3 in HeLa cells by 2-4 h post-infection and stimulates maximal transcription factor activity by 8-12 h.[134] IRF-3 activation depends on the cellular dsRNA sensors RIG-I, Mda-5, and IPS-1.[134-136] Importantly, decreased expression of RIG-I or IPS-1 by RNA interference does not inhibit NF-κB activation,[134] suggesting that reovirus engages the NF-κB pathway via an alternative mechanism. IRF-3 is required for efficient induction of apoptosis in reovirus-infected cells and elaboration of IFN-β in cultured cells and in vivo.[134] However, signals emanating from the type I interferon receptor are dispensable for apoptosis induction, indicating that cell death is mediated through an IRF-3-dependent, interferon-independent mechanism.[137]

CELLULAR GENE EXPRESSION PROFILES FOLLOWING REOVIRUS INFECTION

The role of transcription factors such as NF-κB and c-Jun/AP1 in the inductive proapoptotic signaling pathway elicited by reovirus led several groups to investigate cellular genes regulated by reovirus using oligonucleotide microarrays.[138-140] Strikingly, these studies identified few classical proapoptotic Bcl-2 family members (such as Bid, Bax, or Bak) or components of death receptor-mediated pathways (such as Fas, Fas ligand, or FADD) upregulated in response to reovirus infection. Additionally, few proapoptotic genes regulated by reovirus were noted to be NF-κB-dependent, indicating that the cell-death response downstream of NF-κB may represent secondary or tertiary events initiated by NF-κB activation at early times post-infection.[139] Despite the lack of apoptosis effectors identified in the microarray studies, three major functional categories of genes that may influence apoptotic processes stand out as significantly regulated in response to reovirus infection: the DNA damage response, the endoplasmic reticulum (ER) stress response, and the host innate antiviral immune response. Analysis of these gene networks may provide clues about the mechanisms by which inductive signaling events elicited by reovirus lead to a widespread apoptotic response in host cells.

DNA damage response genes were found to be significantly downregulated by reovirus infection in two microarray studies.[138,139] Reovirus-regulated DNA damage response genes include damage-specific DNA binding protein 2 (DDB2), excision repair cross-complementing group 4 (ERCC4), and fusion involved in t(12; 16) in malignant liposarcoma (FUS). The physiological role of this response in viral replication or cell death is unknown. However, interactions between the NF-κB pathway and the proapoptotic tumor suppressor protein p53 pathway[141,142] may link these responses in reovirus-infected cells. In support of this hypothesis, the NF-κB-dependent genes PLK3[143] and IER3 (IEX-1[128]), both of which are induced by reovirus infection as rapidly as 2 h post-infection,[139] can regulate p53-mediated apoptosis.

Genes involved in the ER stress response are also regulated following reovirus infection. These include growth and DNA-damage-inducible 45 α and β (GADD45α and GADD45β) and heat shock protein 70 (Hsp70), which were identified in all three microarray studies.[138-140] Interestingly, expression of those genes was significantly greater following infection with reovirus strains that potently inhibit host-cell protein synthesis (T3C8 and T3C87) in comparison to a strain that does not (T3D), suggesting that the ER stress response may be potentiated by mechanisms of translation inhibition.[140] One functional effect of ER stress is the activation of PERK,[145] a kinase that phosphorylates translation initiation factor eIF2α to inhibit translation.[146] Despite inhibition of cellular protein synthesis, reovirus replication is enhanced in the presence of PERK, suggesting that reovirus may benefit from the ER stress response, perhaps by preferentially allowing synthesis of viral proteins.[140]

A large number of genes regulated by reovirus infection identified in previous microarray studies belong to the cellular innate immune response to viral infection, particularly IFN-stimulated genes (ISGs) upregulated downstream of NF-κB-dependent production of type I IFN. Secreted type I IFNs activate the Janus-activated kinase (JAK)-signal transducer and activator of transcription (STAT) signal-transduction pathway leading to transcription of ISGs via a STAT1/STAT2/IRF9 complex, known as ISGF3.[147,148] ISGs induce an antiviral state in uninfected cells through a variety of mechanisms including inhibition of protein synthesis.[149] Type I IFNs also greatly sensitize tissue-culture cells to apoptosis in response to dsRNA and influenza virus.[150] Reovirus infection upregulates STAT1, STAT2, and many of the classical ISGs, including 2'5' oligoadenylate synthase (OAS1), double-stranded RNA-activated protein kinase (PKR), ISG15, and myxoma resistance 1 and 2 (Mx1/Mx2) (Table 1). The type I IFN response is significantly enhanced following infection with two strains that are potent apoptosis inducers (T3D and T3C87) in comparison to a strain that is not (T3C8).[140] However, apoptosis following reovirus infection does not require type I IFN signaling,[137] suggesting that other mechanisms dependent on IRF-3 and NF-κB are responsible.

Several proteins involved in the intrinsic, mitochondrial-dependent apoptosis pathway are also regulated by reovirus infection,[138,139] including MCL1,[151] PAWR (Par-4[152]), BNIP3L,[153] and MOAP1.[154] These genes are NF-κB-independent, suggesting that they may function in concert with, or in parallel to, NF-κB-dependent genes to modulate the apoptotic response in reovirus-infected cells. Proteins in the extrinsic, death receptor-mediated apoptosis pathway regulated by reovirus infection as determined by either microarray or RT-PCR include TNF-related apoptosis-inducing ligand (TRAIL), death receptors DR4, DR5, and DR6, and the Fas death domain-associated protein DAXX (refs. 140, 155 and O'Donnell, Holm, and Dermody, unpublished). While TRAIL and DR5 have been shown to be involved in reovirus-induced apoptosis,[155] upregulation of both of these genes was not observed in the same microarray study, indicating that either these genes are under differential temporal regulation or endogenous mRNA or protein levels are sufficient to bring about an apoptotic response. DAXX is a particularly intriguing candidate, as it is a multifunctional protein that associates with the Fas death domain to potentiate Fas-induced apoptosis.[156] DAXX also joins with two other IFN-induced proteins upregulated by reovirus, PML and Sp100, to direct stress responses in the nucleus.[157] Together, these discovery-based approaches have provided a number of potential proapoptotic candidates that warrant further examination.

EFFECTORS OF REOVIRUS-INDUCED APOPTOSIS

Apoptosis induced by reovirus requires both extrinsic death receptor pathways and mitochondrial damage. Activation of the extrinsic pathway following reovirus infection is mediated by TRAIL and its receptors DR4 and DR5.[155] TRAIL is released from cells following reovirus infection with maximal levels detected at 48 h post-infection. DR5 is upregulated in response to reovirus beginning at 4 h and increases to maximal levels at 24 h post-infection.[155] Treatment of cells with soluble TRAIL receptors or TRAIL-specific antibodies decreases apoptosis following reovirus infection of HEK293 cells[155] and neuroblastoma cells.[158] Death receptors engaged by TRAIL mediate apoptotic signaling through the initiator caspase, caspase 8,[159] which is activated following reovirus infection.[160] Expression of a dominant-negative FADD mutant, which blocks the capacity of caspase 8 to engage death receptors,[161] decreases reovirus-induced activation of caspase 3,[155] which serves as the executioner caspase.[159] Mechanisms by which reovirus upregulates the release of TRAIL and DR5 expression are not fully understood. Maximal levels of TRAIL are detected following NF-κB activation, suggesting that TRAIL release is mediated by NF-κB.[91,155] However, TRAIL sensitivity is potentiated by blockade of NF-κB activation observed later in infection,[112] which might be mediated by downregulation of the antiapoptotic protein, cellular FLICE (caspase 8) inhibitory protein (cFLIP).[162] TRAIL and death receptor expression are regulated by type I IFN,[163,164] providing further support for the hypothesis that extrinsic apoptotic pathways initiated by reovirus infection are activated downstream of type I IFN action.

Reovirus infection also activates the intrinsic apoptosis pathway, which is initiated by mitochondrial injury and release of cytochrome c and Smac/DIABLO.[165-168] Release of these mediators leads to activation of caspase 9, which in turn activates caspase 3.[166,167,169] Smac/DIABLO can be detected in the cytoplasm of HEK293 cells approximately 4 h following reovirus infection,[170] which is subsequent to the activation of NF-κB.[91] Identification of Noxa as an NF-κB- and IRF-3-dependent protein that is upregulated following reovirus infection and required for efficient apoptosis induction provides a mechanistic link between the initial signaling events and the intrinsic, mitochondrial-dependent apoptotic pathways.[137] These findings suggest a direct connection between initial signaling events

Table 1. Interferon-inducible genes regulated by reovirus infection in 2 out of 3 microarray analyses[a]

		Reference		
Gene Symbol- Human (Mouse)[b]	Gene Name/Alternate Designation[c]	DeBiasi et al[138]	O'Donnell et al[139]	Smith et al[140]
CXCL10	Interferon-inducible cytokine IP-10, IP-10	x	x	x
CXCL11	Interferon-induced T-cell chemoattractant, I-TAC	x	x	
GBP1	Guanylate binding protein 1, interferon-inducible, 67kDa	x		x
HSPA1A	Heat shock protein 70kD 1A		x	x
HSPA1B	Heat shock protein 70kD 1B		x	x
IFI16 (Ifi204)	Interferon, gamma-inducible protein 16	x	x	x
IFI35	Interferon-induced protein 35	x	x	x
IFI44	Microtubule associated protein, 44 kDa, MTAp44	x	x	
IFIT1	IFI-56k	x	x	x
IFIT2	IFI-54k	x	x	x
IFIT3	Interferon-induced protein with tetratricopeptide repeats 3	x	x	x
IFIT5	Interferon-induced protein with tetratricopeptide repeats 5	x	x	
IFITM1	9-27	x	x	
IFITM3	1-8U	x	x	
IFNB1	Interferon-β	x		x
IRF1	Interferon regulatory factor 1		x	x

continued on next page

Table 1. Continued

		Reference		
Gene Symbol- Human (Mouse)[b]	Gene Name/Alternate Designation[c]	DeBiasi et al[138]	O'Donnell et al[139]	Smith et al[140]
IRF7	Interferon regulatory factor 7	x	x	x
ISG15	ISG15 ubiquitin-like modifier, G1P2	x	x	x
ISG20	Interferon stimulated exonuclease gene, 20kDa		x	x
ISGF3G	Interferon regulatory factor 9, IRF9	x	x	x
MX2	Myxovirus resistance 2	x	x	x
OAS1	2',5'-oligoadenylate synthetase 1, 40/46kDa	x	x	
OAS2	2',5'-oligoadenylate synthetase-like, 59kDa	x	x	
PLAUR	Plasminogen activator, urokinase receptor, CD87		x	x
PSMB9	Proteasome subunit, beta type 9		x	x
RSAD2	Radical S-adenosyl methionine domain containing 2	x		x
SP100	SP100 nuclear antigen		x	x
SP110 (5830484A20Rik)	IFI41, SP110 nuclear body protein	x	x	
TAP1	Transporter 1, ATP-binding cassette, sub-family B	x	x	x

a) Genes were identified as reovirus-regulated and interferon-inducible by the criteria stated in the respective reference. b) Gene symbol is given for human genes. If the symbol differs between human and mouse, the mouse symbol is given in parentheses. c) If different from gene symbol, the full name or common designation is given.

and mitochondrial damage. Another link could be Bid, a proapoptotic Bcl-2 family member that, following an activating cleavage event, translocates to mitochondria and mediates cytochrome c release and activation of caspase 9.[169,172] Bid cleavage can be detected at approximately 10 h post infection of HEK293 cells or 24 h post infection in murine fibroblasts.[160,170,171] Bid is required for apoptosis induction following reovirus infection and potentiates reovirus disease in the central nervous system of newborn mice.[171] Together, these data suggest that mitochondrial factors and the overall stability of mitochondria play important contributing roles in the induction of apoptosis by reovirus.

REOVIRUS-INDUCED APOPTOSIS IN VIVO

Determinants of reovirus apoptosis in vivo show some similarity to those observed in cell-culture models, as well as some surprising differences. Markers of apoptosis, such as caspase 3 activation and cleavage of cellular DNA, are detected in both the brain and heart of newborn mice infected with reovirus.[92-94] Reovirus induces apoptosis in neurons following either intracranial[92] or peroral[93] inoculation. Differences in the capacity of reovirus strains to induce encephalitis are associated with the capacity of these strains to induce neuronal apoptosis.[173] Moreover, as judged from studies using primary neuronal cultures, apoptosis induction may enhance viral replication in those tissues.[173] Accordingly, inhibition of neuronal apoptosis with minocycline delays reovirus encephalitis and reduces virus growth following intracranial inoculation.[174] Similar to its role in cell-culture models, NF-κB is required for apoptosis of neurons following reovirus infection in vivo, as mice deficient in the NF-κB p50 subunit are protected from neuronal injury.[93]

In keeping with the pathogenesis of reovirus encephalitis, myocarditis following reovirus infection also is mediated by apoptosis, which can be attenuated by pharmacologic inhibitors of either caspases[95] or calpain.[94] The innate antiviral immune response plays a key role in determining the extent of myocardial injury following reovirus infection. Nonmyocarditic reovirus strains induce higher levels of type I IFN and are more sensitive to its effects than myocarditic strains.[175] Additionally, the ISGs IRF-1 and PKR play protective roles in reovirus myocarditis.[176,177] Type I IFN production in the heart or in cardiac myocyte cultures is downstream of both IRF-3[178] and NF-κB.[93] Intriguingly, the lack of NF-κB p50 markedly enhances reovirus myocarditis, but this disease manifestation can be attenuated by treatment with IFN-β.[93] These findings demonstrate a tissue-specific role for NF-κB following reovirus infection in vivo: NF-κB is required for reovirus-induced apoptosis in the CNS, whereas it protects the heart from viral damage via activation of type I IFN.

CONCLUSION AND FUTURE DIRECTIONS

Despite the accumulated knowledge about reovirus attachment to cell-surface receptors and internalization into host cells, a precise understanding of the role of the viral attachment and internalization receptors in reovirus disease is not available. Serotype 1 and serotype 3 reovirus strains vary in the types of cell-surface carbohydrate used as coreceptors,[19] but both serotypes bind to JAM-A.[20,105] These observations make it unlikely that JAM-A is the sole determinant of reovirus tropism. It is possible that JAM-A serves as a serotype-independent reovirus receptor at some sites within the host and other as yet unidentified receptors, perhaps carbohydrate in nature, confer serotype-dependent differences in growth in other tissues. It is also possible that reovirus serotypes engage JAM-A with different affinities, which influences tissue tropism in infected animals.

Little is known about the signaling events that are initiated by the binding of reovirus to its cell-surface receptors. JAM-A contains a cytoplasmic domain that is approximately 45 amino acids in length, includes 13 potential phosphorylation sites, and interacts with several PDZ domain-containing proteins, suggesting a role in ligand-induced cell signaling.[179,180] Although signaling through JAM-A is not required for reovirus infection[40,108] or apoptosis,[108] the role of JAM-A signaling, if any, in the reovirus infectious cycle remains unknown. The cytoplasmic domains of the β1 integrin heterodimers that function as reovirus internalization receptors also are involved in a number of signaling pathways.[43] In particular, the β1 integrin cytoplasmic tail is linked to cytoskeletal proteins such as talin[181] and α-actinin,[182] in addition to signaling molecules like paxillin and focal adhesion kinase.[183] Furthermore, the β1 integrin cytoplasmic domain contains two NPXY motifs,[184] which serve as recognition sites for the cellular endocytic machinery[185] and are required for transport of reovirus to late endosomes for viral disassembly.[44,51] The mechanisms by which signaling pathways elicited by β1 integrins promote reovirus infection have not been resolved.

As with most nonenveloped viruses, it is unclear how reovirus overcomes the physiological barrier posed by cell membranes during viral entry. Although structural features of the μ1 protein that contribute to membrane penetration have been identified, the precise role of specific μ1 domains in delivery of the viral core into the cytoplasm is not known. The μ1 protein is a key regulator of reovirus-induced apoptosis,[108,110,111] but it is not clear how the viral disassembly events culminating in μ1-mediated membrane penetration elicit proapoptotic signaling. It is possible that endosomal disruption by μ1 leads to the release of hydrolytic enzymes such as cathepsins, which in turn damage mitochondria and stimulate death signaling.[186-188] Interestingly, mitochondrial injury has been reported as early as 4 h following reovirus adsorption, suggesting the involvement of an early viral replication event.[160,170] It is also possible that release of these enzymes causes apoptosis via their action on death regulators such as Bid.[189] Alternatively, fragments of μ1 produced during proteolytic viral disassembly are known to gain access to the cytoplasm,[85] and peptides derived from the μ1 ϕ domain can destabilize membranes and induce cell death.[190,191,192] These fragments may activate other cellular sensors of viral infection or directly injure mitochondria to activate proapoptotic signaling pathways.

Events in reovirus replication following viral disassembly in endosomes but prior to viral RNA synthesis result in the activation of NF-κB, which is required for apoptosis following reovirus infection.[91] The signaling pathways that connect the μ1-mediated events during viral disassembly to the activation of NF-κB are not known. Furthermore, although NF-κB activation is required for apoptosis induction by reovirus, the activation of this transcription factor alone is not sufficient.[162] These results suggest a role for other cellular signal transducers in the initiation of the apoptotic response. Since MEKK1 and its downstream target JNK also are required for reovirus-induced apoptosis,[122,123] it is possible that the MAP kinase cascade acts together with NF-κB to trigger the apoptotic response. However, neither the mechanism of activation of the MAP kinase cascade nor the relationship between MAP kinases and NF-κB during reovirus infection is understood.

Insights into how the activation of intracellular signaling pathways results in the execution of the cell-death response are also being elucidated. Both classical extrinsic and intrinsic apoptotic pathways are activated following reovirus infection.[160] In some cell types, these pathways appear to be activated by the release of TRAIL from infected cells and the upregulation of TRAIL receptors DR4 and DR5.[155] However, mechanisms by which reovirus infection results in secretion of TRAIL or upregulation of DR4 and DR5 remain unknown. Microarray studies comparing gene expression profiles of cells infected with reovirus strains that differ significantly in apoptotic potential[138] or cells blocked in apoptosis due to functional absence of NF-κB[139] do not demonstrate upregulation of any classical apoptosis effector pathway components. Rather, they show an upregulation of several ISGs in an NF-κB dependent manner,[138,139] suggesting a requirement for the expression of type I IFNs in the apoptotic response to reovirus. However, type I IFNs are not required for the induction of apoptosis following infection by reovirus,[137] in at least in some cell types. Nonetheless, these cytokines are clearly crucial for cell fate decisions as part of the innate immune response[150] and may contribute to reovirus-induced cell death in some tissues.

Studies describing the pathogenesis of reovirus in mice lacking the p50 subunit of NF-κB demonstrate that p50 serves two distinctly different functions in reovirus pathogenesis, inducing apoptosis in the brain, while mediating survival in the heart.[93] At least two nonmutually exclusive possibilities may account for these different outcomes following reovirus-induced NF-κB activation. It is possible that NF-κB activation by reovirus leads to expression of the same constellation of genes regardless of cell type, and proapoptotic or prosurvival signaling pathways are dictated by the cellular response to these expression patterns. Alternatively, reovirus-induced NF-κB activation may activate different signaling pathways depending on the cell type and tissue microenvironment. In the CNS, NF-κB signaling may upregulate expression of apoptosis-inducing genes, whereas in the heart, NF-κB signaling may lead to expression of prosurvival genes. Further studies are required to precisely define the basis for the differences in reovirus virulence in different tissues. Such studies will reveal new mechanisms by which viral attachment and disassembly regulate prodeath signaling responses and extend an understanding of how viruses cause tissue-specific injury.

ACKNOWLEDGMENTS

We thank Karl Boehme and Kristen Ogden for review of the manuscript and members of our laboratories for many helpful discussions. We acknowledge support from Public Health Service awards T32 AI49824 (G.H.H.), F32 AI071440 (G.H.H.), R15 AI94440 (G.H.H.), R01 AI32539 (T.S.D.), R37 AI38296 (T.S.D.), R01 AI50080 (T.S.D.), and R01 AI76983 (T.S. and T.S.D.) and the Elizabeth B. Lamb Center for Pediatric Research.

REFERENCES

1. Dermody TS, Parker JS, Sherry B. Orthoreoviruses. In: Knipe DM, Howley PM, eds. Fields Virology. Sixth Edition. Philadelphia: Lippincott Williams & Wilkins, In press.
2. Virgin HW, Tyler KL, Dermody TS. Reovirus. In: Nathanson N, ed. Viral Pathogenesis. New York: Lippincott-Raven, 1997:669-699.
3. Duncan R, Horne D, Cashdollar LW et al. Identification of conserved domains in the cell attachment proteins of the three serotypes of reovirus. Virology 1990; 174:399-409.
4. Nibert ML, Dermody TS, Fields BN. Structure of the reovirus cell-attachment protein: A model for the domain organization of σ1. J Virol 1990; 64:2976-2989.
5. Tyler KL, McPhee DA, Fields BN. Distinct pathways of viral spread in the host determined by reovirus S1 gene segment. Science 1986; 233:770-774.

6. Weiner HL, Powers ML, Fields BN. Absolute linkage of virulence and central nervous system tropism of reoviruses to viral hemagglutinin. J Infect Dis 1980; 141:609-616.
7. Weiner HL, Drayna D, Averill Jr DR et al. Molecular basis of reovirus virulence: Role of the S1 gene. Proc Natl Acad Sci USA 1977; 74:5744-5748.
8. Morrison LA, Sidman RL, Fields BN. Direct spread of reovirus from the intestinal lumen to the central nervous system through vagal autonomic nerve fibers. Proc Natl Acad Sci USA 1991; 88:3852-3856.
9. Tardieu M, Powers ML, Weiner HL. Age-dependent susceptibility to reovirus type 3 encephalitis: Role of viral and host factors. Ann Neurol 1983; 13:602-607.
10. Dichter MA, Weiner HL. Infection of neuronal cell cultures with reovirus mimics in vitro patterns of neurotropism. Ann Neurol 1984; 16:603-610.
11. Weiner HL, Ault KA, Fields BN. Interaction of reovirus with cell surface receptors. I. Murine and human lymphocytes have a receptor for the hemagglutinin of reovirus type 3. J Immunol 1980; 124:2143-2148.
12. Lee PW, Hayes EC, Joklik WK. Protein σ1 is the reovirus cell attachment protein. Virology 1981; 108:156-163.
13. Furlong DB, Nibert ML, Fields BN. Sigma 1 protein of mammalian reoviruses extends from the surfaces of viral particles. J Virol 1988; 62:246-256.
14. Fraser RDB, Furlong DB, Trus BL et al. Molecular structure of the cell-attachment protein of reovirus: Correlation of computer-processed electron micrographs with sequence-based predictions. J Virol 1990; 64:2990-3000.
15. Gentsch JR, Pacitti AF. Effect of neuraminidase treatment of cells and effect of soluble glycoproteins on type 3 reovirus attachment to murine L cells. J Virol 1985; 56:356-364.
16. Paul RW, Choi AH, Lee PWK. The α-anomeric form of sialic acid is the minimal receptor determinant recognized by reovirus. Virology 1989; 172:382-385.
17. Dermody TS, Nibert ML, Bassel-Duby R et al. Sequence diversity in S1 genes and S1 translation products of 11 serotype 3 reovirus strains. J Virol 1990; 64:4842-4850.
18. Chappell JD, Gunn VL, Wetzel JD et al. Mutations in type 3 reovirus that determine binding to sialic acid are contained in the fibrous tail domain of viral attachment protein sigma1. J Virol 1997; 71:1834-1841.
19. Chappell JD, Duong JL, Wright BW et al. Identification of carbohydrate-binding domains in the attachment proteins of type 1 and type 3 reoviruses. J Virol 2000; 74:8472-8479.
20. Barton ES, Forrest JC, Connolly JL et al. Junction adhesion molecule is a receptor for reovirus. Cell 2001; 104:441-451.
21. Martin-Padura I, Lostaglio S, Schneemann M et al. Junctional adhesion molecule, a novel member of the immunoglobulin superfamily that distributes at intercellular junctions and modulates monocyte transmigration. J Cell Biol 1998; 142:117-127.
22. Williams LA, Martin-Padura I, Dejana E et al. Identification and characterisation of human junctional adhesion molecule (JAM). Mol Immunol 1999; 36:1175-1188.
23. Liu Y, Nusrat A, Schnell FJ et al. Human junction adhesion molecule regulates tight junction resealing in epithelia. J Cell Sci 2000; 113:2363-2374.
24. Dryden KA, Wang G, Yeager M et al. Early steps in reovirus infection are associated with dramatic changes in supramolecular structure and protein conformation: Analysis of virions and subviral particles by cryoelectron microscopy and image reconstruction. J Cell Biol 1993; 122:1023-1041.
25. Chappell JD, Prota A, Dermody TS et al. Crystal structure of reovirus attachment protein σ1 reveals evolutionary relationship to adenovirus fiber. EMBO J 2002; 21:1-11.
26. Reiter DM, Frierson JM, Halvorson EE et al. Crystal structure of reovirus attachment protein σ1 in complex with sialylated oligosaccharides. PLoS Pathog 2011; 7:e1002166.
27. van Raaij MJ, Mitraki A, Lavigne G et al. A triple β-spiral in the adenovirus fibre shaft reveals a new structural motif for a fibrous protein. Nature 1999; 401:935-938.
28. Guardado CP, Fox GC, Hermo Parrado XL et al. Structure of the carboxy-terminal receptor-binding domain of avian reovirus fibre sigmaC. J Mol Biol 2005; 354:137-149.
29. Prota AE, Campbell JA, Schelling P et al. Crystal structure of human junctional adhesion molecule 1: Implications for reovirus binding. Proc Natl Acad Sci USA 2003; 100:5366-5371.
30. Stehle T, Dermody TS. Structural similarities in the cellular receptors used by adenovirus and reovirus. Viral Immunol 2004; 17:129-143.
31. Forrest JC, Campbell JA, Schelling P et al. Structure-function analysis of reovirus binding to junctional adhesion molecule 1. Implications for the mechanism of reovirus attachment. J Biol Chem 2003; 278:48434-48444.
32. Kirchner E, Guglielmi KM, Strauss H et al. Structure of reovirus σ1 in complex with its receptor junctional adhesion molecule-A. PLoS Pathog 2008; 4:e1000235.
33. Bewley MC, Springer K, Zhang YB et al. Structural analysis of the mechanism of adenovirus binding to its human cellular receptor, CAR. Science 1999; 286:1579-1583.
34. van Raaij MJ, Chouin E, van der Zandt H et al. Dimeric structure of the coxsackievirus and adenovirus receptor D1 domain at 1.7 A resolution. Structure 2000; 8:1147-1155.

35. Spear PG. Viral interactions with receptors in cell junctions and effects on junctional stability. Dev Cell 2002; 3:462-464.
36. Compton T. Receptors and immune sensors: The complex entry path of human cytomegalovirus. Trends Cell Biol 2004; 14:5-8.
37. Ugolini S, Mondor I, Sattentau QJ. HIV-1 attachment: Another look. Trends Microbiol 1999; 7:144-149.
38. Berger EA, Murphy PM, Farber JM. Chemokine receptors as HIV-1 coreceptors: Roles in viral entry, tropism, and disease. Annu Rev Immunol 1999; 17:657-700.
39. Barton ES, Connolly JL, Forrest JC et al. Utilization of sialic acid as a coreceptor enhances reovirus attachment by multistep adhesion strengthening. J Biol Chem 2001; 276:2200-2211.
40. Maginnis MS, Forrest JC, Kopecky-Bromberg SA et al. β1 integrin mediates internalization of mammalian reovirus. J Virol 2006; 80:2760-2770.
41. Breun LA, Broering TJ, McCutcheon AM et al. Mammalian reovirus L2 gene and λ2 core spike protein sequences and whole-genome comparisons of reoviruses type 1 Lang, type 2 Jones, and type 3 Dearing. Virology 2001; 287:333-348.
42. Hynes RO. Integrins: Versatility, modulation, and signaling in cell adhesion. Cell 1992; 69:11-25.
43. Hynes R. Integrins: Bidirectional, allosteric signaling machines. Cell 2002; 110:673-687.
44. Maginnis MS, Mainou BA, Derdowski AM et al. NPXY motifs in the β1 integrin cytoplasmic tail are required for functional reovirus entry. J Virol 2008; 82:3181-3191.
45. Sturzenbecker LJ, Nibert ML, Furlong DB et al. Intracellular digestion of reovirus particles requires a low pH and is an essential step in the viral infectious cycle. J Virol 1987; 61:2351-2361.
46. Borsa J, Morash BD, Sargent MD et al. Two modes of entry of reovirus particles into L cells. J Gen Virol 1979; 45:161-170.
47. Borsa J, Sargent MD, Lievaart PA et al. Reovirus: Evidence for a second step in the intracellular uncoating and transcriptase activation process. Virology 1981; 111:191-200.
48. Rubin DH, Weiner DB, Dworkin C et al. Receptor utilization by reovirus type 3: Distinct binding sites on thymoma and fibroblast cell lines result in differential compartmentalization of virions. Microb Pathog 1992; 12:351-365.
49. Ehrlich M, Boll W, Van Oijen A et al. Endocytosis by random initiation and stabilization of clathrin-coated pits. Cell 2004; 118:591-605.
50. Georgi A, Mottola-Hartshorn C, Warner A et al. Detection of individual fluorescently labeled reovirions in living cells. Proc Natl Acad Sci USA 1990; 87:6579-6583.
51. Mainou BA, Dermody TS. Transport to late endosomes is required for efficient reovirus infection. J Virol 2012; 86:8346-8358.
52. Canning WM, Fields BN. Ammonium chloride prevents lytic growth of reovirus and helps to establish persistent infection in mouse L cells. Science 1983; 219:987-988.
53. Maratos-Flier E, Goodman MJ, Murray AH et al. Ammonium inhibits processing and cytotoxicity of reovirus, a nonenveloped virus. J Clin Invest 1986; 78:617-625.
54. Maxfield FR. Weak bases and ionophores rapidly and reversibly raise the pH in endocytic vesicles in cultured mouse fibroblasts. J Cell Biol 1982; 95:676-681.
55. Ohkuma S, Poole B. Fluorescence probe measurement of the intralysosomal pH in living cells and the perturbation of pH by various agents. Proc Natl Acad Sci USA 1978; 75:3327-3331.
56. Barrett AJ, Kembhavi AA, Brown MA et al. L-trans-Epoxysuccinyl-leucylamido(4-guanidino)butane (E-64) and its analogues as inhibitors of cysteine proteinases including cathepsins B, H and L. Biochem J 1982; 201:189-198.
57. Baer GS, Dermody TS. Mutations in reovirus outer-capsid protein σ3 selected during persistent infections of L cells confer resistance to protease inhibitor E64. J Virol 1997; 71:4921-4928.
58. Chandran K, Nibert ML. Protease cleavage of reovirus capsid protein μ1/μ1C is blocked by alkyl sulfate detergents, yielding a new type of infectious subvirion particle. J Virol 1998; 762:467-475.
59. Ebert DH, Wetzel JD, Brumbaugh DE et al. Adaptation of reovirus to growth in the presence of protease inhibitor E64 segregates with a mutation in the carboxy terminus of viral outer-capsid protein σ3. J Virol 2001; 75:3197-3206.
60. Jané-Valbuena J, Nibert ML, Spencer SM et al. Reovirus virion-like particles obtained by recoating infectious subvirion particles with baculovirus-expressed σ3 protein: An approach for analyzing σ3 functions during virus entry. J Virol 1999; 73:2963-2973.
61. Bond JS, Butler PE. Intracellular proteases. Annu Rev Biochem 1987; 56:333-364.
62. Gal S, Gottesman MM. The major excreted protein (MEP) of transformed mouse cells and cathepsin L have similar protease specificity. Biochem Biophys Res Commun 1986; 139:156-162.
63. Gottesman MM, Sobel ME. Tumor promoters and Kirsten sarcoma virus increase synthesis of a secreted glycoprotein by regulating levels of translatable mRNA. Cell 1980; 19:449-455.
64. Kirschke H, Langner J, Wiederanders B et al. Cathepsin L. A new proteinase from rat-liver lysosomes. Eur J Biochem 1977; 74:293-301.

65. Ebert DH, Deussing J, Peters C et al. Cathepsin L and cathepsin B mediate reovirus disassembly in murine fibroblast cells. J Biol Chem 2002; 277:24609-24617.
66. Riese RJ, Wolf PR, Bromme D et al. Essential role for cathepsin S in MHC class II-associated invariant chain processing and peptide loading. Immunity 1996; 4:357-366.
67. Golden JW, Bahe JA, Lucas WT et al. Cathepsin S supports acid-independent infection by some reoviruses. J Biol Chem 2004; 279:8547-8557.
68. Johnson EM, Wetzel JD, Doyle JD et al. Genetic and pharmacologic alteration of cathepsin expression influences reovirus pathogenesis. J Virol 2009; 83:9630-9640.
69. Nygaard RM, Golden JW, Schiff LA. Impact of host proteases on reovirus infection in the respiratory tract. J Virol 2012; 86:1238-1243.
70. Tosteson MT, Nibert ML, Fields BN. Ion channels induced in lipid bilayers by subvirion particles of the nonenveloped mammalian reoviruses. Proc Natl Acad Sci USA 1993; 90:10549-10552.
71. Lucia-Jandris P, Hooper JW, Fields BN. Reovirus M2 gene is associated with chromium release from mouse L cells. J Virol 1993; 67:5339-5345.
72. Hooper JW, Fields BN. Role of the μ1 protein in reovirus stability and capacity to cause chromium release from host cells. J Virol 1996; 70:459-467.
73. Olland AM, Jané-Valbuena J, Schiff LA et al. Structure of the reovirus outer capsid and dsRNA-binding protein σ3 at 1.8 Å resolution. EMBO J 2001; 20:979-989.
74. Nason E, Wetzel J, Mukherjee S et al. A monoclonal antibody specific for reovirus outer-capsid protein σ3 inhibits σ1-mediated hemagglutination by steric hindrance. J Virol 2001; 75:6625-6634.
75. Wetzel JD, Wilson GJ, Baer GS et al. Reovirus variants selected during persistent infections of L cells contain mutations in the viral S1 and S4 genes and are altered in viral disassembly. J Virol 1997; 71:1362-1369.
76. Clark KM, Wetzel JD, Bayley J et al. Reovirus variants selected for resistance to ammonium chloride have mutations in viral outer-capsid protein σ3. J Virol 2006; 80:671-681.
77. Doyle JD, Danthi P, Kendall EA et al. Molecular determinants of proteolytic disassembly of the reovirus outer capsid. J Biol Chem 2012; 287:8029-8038.
78. Nibert ML, Schiff LA, Fields BN. Mammalian reoviruses contain a myristoylated structural protein. J Virol 1991; 65:1960-1967.
79. Smith RE, Zweerink HJ, Joklik WK. Polypeptide components of virions, top component and cores of reovirus type 3. Virology 1969; 39:791-810.
80. Odegard AL, Chandran K, Zhang X et al. Putative autocleavage of outer capsid protein μ1, allowing release of myristoylated peptide μ1N during particle uncoating, is critical for cell entry by reovirus. J Virol 2004; 78:8732-8745.
81. Nibert ML, Odegard AL, Agosto MA et al. Putative autocleavage of reovirus μ1 protein in concert with outer-capsid disassembly and activation for membrane permeabilization. J Mol Biol 2005; 345:461-474.
82. Bodkin DK, Nibert ML, Fields BN. Proteolytic digestion of reovirus in the intestinal lumens of neonatal mice. J Virol 1989; 63:4676-4681.
83. Nibert ML, Fields BN. A carboxy-terminal fragment of protein μ1/μ1C is present in infectious subvirion particles of mammalian reoviruses and is proposed to have a role in penetration. J Virol 1992; 66:6408-6418.
84. Chandran K, Walker SB, Chen Y et al. In vitro recoating of reovirus cores with baculovirus-expressed outer-capsid proteins μ1 and σ3. J Virol 1999; 73:3941-3950.
85. Chandran K, Parker JS, Ehrlich M et al. The delta region of outer-capsid protein μ1 undergoes conformational change and release from reovirus particles during cell entry. J Virol 2003; 77:13361-13375.
86. Ivanovic T, Agosto MA, Zhang L et al. Peptides released from reovirus outer capsid form membrane pores that recruit virus particles. EMBO J 2008; 27:1289-1298.
87. Chandran K, Farsetta DL, Nibert ML. Strategy for nonenveloped virus entry: A hydrophobic conformer of the reovirus membrane penetration protein μ1 mediates membrane disruption. J Virol 2002; 76:9920-9933.
88. Liemann S, Chandran K, Baker TS et al. Structure of the reovirus membrane-penetration protein, μ1, in a complex with its protector protein, σ3. Cell 2002; 108:283-295.
89. Tyler KL, Squier MK, Rodgers SE et al. Differences in the capacity of reovirus strains to induce apoptosis are determined by the viral attachment protein σ1. J Virol 1995; 69:6972-6979.
90. Rodgers SE, Barton ES, Oberhaus SM et al. Reovirus-induced apoptosis of MDCK cells is not linked to viral yield and is blocked by Bcl-2. J Virol 1997; 71:2540-2546.
91. Connolly JL, Rodgers SE, Clarke P et al. Reovirus-induced apoptosis requires activation of transcription factor NF-κB. J Virol 2000; 74:2981-2989.
92. Oberhaus SM, Smith RL, Clayton GH et al. Reovirus infection and tissue injury in the mouse central nervous system are associated with apoptosis. J Virol 1997; 71:2100-2106.
93. O'Donnell SM, Hansberger MW, Connolly JL et al. Organ-specific roles for transcription factor NF-κB in reovirus-induced apoptosis and disease. J Clin Invest 2005; 115:2341-2350.
94. DeBiasi R, Edelstein C, Sherry B et al. Calpain inhibition protects against virus-induced apoptotic myocardial injury. J Virol 2001; 75:351-361.

95. DeBiasi RL, Robinson BA, Sherry B et al. Caspase inhibition protects against reovirus-induced myocardial injury in vitro and in vivo. J Virol 2004; 78:11040-11050.
96. Connolly JL, Barton ES, Dermody TS. Reovirus binding to cell surface sialic acid potentiates virus-induced apoptosis. J Virol 2001; 75:4029-4039.
97. Tyler KL, Squier MKT, Brown AL et al. Linkage between reovirus-induced apoptosis and inhibition of cellular DNA synthesis: Role of the S1 and M2 genes. J Virol 1996; 70:7984-7991.
98. Ernst H, Shatkin AJ. Reovirus hemagglutinin mRNA codes for two polypeptides in overlapping reading frames. Proc Natl Acad Sci USA 1985; 82:48-52.
99. Jacobs BL, Atwater JA, Munemitsu SM et al. Biosynthesis of reovirus-specified polypeptides. The S1 mRNA synthesized in vivo is structurally and functionally indistinguishable from in vitro-synthesized S1 mRNA and encodes two polypeptides, σ1a and σ1bNS. Virology 1985; 147:9-18.
100. Sarkar G, Pelletier J, Bassel-Duby R et al. Identification of a new polypeptide coded by reovirus gene S1. J Virol 1985; 54:720-725.
101. Rodgers SE, Connolly JL, Chappell JD et al. Reovirus growth in cell culture does not require the full complement of viral proteins: Identification of a σ1s-null mutant. J Virol 1998; 72:8597-8604.
102. Hoyt CC, Richardson-Burns SM, Goody RJ et al. Nonstructural protein sigma1s is a determinant of reovirus virulence and influences the kinetics and severity of apoptosis induction in the heart and central nervous system. J Virol 2005; 79:2743-2753.
103. Boehme KW, Guglielmi KM, Dermody TS. Reovirus nonstructural protein σ1s is required for establishment of viremia and systemic dissemination. Proc Natl Acad Sci USA 2009; 106:19986-19991.
104. Boehme KW, Frierson JM, Konopka JL et al. The reovirus σ1s protein is a determinant of hematogenous but not neural viral dissemination in mice. J Virol 2011; 85:11781-11790.
105. Campbell JA, Shelling P, Wetzel JD et al. Junctional adhesion molecule-A serves as a receptor for prototype and field-isolate strains of mammalian reovirus. J Virol 2005; 79:7967-7978.
106. Dermody TS, Nibert ML, Bassel-Duby R et al. A sigma 1 region important for hemagglutination by serotype 3 reovirus strains. J Virol 1990; 64:5173-5176.
107. Connolly JL, Dermody TS. Virion disassembly is required for apoptosis induced by reovirus. J Virol 2002; 76:1632-1641.
108. Danthi P, Hansberger MW, Campbell JA et al. JAM-A-independent, antibody-mediated uptake of reovirus into cells leads to apoptosis. J Virol 2006; 80:1261-1270.
109. Hazelton PR, Coombs KM. The reovirus mutant tsA279 has temperature-sensitive lesions in the M2 and L2 genes: The M2 gene is associated with decreased viral protein production and blockade in transmembrane transport. Virology 1995; 207:46-58.
110. Danthi P, Kobayashi T, Holm GH et al. Reovirus apoptosis and virulence are regulated by host cell membrane penetration efficiency. J Virol 2008; 82:161-172.
111. Danthi P, Kobayashi T, Coffey CM et al. Independent regulation of reovirus membrane penetration and apoptosis by the μ1 f domain. PLoS Pathog 2008; 4:e1000248.
112. Clarke P, Meintzer SM, Moffitt LA et al. Two distinct phases of virus-induced nuclear factor kappa B regulation enhance tumor necrosis factor-related apoptosis-inducing ligand-mediated apoptosis in virus-infected cells. J Biol Chem 2003; 278:18092-18100.
113. Beg A, Finco T, Nantermet P et al. Tumor necrosis factor and interleukin-1 lead to phosphorylation and loss of IκBα: A mechanism for NF-κB activation. Mol Cell Biol 1993; 13:3301-3310.
114. Cahir McFarland ED, Izumi KM, Mosialos G. Epstein-barr virus transformation: Involvement of latent membrane protein 1-mediated activation of NF-kappaB. Oncogene 1999; 18:6959-6964.
115. McKinsey TA, Brockman JA, Scherer DC et al. Inactivation of IkappaBbeta by the tax protein of human T-cell leukemia virus type 1: A potential mechanism for constitutive induction of NF-kappaB. Mol Cell Biol 1996; 16:2083-2090.
116. Abbadie C, Kabrun N, Bouali F et al. High levels of c-rel expression are associated with programmed cell death in the developing avian embryo and in bone marrow cells in vitro. Cell 1993; 75:899-912.
117. Grimm S, Bauer MKA, Baeuerle PA et al. Bcl-2 down-regulates the activity of transcription factor NF-κB induced upon apoptosis. J Cell Biol 1996; 134:13-23.
118. Jung M, Zhang Y, Lee S et al. Correction of radiation sensitivity in ataxia telangiectasia cells by a truncated IκB-α. Science 1995; 268:1619-1621.
119. Beg A, Baltimore D. An essential role for NF-κB in preventing TNF-α-induced cell death. Science 1996; 274:782-784.
120. Liu ZG, Hsu H, Goeddel D et al. Dissection of TNF receptor 1 effector functions: JNK activation is not linked to apoptosis while NF-κB activation prevents cell death. Cell 1996; 87:565-576.
121. Van Antwerp D, Martin S, Kafri T et al. Suppression of TNF-α-induced apoptosis by NF-κB. Science 1996; 274:787-789.
122. Clarke P, Meintzer SM, Widmann C et al. Reovirus infection activates JNK and the JNK-dependent transcription factor c-Jun. J Virol 2001; 75:11275-11283.

123. Clarke P, Meintzer SM, Wang Y et al. JNK regulates the release of proapoptotic mitochondrial factors in reovirus-infected cells. J Virol 2004; 78:13132-13138.
124. Norman KL, Hirasawa K, Yang AD et al. Reovirus oncolysis: The Ras/RalGEF/p38 pathway dictates host cell permissiveness to reovirus infection. Proc Natl Acad Sci USA 2004; 101:11099-11104.
125. Meusel TR, Imani F. Viral induction of inflammatory cytokines in human epithelial cells follows a p38 mitogen-activated protein kinase-dependent but NF-kappa B-independent pathway. J Immunol 2003; 171:3768-3774.
126. Duncan MR, Stanish SM, Cox DC. Differential sensitivity of normal and transformed human cells to reovirus infection. J Virol 1978; 28:444-449.
127. Strong JE, Coffey MC, Tang D et al. The molecular basis of viral oncolysis: Usurpation of the Ras signaling pathway by reovirus. EMBO J 1998; 17:3351-3362.
128. Strong JE, Lee PW. The v-erbB oncogene confers enhanced cellular susceptibility to reovirus infection. J Virol 1996; 70:612-616.
129. Coffey MC, Strong JE, Forsyth PA et al. Reovirus therapy of tumors with activated Ras pathway. Science 1998; 282:1332-1334.
130. Mundschau LJ, Faller DV. Oncogenic ras induces an inhibitor of double-stranded RNA-dependent eukaryotic initiation factor 2 alpha-kinase activation. J Biol Chem 1992; 267:23092-23098.
131. Williams ME, Cox DC, Stevenson JR. Rejection of reovirus-treated L1210 leukemia cells by mice. Cancer Immunol Immunother 1986; 23:87-92.
132. Wilcox ME, Yang W, Senger D et al. Reovirus as an oncolytic agent against experimental human malignant gliomas. J Natl Cancer Inst 2001; 93:903-912.
133. Norman KL, Coffey MC, Hirasawa K et al. Reovirus oncolysis of human breast cancer. Hum Gene Ther 2002; 13:641-652.
134. Holm GH, Zurney J, Tumilasci V et al. Retinoic acid-inducible gene-I and interferon-β promoter stimulator-1 augment proapoptotic responses following mammalian reovirus infection via interferon regulatory factor-3. J Biol Chem 2007; 282:21953-21961.
135. Kato H, Takeuchi O, Mikamo-Satoh E et al. Length-dependent recognition of double-stranded ribonucleic acids by retinoic acid-inducible gene-I and melanoma differentiation-associated gene 5. J Exp Med 2008; 205:1601-1610
136. Loo YM, Fornek J, Crochet N et al. Distinct RIG-I and MDA5 signaling by RNA viruses in innate immunity. J Virol 2008; 82:335-345.
137. Knowlton JJ, Dermody TS, Holm GH. Apoptosis induced by mammalian reovirus is interferon-β-independent and enhanced by IRF-3- and NF-κB-dependent expression of Noxa. J. Virol 2012; 86:1650-1660.
138. DeBiasi RL, Clarke P, Meintzer SM et al. Reovirus-induced alteration in expression of apoptosis and DNA repair genes with potential roles in viral pathogenesis. J Virol 2003; 77:8934-8947.
139. O'Donnell SM, Holm GH, Pierce JM et al. Identification of an NF-κB-dependent gene network in cells infected by mammalian reovirus. J Virol 2006; 80:1077-1086.
140. Smith JA, Schmechel SC, Raghavan A et al. Reovirus induces and benefits from an integrated cellular stress response. J Virol 2006; 80:2019-2033.
141. Webster GA, Perkins ND. Transcriptional cross talk between NF-κ B and p53. Mol Cell Biol 1999; 19:3485-3495.
142. Dreyfus D, Nagasawa M, Gelfand E et al. Modulation of p53 activity by IκBα: Evidence suggesting a common phylogeny between NF-κB and p53 transcription factors. BMC Immunol 2005; 6:12.
143. Li Z, Niu J, Uwagawa T et al. Function of polo-like kinase 3 in NF-κB-mediated proapoptotic response. J Biol Chem 2005; 280:16843-16850.
144. Huang YH, Wu JY, Zhang Y et al. Synergistic and opposing regulation of the stress-responsive gene IEX-1 by p53, c-Myc, and multiple NF-kappaB/rel complexes. Oncogene 2002; 21:6819-6828.
145. Liu CY, Schröder M, Kaufman RJ. Ligand-independent dimerization activates the stress response kinases IRE1 and PERK in the lumen of the endoplasmic reticulum. J Biol Chem 2000; 275:24881-24885.
146. Harding HP, Zhang Y, Bertolotti A et al. Perk is essential for translational regulation and cell survival during the unfolded protein response. Mol Cell 2000; 5:897-904.
147. Sen GC. Viruses and interferons. Annu Rev Microbiol 2001; 55:255-281.
148. Stark GR, Kerr IM, Williams BR et al. How cells respond to interferons. Annu Rev Biochem 1998; 67:227-264.
149. Kunzi MS, Pitha PM. Interferon targeted genes in host defense. Autoimmunity 2003; 36:457-461.
150. Tanaka N, Sato M, Lamphier MS et al. Type I interferons are essential mediators of apoptotic death in virally infected cells. Genes Cells 1998; 3:29-37.
151. Bingle CD, Craig RW, Swales BM et al. Exon skipping in Mcl-1 results in a Bcl-2 homology domain 3 only gene product that promotes cell death. J Biol Chem 2000; 275:22136-22146.
152. Gurumurthy S, Goswami A, Vasudevan KM et al. Phosphorylation of Par-4 by protein kinase A is critical for apoptosis. Mol Cell Biol 2005; 25:1146-1161.

153. Imazu T, Shimizu S, Tagami S et al. Bcl-2/E1B 19 kDa-interacting protein 3-like protein (Bnip3L) interacts with Bcl-2/Bcl-xL and induces apoptosis by altering mitochondrial membrane permeability. Oncogene 1999; 18:4523-4529.
154. Tan KO, Tan KML, Chan SL et al. MAP-1, a novel proapoptotic protein containing a BH3-like motif that associates with bax through its Bcl-2 homology domains. J Biol Chem 2001; 276:2802-2807.
155. Clarke P, Meintzer SM, Gibson S et al. Reovirus-induced apoptosis is mediated by TRAIL. J Virol 2000; 74:8135-8139.
156. Torii S, Egan DA, Evans RA et al. Human Daxx regulates Fas-induced apoptosis from nuclear PML oncogenic domains (PODs). EMBO J 1999; 18:6037-6049.
157. Takahashi Y, Lallemand-Breitenbach V, Zhu J et al. PML nuclear bodies and apoptosis. Oncogene 2004; 23:2819-2824.
158. Richardson-Burns SM, Kominsky DJ, Tyler KL. Reovirus-induced neuronal apoptosis is mediated by caspase 3 and is associated with the activation of death receptors. J Neurovirol 2002; 8:365-380.
159. Blatt NB, Glick GD. Signaling pathways and effector mechanisms preprogrammed cell death. Bioorg Med Chem 2001; 9:1371-1384.
160. Kominsky DJ, Bickel RJ, Tyler KL. Reovirus-induced apoptosis requires both death receptor- and mitochondrial-mediated caspase-dependent pathways of cell death. Cell Death Differ 2002; 9:926-933.
161. Wajant H, Johannes FJ, Haas E et al. Dominant-negative FADD inhibits TNFR60-, Fas/Apo1- and TRAIL-R/Apo2-mediated cell death but not gene induction. Curr Biol 1998; 8:113-116.
162. Clarke P, Debiasi RL, Meintzer SM et al. Inhibition of NF-kappa B activity and cFLIP expression contribute to viral-induced apoptosis. Apoptosis 2005; 10:513-524.
163. Chawla-Sarkar M, Lindner DJ, Liu YF et al. Apoptosis and interferons: Role of interferon-stimulated genes as mediators of apoptosis. Apoptosis 2003; 8:237-249.
164. Shigeno M, Nakao K, Ichikawa T et al. Interferon-alpha sensitizes human hepatoma cells to TRAIL-induced apoptosis through DR5 upregulation and NF-kappa B inactivation. Oncogene 2003; 22:1653-1662.
165. Li P, Nijhawan D, Budhardjo I et al. Cytochrome c and dATP-dependent formation of Apaf-1/caspase-9 complex initiates an apoptotic protease cascade. Cell 1997; 91:479-489.
166. Verhagen AM, Ekert PG, Pakusch M et al. Identification of DIABLO, a mammalian protein that promotes apoptosis by binding to and antagonizing IAP proteins. Cell 2000; 102:43-53.
167. Du C, Fang M, Li Y et al. Smac, a mitochondrial protein that promotes cytochrome c-dependent caspase activation by eliminating IAP inhibition. Cell 2000; 102:33-42.
168. Joza N, Susin SA, Daugas E et al. Essential role of the mitochondrial apoptosis-inducing factor in programmed cell death. Nature 2001; 410:549-554.
169. Li H, Zhu H, Xu CJ et al. Cleavage of BID by caspase 8 mediates the mitochondrial damage in the Fas pathway of apoptosis. Cell 1998; 94:491-501.
170. Kominsky DJ, Bickel RJ, Tyler KL. Reovirus-induced apoptosis requires mitochondrial release of Smac/DIABLO and involves reduction of cellular inhibitor of apoptosis protein levels. J Virol 2002; 76:11414-11424.
171. Danthi P, Pruijssers AJ, Berger AK et al. Bid regulates the pathogenesis of neurotropic reovirus. PLoS Pathog 2010; 6:e1000980.
172. Luo X, Budihardjo I, Zou H et al. Bid, a Bcl2 interacting protein, mediates cytochrome c release from mitochondria in response to activation of cell surface death receptors. Cell 1998; 94:481-490.
173. Richardson-Burns SM, Tyler KL. Regional differences in viral growth and central nervous system injury correlate with apoptosis. J Virol 2004; 78:5466-5475.
174. Richardson-Burns SM, Tyler KL. Minocycline delays disease onset and mortality in reovirus encephalitis. Exp Neurol 2005; 192:331-339.
175. Sherry B, Torres J, Blum MA. Reovirus induction of and sensitivity to beta interferon in cardiac myocyte cultures correlate with induction of myocarditis and are determined by viral core proteins. J Virol 1998; 72:1314-1323.
176. Azzam-Smoak K, Noah DL, Stewart MJ et al. Interferon regulatory factor-1, interferon-beta, and reovirus-induced myocarditis. Virology 2002; 298:20-29.
177. Stewart MJ, Blum MA, Sherry B. PKR's protective role in viral myocarditis. Virology 2003; 314:92-100.
178. Noah DL, Blum MA, Sherry B. Interferon regulatory factor 3 is required for viral induction of beta interferon in primary cardiac myocyte cultures. J Virol 1999; 73:10208-10213.
179. Bazzoni G, Martinez-Estrada OM, Orsenigo F et al. Interaction of junctional adhesion molecule with the tight junction components ZO-1, cingulin, and occludin. J Biol Chem 2000; 275:20520-20526.
180. Ebnet K, Schulz CU, Meyer Zu Brickwedde MK et al. Junctional adhesion molecule interacts with the PDZ domain-containing proteins AF-6 and ZO-1. J Biol Chem 2000; 275:27979-27988.
181. Pfaff M, Liu S, Erle DJ et al. Integrin beta cytoplasmic domains differentially bind to cytoskeletal proteins. J Biol Chem 1998; 273:6104-6109.

182. Otey CA, Pavalko FM, Burridge K. An interaction between alpha-actinin and the beta 1 integrin subunit in vitro. J Cell Biol 1990; 111:721-729.
183. Schaller MD, Otey CA, Hildebrand JD et al. Focal adhesion kinase and paxillin bind to peptides mimicking beta integrin cytoplasmic domains. J Cell Biol 1995; 130:1181-1187.
184. Reszka AA, Hayashi Y, Horwitz AF. Identification of amino acid sequences in the integrin beta 1 cytoplasmic domain implicated in cytoskeletal association. J Cell Biol 1992; 117:1321-1330.
185. Chen WJ, Goldstein JL, Brown MS. NPXY, a sequence often found in cytoplasmic tails, is required for coated pit-mediated internalization of the low density lipoprotein receptor. J Biol Chem 1990; 265:3116-3123.
186. Deiss LP, Galinka H, Berissi H et al. Cathepsin D protease mediates programmed cell death induced by interferon-gamma, Fas/APO-1 and TNF-alpha. EMBO J 1996; 15:3861-3870.
187. Guicciardi ME, Deussing J, Miyoshi H et al. Cathepsin B contributes to TNF-alpha-mediated hepatocyte apoptosis by promoting mitochondrial release of cytochrome c. J Clin Invest 2000; 106:1127-1137.
188. Roberg K. Relocalization of cathepsin D and cytochrome c early in apoptosis revealed by immunoelectron microscopy. Lab Invest 2001; 81:149-158.
189. Stoka V, Turk B, Schendel SL et al. Lysosomal protease pathways to apoptosis. Cleavage of bid, not pro-caspases, is the most likely route. J Biol Chem 2001; 276:3149-3157.
190. Coffey CM, Sheh A, Kim IS et al. Reovirus outer capsid protein μ1 induces apoptosis and associates with lipid droplets, endoplasmic reticulum, and mitochondria. J. Virol. 80:8422-38, 2006.
191. Kim JW, Lyi SM, Parrish CR, Parker JS. A proapoptotic peptide derived from reovirus outer capsid protein μ1 has membrane-destabilizing activity. J Virol 2011; 85:1507-16.
192. Wisniewski ML, Werner BG, Hom LG et al. Reovirus infection or ectopic expression of outer capsid protein μ1 induces apoptosis independently of the cellular proapoptotic proteins Bax and Bak. J Virol 2011; 85:296-304.

CHAPTER 4

ENTRY OF INFLUENZA VIRUS

Xiangjie Sun and Gary R. Whittaker*
Department of Microbiology and Immunology, Cornell University, Ithaca, New York, USA
**Corresponding Author: Gary Whittaker—Email: grw7@cornell.edu*

Abstract: As a major pathogen of human and certain animal species, influenza virus causes wide spread and potentially devastating disease. To initiate infection, the virus first binds to cellular receptors comprising either -(2,3) or -(2,6) linked sialic acid. Recent advances in our understanding of the influenza virus receptor and viral host species involved have shed light on the molecular mechanism of how influenza virus transmits across species and adapts to a new host. Following receptor binding, influenza viruses are internalized through multiple endocytic pathways, including both clathrin- and non-clathrin-dependent routes, which have recently been visualized at single viral particle level. The viral envelope then fuses with the endosomal membrane in a low pH-dependent manner and the viral genome is released into the cytosol, followed by further transport to the nucleus where genome replication occurs.

INTRODUCTION

Influenza A viruses are major pathogens of humans and other animal species and share the common properties of possessing a segmented single-strand negative sense RNA genome encapisdated into ribonucleoproteins (RNPs) that are further packaged into enveloped virions.[1] They are now established to enter cells via pH-dependent endocytosis and replicate their genomes in the nucleus. This chapter aims to summarize our current knowledge of how influenza A virus binds to cells and is internalized into low pH endosomes, and how these processes are coordinated with membrane fusion, virus uncoating and nuclear import. This process is summarized in Figure 1.

Viral Entry into Host Cells, edited by Stefan Pöhlmann and Graham Simmons.

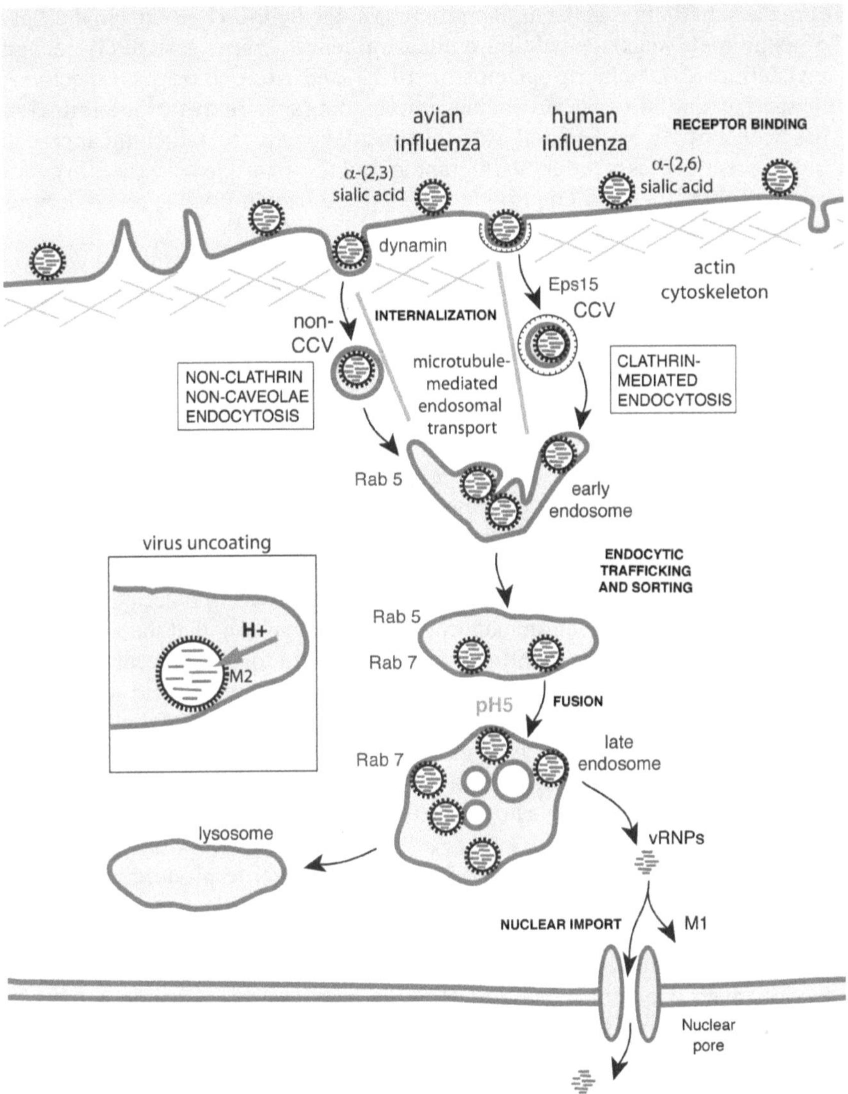

Figure 1. Routes of influenza virus entry into host cells. Avian and human influenza virus binds to α-(2-3) or α-(2-6)-linked cell surface sialic acid. Viruses are then internalized by clathrin-dependent or -independent endocytosis, in a cytoskeleton-dependent manner. Virions are trafficked through Rab 5 and Rab 7-positive early and late endosomes, where fusion occurs in a low pH-dependent manner. For virus uncoating the M2 ion channel allows acidification of the virus interior and release of M1 before the genomic RNPs enter the nucleus via nuclear pores.

INFLUENZA VIRUS RECEPTORS

It has been recognized for many years that cell surface sialic acids from either glycoprotein or glycolipid are the receptors for influenza viruses.[2] Although there is considerable variety in the terminal sialic acids expressed on the cell surface with regard to the substitution at the amino group or at the hydroxyl group and the linkage to the penultimate sugar, avian and human influenza viruses selectively recognize N-acetylsialic acids attached to galactose in α-(2,3) and α-(2,6) linkages respectively.[3-5] The question of how influenza virus achieves its binding specificity for the terminal sialic acid has been partially understood by biochemical, genetic and structural approaches; especially recent studies of the crystallographic structure of influenza viruses in complex with sialic acid analogs, which provide an extensive insight into the receptor binding properties of influenza virus.[6-9]

Binding of influenza virus to its sialic acid receptor is mediated by the viral glycoprotein hemagglutinin (HA), which is a trimeric molecule present on the viral surface. Based on protein antigenicity there are currently 16 distinct HA subtypes, with avian species (notably water fowl and shore birds) harboring all of the HA subtypes.[10,11] These species are therefore regarded as the natural reservoir of influenza virus. In humans, influenza H1, H2 and H3 viruses have caused pandemics in the last century, and more recently H5, H7 and H9 viruses transmitted from birds have resulted in sporadic human influenza outbreaks.[12-14] Only a limited number of influenza subtypes are seen in other animals (including domestic poultry).[11] The receptor binding specificity of the virus from a given species correlates with the abundance of cellular receptor in its host. Human lung and upper respiratory tracts have abundant α-(2,6) linked sialic acid, whereas sialic acid in α-(2,3) linkage is predominant in avian enteric tracts, giving rise to a simple model of avian-human tropism and suggesting that the availability of cellular receptor provides the positive selective pressure for viral receptor specificity and restricts influenza host range.[15-17] However, α-(2,3) linked sialic acid is certainly not excluded from human respiratory tracts. Residual expression of α-(2,3) linked sialic acid in human airway ciliated cells, was originally proposed to explain the initial infection of a highly virulent avian virus in humans without any changes at viral receptor binding sites.[18,19] More recently, immunohistochemistry of the human respiratory tract has shown that whereas α-(2,6) linked sialic acid is expressed extensively in the upper regions, α-(2,3) linked sialic acid can be quite abundant in the lower respiratory tract therefore allowing the highly pathogenic avian influenza (HPAI) virus A/Vietnam/1194/04 (H5N1) attachment to type II pneumocytes, alveolar macrophages, and nonciliated cuboidal epithelial cells from the lower respiratory tract; in part explaining the possibility of extensive primary pneumonia of H5N1 HPAI in humans.[20,21]

Our understanding of the molecular mechanism of virus binding to cellular receptor with strong specificity has been greatly improved by structural studies of HA at the atomic level. Most notably, the crystallographic structures of the complexes formed by sialic acid analogs and the hemagglutinin proteins of the H1, H3, H5, H7 and H9 subtypes have been resolved and a model for interpreting the interactions between viral HA proteins and receptors has been established.[9,22-24] In this model, it was suggested that the viral receptor bound to a membrane-distal pocket, in which the 190 helix (residues 190 to 198), the 130 loop (residues 135-138) and the 220 loop (residues 221-228) form the three sides of the pocket, with the conserved residues Tyr 98, Trp 153, His183 and Tyr 195 located at the base.[2,22] The amino acid residues constituting the receptor-binding pocket of HA interact with sialic acid by extensive hydrogen bonding and van der Waals interactions. Later, a second receptor binding site located at the interface between HA1 and HA2 was also identified, but its significance remains unclear due a weaker binding affinity for α-(2,3) linked sialic acid; α-(2,6) linked sialic acid fails to bind to this site.[8]

The exact residues in the receptor-binding site that contribute to receptor specificity may be different in influenza viruses across species. A comparison of HA sequences from avian influenza H1-H13 subtypes revealed six conserved residues, including 138A, 190E, 194L, 225G, 226Q and 228G.[25] Human H2 and H3 viruses with a preference for α-(2,6) linked sialic acid have conserved 226L and 228S residues, indicative of the importance of these two residues in receptor binding specificities.[26] In contrast, human H1 viruses have an avian consensus sequence at residues 226Q and 228G, but instead have substitutions at residues 190 and 225. Interestingly, HA from the 1918 influenza H1 pandemic virus, A/New York/1/18, had only one mutation at residue 190 compared to the avian consensus sequence and could bind to both α-(2,6) and α-(2,3) sialic acid, whereas another 1918 isolate A/South Carolina/1/18 had both the expected H1 substitutions at position 190 and 225, and preferentially bound to α-(2,6)-linked sialic acid.[27] As a whole, we can conclude that the amino acid residues at the primary cellular receptor-binding pocket are the major determinant of viral receptor specificity and that mutation(s) at this site can confer the ability to bind to a different cellular receptor.

Understanding the correlation between unique sequences at the HA receptor binding site and influenza virus binding specificity may help predict at which position(s) a virus needs to mutate for adaptation to a new host. As demonstrated for the H5N1 influenza virus A/Vietnam/1203/2004, which was transmitted from birds and caused mortality in humans, viral receptor binding specificity can be modulated by altering the consensus residues that have proven to be important for H1, H2 and H3 subtypes. For instance, mutations at the residues 226 and 228 that can convert avian H2 and H3 HAs to human receptor specificity, when introduced into the A/Vietnam/1203/2004 H5 HA background, showed a reduced binding affinity for α-2,3 linked sialic acid and could bind to a natural human α-(2,6) glycan, indicating a possible path for this avian H5 virus to alter receptor specificity and gain the ability to spread in human population.[28] Host-dependent glycosylation adjacent to the receptor binding site of HA molecule also affects its receptor binding properties. It was shown that the mutations in the HPAI fowl plaque virus (FPV) HA that eliminated the oligosaccharides at position Asn 123 and Asn 149 showed stronger binding affinity for receptor.[29] The reduced receptor binding affinity modulated by the presence of glycosylation of HA protein is believed to be important during virus release from host cells.[29,30] Moreover, mutations at glycosylation sites close to the receptor binding pocket are often seen during viral adaptation to a new host, e.g., for human influenza grown in embryonated eggs or adapted to mice, suggesting that the host dependent glycosylation may be involved in the regulation of HA molecule binding to specific receptors encountered in different hosts.[31,32]

Although cellular sialic acid has been regarded as the sole receptor for influenza virus, there is evidence supporting the existence of a coreceptor, which may be essential for viral internalization and subsequent infection. It was shown that a mutation in A/X-31virus at residue 98 (Y98F) bound only weakly with erythrocytes, yet the virus replicated in MDCK cells and embryonated eggs as efficiently as parental virus.[7] These data may be explained by differences in the affinity of the Y98F virus for sialic acid, or could be due to a sialic acid-independent pathway. In a more recent study, it was found that influenza virus could infect MDCK cells that had been enzymatically treated to remove all accessible sialic acid residues.[33] Moreover, influenza is unable to infect Lec1 cells, which are deficient in terminal N-linked glycosylation, despite the fact that these cells express abundant glycosphingolipids at the cell surface.[34] Conversely cells deficient in glycosphingolipids, but containing normal N-linked glycoprotein, showed no defects in virus entry and infection.[35,36] Overall, these results suggest that a specific coreceptor may be required for viral internalization in addition to initial sialic acid binding, or that the critcal sialic acid is on a specific (or selection of specific) protein(s).

Conventional studies of influenza virus receptor binding specificity depend heavily on agglutination assays using resialylated erythrocytes, which express specific sialic acid determinants by linkage-specific sialyltransferases, following enzymatic removal of endogenous sialic acid from red blood cells.[37] However, this method suffers from great variation in the preparation of erythrocytes and has limited sensitivity. The recent application of high throughput glycan microarrays to study of influenza virus receptor specificity provides us with a more powerful tool and makes it possible to perform quick and accurate evaluation of viral receptor binding specificity.[38] We hope that in the near future we will have a better understanding of influenza virus receptor binding properties, especially for viruses newly adapted to a different host.

VIRAL INTERNALIZATION AND ENDOCYTIC TRAFFICKING

Upon viral receptor binding, influenza viruses are endocytosed, leading to their residency in intracellular vesicles. Earlier studies, heavily dependent on morphological observation by electron microscopy, suggested that influenza virus could get internalized by clathrin-mediated endocytosis.[39,40] More recent studies, taking advantage of recent advances in cell biology and molecular biology, have provided us with new findings regarding viral endocytic pathway(s). By using dominant-negative mutants involved in clathrin- and caveolin-mediated endocytosis (Eps15 and caveolin-1 respectively), as well as siRNA targeting of the clathrin heavy chain, it was found that influenza infectivity was not affected significantly in HeLa cells, indicating that influenza virus could be internalized, and cells infected, by additional endocytic pathways beside clathrin-dependent endocytosis.[41,42] The involvement of multiple endocytic pathways in influenza virus entry was further confirmed by individually tracking viral particles in real time, providing more quantitative analyses. Live cell microscopy of BSC-1 cells revealed that about 65% of receptor-bound viruses were internalized by clathrin-mediated endocytosis.[43] The remaining one-third of the viruses could enter cells by clathrin, caveolae-independent endocytosis, suggesting that clathrin-independent endocytosis occurs in live cells and is a functional route contributing to influenza virus infection. Moreover, Zhuang and coworkers have suggested that the virus using clathrin-mediated endocytosis tended to be internalized through the de novo formation of clathrin-coated pit at the virus binding sites.[43] The molecular mechanism of how influenza virus might induce de novo clathrin coated pit formation remains unclear.

After being recruited into either clathrin-coated or noncoated pits, virus-containing pits can be pinched off to form intracellular vesicles by dynamin GTPases.[44] Like other vesicles derived from the plasma membrane, virus-containing vesicles mature into early endosomes and subsequently late endosomes, from which viral fusion and uncoating occurs. In addition to biochemical evidence, relying mainly on low pH-induced fusion of viruses with erythrocytes or liposomes,[45] viral fusion with endosomal membranes can also be visualized in real time by using either a single fluorescent dye (e.g., DiD) or with dual-labeled viruses (DiO and octadecyl rhodamine, R18).[46,47] Fusion events are indicated by increased fluorescence intensity due to fluorescence dequenching in single fluorescence labeling or by shifted fluorescence wavelength upon both dequenching and fluorescent resonance energy transfer (FRET) in dual-fluorescence labeling. Dual-wavelength imaging appears to offer some advantages in tracking numerous particles simultaneously because the fusion event is revealed by a change in the fluorescence wavelength, which may be more evident than that revealed by the change in a single fluorescence intensity. Notably, fusion in HeLa cells, which was not observed in single-fluorescence imaging, could be demonstrated by dual-wavelength imaging.[46,47]

Rab5 and Rab7 GTPase proteins, as well as specific isoforms of protein kinase C and the cellular vacuolar protein sorting (VPS) pathway, have been demonstrated to be involved in the maturation of virus-containing endosomes.[48-50] By live cell imaging in BSC-1 cells, most fusion events seem to occur from endosomes containing both Rab 5 and Rab 7,[51] and as such would not be considered true "late endosomes", but rather intermediates between early and late endosomes. One characteristic of influenza virus entry is that the early endosome-enclosed influenza viruses were transited rapidly along microtubules, compared to transferrin-containing early endosomes, which moved much more slowly and were microtubule independent.[46,51] The sorting of influenza virus into dynamic, rapidly maturing endosomes occurs at the plasma membrane and is believed to be determined by the specific adaptor protein(s) binding to the cytoplasmic tails of the influenza receptor. This finding might explain the requirement for sialated glycoprotein rather than glycospingolipid for receptor binding. Unexpectedly, influenza entry was found to be independent of the clathrin-adaptor protein AP-2.[51] However as mentioned above, due to the lack of specificity of the influenza receptor, specific adaptor proteins that mediate viral internalization remain undefined.

When influenza resides in endosomes, transport toward to the nucleus can be facilitated by intracellular actin and microtubule motor proteins. Viral trajectories in nonpolarized CHO cells could be assigned into three stages, from the initial actin-dependent movement in the cell periphery, to a rapid, dynein-directed transport on microtubules, followed by an intermittent movement involving both plus- and minus-end-directed microtubule-based motilities in the perinuclear region.[46] Although viral movement could be facilitated by the existence of intracellular motor proteins, it was not absolutely necessary for infection in nonpolarized cells.[52] However, in polarized epithelial cells, which influenza encounters during in vivo infection, intact actin filaments are essential and actin filament disruption treatment leads to failed viral infection from the apical surface.[52] Considering the existence of dense actin meshwork at polarized apical side, influenza virus transport in polarized cells seems to require distinct actin motor proteins. Myosin VI, the only actin motor protein traveling toward the minus ends of actin filaments, has recently been shown to have a role in transporting endocytosed virus across the actin meshwork of polarized MDCK cells (X. Sun and G. Whittaker, unpublished data).

All of the above mentioned viral endocytic trafficking studies have used tissue culture-adapted influenza viruses, which are spherical in nature. However, viruses from clinical samples can be present in long filamentous forms up to several micrometers in length.[53,54] These viruses are infectious, indicating their importance in viral pathogenesis and transmission in vivo. The questions of how the filamentous viruses get into the cells and initiate infection largely remains unclear, in part due to the technical difficulty of purifying the filamentous form of the virus. Initial studies have suggested that filamentous influenza could be endocytosed in a delayed, dynamin-independent manner, compared to rapid, dynamin-dependent endocytosis of spherical virions.[55]

MEMBRANE FUSION

In addition to receptor binding, a second major function of the influenza hemagglutinin (HA) is to mediate fusion of virus and cell membranes.[2] Such a fusion process is essential to deliver the genome of any enveloped virus into the cell, and influenza HA has proven to be a paradigm of virus-cell fusion, as well as providing a mechanistic framework for other fusion events; e.g., those mediated by cellular SNARE proteins. The role of HA in influenza virus entry has been reviewed extensively elsewhere and the reader is referred here for more detail.[2]

Different influenza HA subtypes have been shown to have distinct properties regarding fusion activity. In contrast to the H3 subtype, the HA protein from a H2 subtype reacts slowly upon low pH-induced conformational change.[56,57] The refolding of HA from its initial form to a fusion competent state may involve reversible intermediates. The transition from a reversible to irreversible conformation is not only affected by pH and the presence of target membrane, but also by the presence of adjacent HA trimers. The existence of adjacent HA trimers could facilitate the transition from a reversible intermediate state to an irreversible conformation, allowing multiple copies of HA to initiate fusion simultaneously.[56,58]

In the influenza virion, HA exists in a metastable state, with the hydrophobic fusion peptide hidden towards the base of the molecule. Fusion is triggered in vivo by exposure to the low pH environment of the endosome. The HA molecule has a high degree of alpha-helical secondary structure, and in the process of fusion it undergoes a major conformational change, accompanied by the formation of a "coiled coil" of alpha helices,[59] which reorientates the fusion peptide to the outermost part of the HA molecule and initiates the fusion event. HA shares many features with fusion proteins of other viruses (e.g., retroviruses and paramyxovirus) and is a founding member of the Class I family of viral fusion proteins,[60] which have common structural features but differing activation requirements. Whereas human immunodeficiency virus (HIV) Env and simian virus 5 (SV5) F proteins have extensive "six-helix bundles" in their active fusogenic state, influenza HA has a much smaller six-helix bundle and has been proposed to employ a "leash in the groove" mechanism to bring the virus envelope and endosomal membrane into close enough proximity to initiate fusion.[61] Fusion is initiated by insertion of the kinked, amphipathic fusion peptide into the outer leaflet of the cell membrane,[62] followed by a hemifusion event and finally fusion pore formation and expansion.[63]

Influenza virus fusion has been extensively studied by biophysical techniques. The most common method involves labeling of the virus envelope with a fluorescent probe and binding of the labeled virus to the surface of erythrocytes (or alternatively liposomes or tissue culture cells), followed by induction of fusion by artificially dropping the external pH. For probes such as octadecyl rhodamine (R18), fusion is monitored by dequenching of R18 as the probe dilutes into the target membrane. Under these conditions, fusion occurs with rapid kinetics ($T^{1/2}$ of approximately 50 seconds or less). Based on in vitro studies, influenza virus fusion is generally considered to have an optimum pH of approximately 5.0, with fusion effectively occurring only between pH 4.5 and pH 5.5.[45] Such a pH requirement fits well with in vivo data showing selective entry of influenza virus through vesicles with properties of "intermediate"/late endosomes.[48,51] This is in contrast to many other pH-dependent enveloped viruses, which can fuse in the range of pH 6.2-6.5, i.e., clearly within early or recycling endosomes.[64]

Virus fusion relies on several features of HA that are initiated during virus assembly in the previous replication cycle. First, HA is cleaved immediately N-terminal to the fusion peptide. This cleavage of HA0 to HA1 and HA2 is essential for the subsequent conformational change to occur in the endosome and allow fusion peptide exposure.[2] Second, the HA trimerizes, with HA trimers functioning in a cooperative manner during fusion.[65] Third, the partitioning of HA into lipid microdomains during budding is essential for subsequent fusogenic activity.[66] In this scenario, the HA (present in a trimeric form in the virus envelope) must be organized cooperatively, possibly in a specific lipid environment, for fusion to occur. The influenza virus envelope contains high levels of cholesterol, depletion of which severely affects the ability of virions to infect cells.[66,67] Such depletion may affect virus fusion directly[67] and/or other early events in virus replication.

VIRUS UNCOATING AND NUCLEAR IMPORT OF VIRAL COMPONENTS

Once fusion has occurred from the endosomal compartment, the uncoated virus is released into the cytoplasm.[68-70] In addition to being a trigger for fusion, endosome acidification is essential for a second event during virus entry-virus uncoating. Within the endosome, H+ ions are transferred into the interior of the virion via the M2 ion channel present in the virus envelope.[71,72] This enables the release of the influenza matrix protein (M1) from the genomic ribonucleoproteins (RNPs) during virus uncoating.[68] The anti-influenza drug amantadine, inhibits the M2 ion channel, thus preventing the interior of the virus from encountering low pH.[69,73,74] Hence, vRNPs are not released and thus do not enter the nucleus and replication cannot occur.

RNPs from the infecting virus particle must enter the nucleus before viral RNA synthesis can occur and because of the large size of the RNPs, an active nuclear transport mechanism is necessary. Nuclear localization signals (NLSs) have been identified in all three polymerase proteins as well as in the nucleoprotein NP. The localization signals in influenza A virus NP have been by far the most intensively studied. These signals have been reviewed recently and the reader is referred here for more detail.[75] In summary, NP and the RNPs contain multiple NLSs that may be redundant in nature, with a nonconventional NLS toward the N-terminus of NP likely being the dominant import signal.[76] Reconstitution experiments have shown that NP is necessary and sufficient to direct nuclear import of a viral RNA segment.[77] In addition, the import function of the multiple NLS signals present on RNPs are clearly regulated by the viral M1 protein, as an amantadine-induced failure of M1 to dissociate from RNPs during virus uncoating inhibits RNP nuclear import;[69,70,78] a block that can be reversed by acidification of the cytoplasm to cause M1-RNP dissociation.[68] Following purification of RNPs from influenza virions and microinjection into live cells, single-particle trajectories show that RNPs are transported to the nuclear envelope by diffusion.[79] Overall the influenza M1 protein, appears to down-regulate the nuclear import of RNPs by inhibiting the interactions between RNPs and nuclear pore complexes, but has no significant effect on the transport properties of RNPs themselves.[78,79] Equally there is no known role for direct interactions of RNPs with the cytoskeleton.[70]

CONCLUSION AND FUTURE PERSPECTIVES

Given the recent interest in pandemic influenza in humans, the entry of both avian and human influenza into cells is of fundamental importance for our understanding of influenza pathogenesis. The viral HA plays pivotal roles in entry, being responsible for both binding and fusion. Detailed structural information on a variety of both avian and human HA subtypes is now available[80] and, combined with new techniques involving glycan microarrays, our knowledge of sialic acid-mediated binding is becoming more sophisticated. In terms of virus fusion, much existing data exists, but this is generally confined to the human H1 and H3 subtypes. Future work focused on connecting the virus binding and fusion events with the route of endocytosis and the relevant cell signaling pathways involved both in virus entry and down-stream events in virus replication promises to be a productive area, especially combined with high-throughput screening and live cell imaging techniques.

REFERENCES

1. Lamb RA, Krug RM. Orthomyxoviridae: The viruses and their replication. In: Knipe DM, Howley PM, eds. Fields Virology. Philadelphia: Lippincott Williams and Wilkins, 2001:1487-1531.
2. Skehel JJ, Wiley DC. Receptor binding and membrane fusion in virus entry: The influenza hemagglutinin. Annu Rev Biochem 2000; 69:531-569.
3. Connor RJ, Kawaoka Y, Webster RG et al. Receptor specificity in human, avian, and equine H2 and H3 influenza virus isolates. Virology 1994; 205(1):17-23.
4. Matrosovich MN, Gambaryan AS, Teneberg S et al. Avian influenza A viruses differ from human viruses by recognition of sialyloligosaccharides and gangliosides and by a higher conservation of the HA receptor-binding site. Virology 1997; 233(1):224-234.
5. Rogers GN, D'Souza BL. Receptor binding properties of human and animal H1 influenza virus isolates. Virology 1989; 173(1):317-322.
6. Carroll SM, Higa HH, Paulson JC. Different cell-surface receptor determinants of antigenically similar influenza virus hemagglutinins. J Biol Chem 1981; 256(16):8357-8363.
7. Martin J, Wharton SA, Lin YP et al. Studies of the binding properties of influenza hemagglutinin receptor-site mutants. Virology 1998; 241(1):101-111.
8. Sauter NK, Glick GD, Crowther RL et al. Crystallographic detection of a second ligand binding site in influenza virus hemagglutinin. Proc Natl Acad Sci USA 1992; 89(1):324-328.
9. Wilson IA, Skehel JJ, Wiley DC. Structure of the haemagglutinin membrane glycoprotein of influenza virus at 3 A resolution. Nature 1981; 289(5796):366-373.
10. Fouchier RA, Munster V, Wallensten A et al. Characterization of a novel influenza A virus hemagglutinin subtype (H16) obtained from black-headed gulls. J Virol 2005; 79(5):2814-2822.
11. Webster RG, Bean WJ, Gorman OT et al. Evolution and ecology of influenza virus. Microbiol Rev 1992; 56:152-179.
12. Claas EC, Osterhaus AD, van Beek R et al. Human influenza A H5N1 virus related to a highly pathogenic avian influenza virus. Lancet 1998; 351(9101):472-477.
13. Koopmans M, Wilbrink B, Conyn M et al. Transmission of H7N7 avian influenza A virus to human beings during a large outbreak in commercial poultry farms in the Netherlands. Lancet 2004; 363(9409):587-593.
14. Lin YP, Shaw M, Gregory V et al. Avian-to-human transmission of H9N2 subtype influenza A viruses: Relationship between H9N2 and H5N1 human isolates. Proc Natl Acad Sci USA 2000; 97(17):9654-9658.
15. Baum LG, Paulson JC. Sialyloligosaccharides of the respiratory epithelium in the selection of human influenza virus receptor specificity. Acta Histochem Suppl 1990; 40:35-38.
16. Ito T, Suzuki Y, Suzuki T et al. Recognition of N-glycolylneuraminic acid linked to galactose by the alpha2,3 linkage is associated with intestinal replication of influenza A virus in ducks. J Virol 2000; 74(19):9300-9305.
17. Suzuki Y, Ito T, Suzuki T et al. Sialic acid species as a determinant of the host range of influenza A viruses. J Virol 2000; 74(24):11825-11831.
18. Matrosovich M, Zhou N, Kawaoka Y et al. The surface glycoproteins of H5 influenza viruses isolated from humans, chickens, and wild aquatic birds have distinguishable properties. J Virol 1999; 73(2):1146-1155.
19. Matrosovich MN, Matrosovich TY, Gray T et al. Human and avian influenza viruses target different cell types in cultures of human airway epithelium. Proc Natl Acad Sci USA 2004; 101(13):4620-4624.
20. Shinya K, Ebina M, Yamada S et al. Avian flu: Influenza virus receptors in the human airway. Nature 2006; 440(7083):435-436.
21. van Riel D, Munster VJ, de Wit E et al. H5N1 virus attachment to lower respiratory tract. Science 2006.
22. Gamblin SJ, Haire LF, Russell RJ et al. The structure and receptor binding properties of the 1918 influenza hemagglutinin. Science 2004; 303(5665):1838-1842.

23. Ha Y, Stevens DJ, Skehel JJ et al. X-ray structures of H5 avian and H9 swine influenza virus hemagglutinins bound to avian and human receptor analogs. Proc Natl Acad Sci USA 2001; 98(20):11181-11186.
24. Ha Y, Stevens DJ, Skehel JJ et al. H5 avian and H9 swine influenza virus haemagglutinin structures: Possible origin of influenza subtypes. EMBO J 2002; 21(5):865-875.
25. Nobusawa E, Aoyama T, Kato H et al. Comparison of complete amino acid sequences and receptor-binding properties among 13 serotypes of hemagglutinins of influenza A viruses. Virology 1991; 182(2):475-485.
26. Vines A, Wells K, Matrosovich M et al. The role of influenza A virus hemagglutinin residues 226 and 228 in receptor specificity and host range restriction. J Virol 1998; 72(9):7626-7631.
27. Glaser L, Stevens J, Zamarin D et al. A single amino acid substitution in 1918 influenza virus hemagglutinin changes receptor binding specificity. J Virol 2005; 79(17):11533-11536.
28. Stevens J, Blixt O, Tumpey TM et al. Structure and receptor specificity of the hemagglutinin from an H5N1 influenza virus. Science 2006.
29. Ohuchi M, Ohuchi R, Feldmann A et al. Regulation of receptor binding affinity of influenza virus hemagglutinin by its carbohydrate moiety. J Virol 1997; 71(11):8377-8384.
30. Mir-Shekari SY, Ashford DA, Harvey DJ et al. The glycosylation of the influenza A virus hemagglutinin by mammalian cells. A site-specific study. J Biol Chem 1997; 272(7):4027-4036.
31. Deom CM, Caton AJ, Schulze IT. Host cell-mediated selection of a mutant influenza A virus that has lost a complex oligosaccharide from the tip of the hemagglutinin. Proc Natl Acad Sci USA 1986; 83(11):3771-3775.
32. Gambaryan AS, Marinina VP, Tuzikov AB et al. Effects of host-dependent glycosylation of hemagglutinin on receptor-binding properties on H1N1 human influenza A virus grown in MDCK cells and in embryonated eggs. Virology 1998; 247(2):170-177.
33. Stray SJ, Cummings RD, Air GM. Influenza virus infection of desialylated cells. Glycobiology 2000; 10(7):649-658.
34. Chu VC, Whittaker GR. Influenza virus entry and infection require host cell N-linked glycoprotein. Proc Natl Acad Sci USA 2004; 101(52):18153-18158.
35. Ablan S, Rawat SS, Blumenthal R et al. Entry of influenza virus into a glycosphingolipid-deficient mouse skin fibroblast cell line. Arch Virol 2001; 146(11):2227-2238.
36. Matrosovich M, Suzuki T, Hirabayashi Y et al. Gangliosides are not essential for influenza virus infection. Glycoconj J 2006; 23(1-2):107-113.
37. Paulson JC, Rogers GN. Resialylated erythrocytes for assessment of the specificity of sialyloligosaccharide binding proteins. Methods Enzymol 1987; 138:162-168.
38. Stevens J, Blixt O, Glaser L et al. Glycan microarray analysis of the hemagglutinins from modern and pandemic influenza viruses reveals different receptor specificities. J Mol Biol 2006; 355(5):1143-1155.
39. Dourmashkin RR, Tyrrell DA. Electron microscopic observations on the entry of influenza virus into susceptible cells. J Gen Virol 1974; 24(1):129-141.
40. Matlin KS, Reggio H, Helenius A et al. Infectious entry pathway of influenza virus in a canine kidney cell line. J Cell Biol 1981; 91(3 Pt 1):601-613.
41. Sieczkarski SB, Whittaker GR. Influenza virus can enter and infect cells in the absence of clathrin-mediated endocytosis. J Virol 2002; 76(20):10455-10464.
42. Sun X, Yau VK, Briggs BJ et al. Role of clathrin-mediated endocytosis during vesicular stomatitis virus entry into host cells. Virology 2005; 338(1):53-60.
43. Rust MJ, Lakadamyali M, Zhang F et al. Assembly of endocytic machinery around individual influenza viruses during viral entry. Nat Struct Mol Biol 2004; 11(6):567-573.
44. Roy AM, Parker JS, Parrish CR et al. Early stages of influenza virus entry into Mv-1 lung cells: Involvement of dynamin. Virology 2000; 267(1):17-28.
45. Hoekstra D, Klappe K. Fluorescence assays to monitor fusion of enveloped viruses. In: Duzgunes N, ed. Methods in Enzymology, Vol. 220: Membrane Fusion Techniques Part A. San Diego: Academic Press, 1993:261-276.
46. Lakadamyali M, Rust MJ, Babcock HP et al. Visualizing infection of individual influenza viruses. Proc Natl Acad Sci USA 2003; 100(16):9280-9285.
47. Sakai T, Ohuchi M, Imai M et al. Dual wavelength imaging allows analysis of membrane fusion of influenza virus inside cells. J Virol 2006; 80(4):2013-2018.
48. Sieczkarski SB, Whittaker GR. Differential requirements of Rab5 and Rab7 for endocytosis of influenza and other enveloped viruses. Traffic 2003; 4(5):333-343.
49. Sieczkarski SB, Brown HA, Whittaker GR. The role of protein kinase C bII in influenza virus entry via late endosomes. J Virol 2003; 77:460-469.
50. Khor R, McElroy LJ, Whittaker GR. The ubiquitin-vacuolar protein sorting system is selectively required during entry of influenza virus into host cells. Traffic 2003; 4(12):857-868.
51. Lakadamyali M, Rust MJ, Zhuang X. Ligands for clathrin-mediated endocytosis are differentially sorted into distinct populations of early endosomes. Cell 2006; 124(5):997-1009.

52. Gottlieb TA, Ivanov IE, Adesnik M et al. Actin microfilaments play a critical role in endocytosis at the apical but not the basolateral surface of polarized epithelial cells. J Cell Biol 1993; 120(3):695-710.
53. Choppin PW. Multiplication of two kinds of influenza A2 virus particles in monkey kidney cells. Virology 1963; 21:342-352.
54. Smirnov Yu A, Kuznetsova MA, Kaverin NV. The genetic aspects of influenza virus filamentous particle formation. Arch Virol 1991; 118(3-4):279-284.
55. Sieczkarski SB, Whittaker GR. Characterization of the host cell entry of filamentous influenza virus. Arch Virol 2005; 150(9):1783-1796.
56. Leikina E, Ramos C, Markovic I et al. Reversible stages of the low-pH-triggered conformational change in influenza virus hemagglutinin. EMBO J 2002; 21(21):5701-5710.
57. Puri A, Booy FP, Doms RW et al. Conformational changes and fusion activity of influenza virus hemagglutinin of the H2 and H3 subtypes: Effects of acid pretreatment. J Virol 1990; 64(8):3824-3832.
58. Markovic I, Leikina E, Zhukovsky M et al. Synchronized activation and refolding of influenza hemagglutinin in multimeric fusion machines. J Cell Biol 2001; 155(5):833-844.
59. Carr CM, Kim PS. A spring-loaded mechanism for the conformational change of influenza hemagglutinin. Cell 1993; 73(4):823-832.
60. Colman PM, Lawrence MC. The structural biology of type I viral membrane fusion. Nat Rev Mol Cell Biol 2003; 4(4):309-319.
61. Park HE, Gruenke JA, White JM. Leash in the groove mechanism of membrane fusion. Nat Struct Biol 2003; 10(12):1048-1053.
62. Han X, Bushweller JH, Cafiso DS et al. Membrane structure and fusion-triggering conformational change of the fusion domain from influenza hemagglutinin. Nat Struct Biol 2001; 8(8):715-720.
63. Hernandez LD, Hoffman LR, Wolfsberg TG et al. Virus-cell and cell-cell fusion. Ann Rev Cell Dev Biol 1996; 12:627-661.
64. Russell DG, Marsh M. Endocytosis in pathogen entry and replication. In: Marsh M, ed. Endocytosis. Oxford: Oxford University Press, 2001:247-280.
65. Danieli T, Pelletier SL, Henis YI et al. Membrane fusion mediated by the influenza virus hemagglutinin requires the concerted action of at least three hemagglutinin trimers. J Cell Biol 1996; 133(3):559-569.
66. Takeda M, Leser GP, Russell CJ et al. Influenza virus hemagglutinin concentrates in lipid raft microdomains for efficient viral fusion. Proc Natl Acad Sci USA 2003; 100(25):14610-14617.
67. Sun X, Whittaker GR. Role for influenza virus envelope cholesterol in virus entry and infection. J Virol 2003; 77(23):12543-12551.
68. Bui M, Whittaker G, Helenius A. Effect of M1 protein and low pH on nuclear transport of influenza virus ribonucleoproteins. J Virol 1996; 70:8391-8401.
69. Martin K, Helenius A. Nuclear transport of influenza virus ribonucleoproteins: The viral matrix protein (M1) promotes export and inhibits import. Cell 1991; 67:117-130.
70. Martin K, Helenius A. Transport of incoming influenza virus nucleocapsids into the nucleus. J Virol 1991; 65:232-244.
71. Pinto LH, Holsinger LJ, Lamb RA. Influenza virus M2 protein has ion channel activity. Cell 1992; 69:517-528.
72. Sugrue RJ, Hay AJ. Structural characteristics of the M2 protein of influenza A viruses: Evidence that it forms a tetrameric channel. Virology 1991; 180:617-624.
73. Bukrinskaya AG, Vorkounova NK, Kornilayeva GV et al. Influenza virus uncoating in infected cells and effect of rimantadine. J Gen Virol 1982; 60:49-59.
74. Hay AJ, Wolstenholme AJ, Skehel JJ et al. The molecular basis of the specific anti-influenza action of amantadine. EMBO J 1985; 4:3021-3024.
75. Whittaker GR, Digard P. Entry and transport of influenza virus. In: Kawaoka Y, ed. Influenza Virology. Current Topics. Wymondham, UK: Caister Academic press, 2006:37-64.
76. Cros JF, Garcia-Sastre A, Palese P. An unconventional NLS is critical for the nuclear import of the influenza A virus nucleoprotein and ribonucleoprotein. Traffic 2005; 6(3):205-213.
77. O'Neill RE, Jaskunas R, Blobel G et al. Nuclear import of influenza virus RNA can be mediated by viral nucleoprotein and transport factors required for protein import. J Biol Chem 1995; 270:22701-22704.
78. Kemler I, Whittaker G, Helenius A. Nuclear import of microinjected influenza virus ribonucleoproteins. Virology 1994; 202:1028-1033.
79. Babcock HP, Chen C, Zhuang X. Using single-particle tracking to study nuclear trafficking of viral genes. Biophys J 2004; 87(4):2749-2758.
80. Russell RJ, Gamblin SJ, Haire LF et al. H1 and H7 influenza haemagglutinin structures extend a structural classification of haemagglutinin subtypes. Virology 2004; 325(2):287-296.

CHAPTER 5

FILOVIRUS ENTRY

Graham Simmons
Blood Systems Research Institute; Department of Laboratory Medicine, University of California, San Francisco, San Francisco, California, USA
Emails: gsimmons@bloodsystems.org; graham.simmons@ucsf.edu.

Abstract: A number of advances in recent years have significantly furthered our understanding of filovirus attachment and cellular tropism. For example, several cell-surface molecules have been identified as attachment factors with the potential to facilitate the in vivo targeting of particular cell types such as macrophages and hepatic cells. Furthermore, our knowledge of internalization and subsequent events during filovirus entry has also been widened, adding new variations to the paradigms for viral entry established for HIV and influenza. In particular, host cell factors such as endosomal proteases and the intracellular receptor Niemann-Pick C1 are now known to play a vital role in activating the membrane fusion potential of filovirus glycoproteins.

INTRODUCTION

The family filoviridae consists of just two members, Marburg virus (MARV) and Ebola virus, first identified in 1967 and 1976, respectively. The Ebola viruses comprise five distinct species: the prototypical Ebola virus (EBOV, *Zaire ebolavirus*); *Sudan ebolavirus* (SUDV); *Bundibugyo ebolavirus* (BDBV); *Taï Forest ebolavirus* (TAFV, formerly known as Côte d'Ivoire) and *Reston ebolavirus* (RESTV). Filoviruses are responsible for sporadic, highly lethal outbreaks of severe hemorrhagic fever in both humans and apes in sub-Saharan Africa. While human cases of filovirus infection are rare, with less than 1700 deaths and 2500 total cases between their discovery in 1967 and 2010,[1] recent years have seen a more sustained level of outbreaks, including multiple introductions of filoviruses into the human population. Although the primary animal host for the filoviruses is still unclear, as with other tropical viral diseases, bats have been strongly implicated as a possible reservoir.[2,3] In addition, the less pathogenic RESTV has been identified in populations of pigs in Asia.[4]

Viral Entry into Host Cells, edited by Stefan Pöhlmann and Graham Simmons.

FILOVIRUS GLYCOPROTEIN

Filoviruses encode a single membrane-bound surface glycoprotein, termed GP, responsible for mediating viral attachment and entry into cells. While MARV encodes GP within a single open-reading frame, the equivalent open-reading frame for ebola virus encodes a truncated, non-structural protein, termed sGP, secreted by infected cells. However, in approximately 20% of GP transcripts, a cotranscriptional editing event introduces a frame-shifting adenine, leading to production of full-length GP.[5] Given that sGP is easily detectable in the blood of infected individuals,[6] a number of roles have been proposed, including sGP synthesis as a mechanism for controlling membrane-bound GP expression levels.[7] The virus may be required to limit GP expression due to excessive levels of membrane-bound GP leading to a loss of cell adhesion, resulting from a downregulation of adhesion molecules.[8,9] In addition, viral particles can tolerate very inefficient levels of GP incorporation without a resultant loss of infectivity,[10] thus methods to reduce GP expression, in particular from the surface of infected cells, may help blunt the immune response mounted against EBOV.

As with many other viral fusion proteins, full-length GP forms a trimer on the virion surface, with each monomer proteolytically processed into two subunits during transit through the trans golgi by the proprotein convertase furin.[11,12] The resulting proteins, a surface subunit, GP_1, and a transmembrane-bound subunit, GP_2, are linked by a single disulfide bond.[13] Unlike most viral fusion proteins,[14] proteolysis of EBOV-GP at the furin cleavage site is not absolutely required for GP to mediate membrane fusion and infection, both in tissue culture and in vivo.[15,16] However, due to the generally conserved nature of the cleavage site between the different strains of EBOV and MARV, it seems likely that cleavage plays a role during natural infection.

GP contains many of the hallmarks of a class I viral fusion protein, most notably the presence of two heptad repeats within GP_2 that associate to form an antiparallel six-helix bundle during membrane fusion.[17] In addition, a highly hydrophobic fusion peptide is positioned close to the N terminus of GP_2. As with a subset of other class I viral fusion proteins, the fusion peptide is not at the very amino terminus of GP_2, but rather forms an internal loop created by two flanking cysteine residues linked by a disulfide-bond. This may explain why furin-mediated cleavage of GP is not absolutely required for function, as viruses without fusion peptides bounded by two cysteines require the cleavage event itself in order to free the fusion peptide.[18]

Filovirus GP is extensively glycosylated,[19] as simply demonstrated by a dramatic increase in mobility by SDS-PAGE following treatment with glycosidases such as PNGase F. Both N- and O-linked carbohydrate moieties are present, with a particularly high density situated within a serine and threonine-rich region in the C-terminal half of GP_1. This region is highly variable between the four characterized strains of EBOV, but due to the high density of O-linked glycans it shows homology to mucins. This region is also markedly hydrophilic, and due to the presence of a disulfide bond linking the more hydrophobic N-terminal region of GP_1 to GP_2, the C-terminus of GP_1 is thought to project out into aqueous milieu.[13,20] Surprisingly, deletion of the mucin-like domain from GP_1 does not impact infectivity mediated by EBOV-GP,[8,9] but rather actually enhances infection.[21] Thus, it is unlikely that the determinants for receptor binding lie within this region of the glycoprotein. However, due to its highly glycosylated nature, and its position distal to the viral membrane, the mucin-like domain likely does contribute to relatively nonspecific viral attachment to some cellular lectins (see below). Further mutagenic mapping of GP_1 highlights the N-terminal 150-160 amino-acids as critical for both EBOV- and MARV-GP mediated entry, suggesting that in fact the amino terminal region may be the site of the receptor-binding domain.[22,23] Phenylalanines at positions 88 and 159, in particular, appear to be critical for GP function.[24]

CELLULAR TROPISM

Studies using filovirus GP pseudotyped onto retroviral or rhabdovirus cores reveal a very broad range of cell tropism in vitro.[25-28] In addition to the transduction of a diverse set of cell types, cell lines from a range of mammalian and avian sources were found to be valid targets for filovirus GP mediated entry. The only consistent exception to the pan-tropism of filoviruses are cell lines of lymphoid origin, which are totally refractory to entry mediated by EBOV-GP.[26] These in vitro findings mirror those seen in infected patients and experimentally challenged animals, where little or no viral RNA or protein has been observed in lymphocytes.[29] Thus, the step of viral entry appears to be the major determinant for filovirus tropism. Cells of the mononuclear phagocytic system are sites of early and sustained viral replication (reviewed in ref. 30). It is likely that organ-specific macrophages act as initial targets for viral replication within specific organs,[31] while blood monocyte/macrophages play a role in dissemination of the virus to tissues,[32] although in vitro assays suggest monocytes may be less infectible than fully mature macrophages.[33] Dendritic cells are also a target for filovirus replication,[34] and may also disseminate virus. Later in infection, other cell types, in addition to macrophages, demonstrate high levels of viral replication, including hepatocytes and endothelial cells, as well as a range of other parenchymal cells. All organs participate in viral replication, with the liver, lungs and spleen appearing to be central sites for viral production and viral induced damage.

RECEPTORS FOR FILOVIRUS ENTRY

Generally, in order to gain entry into target cells, enveloped viruses such as the filoviruses must first engage specific cellular receptors. The presence of these receptors on target cells is thus a major determinant of cellular and tissue tropism. Detailed studies of entry mediated by EBOV and MARV glycoproteins, established that for EBOV at least, a proteinaceous component, with a requirement for glycosylation, is necessary for entry.[25,27] Interestingly, neither treatment with pronase, nor disrupting carbohydrate addition, inhibited entry mediated by MARV GP,[27] suggesting that either the two filoviruses use different receptors, or there are functional differences in the way the same receptor is utilized. The fact that pronase treatment did not negatively impact MARV entry does not exclude the role of proteins in infection as other viruses known to use proteinaceous receptors are also insensitive to pronase.[27] In support of this, soluble versions of EBOV GP can block MARV infection, and vice versa, suggesting a shared receptor.[23]

Recently, a bioinformatics approach was used to identify T-cell immunoglobulin and mucin-domain containing protein-1 (TIM-1) as a potential receptor for EBOV on mucosal epithelia.[35] It remains to be seen whether TIM-1 acts as a true receptor, or merely as a tissue-specific attachment factor, particularly as TIM-1 is also expressed on subsets of activated T cells which are refractory to infection. Either way, use of TIM-1 could explain the high levels of filovirus replication seen in lung tissue. Three members of the Tyro3 receptor tyrosine kinase family have been demonstrated to facilitate filovirus infection.[36] Stable expression of any of Axl, Dtk or Mer in refractory lymphocyte cell lines permitted infection with either pseudovirions or live virus. In turn, antibodies, Gas6 ligand and soluble versions of the ectodomains from each Tyro3 family member were able to prevent the enhancement of infection seen on Tyro3-expressing lymphocytes.[36] Axl, but not Dtk or Mer, is readily detectable on many EBOV sensitive cell lines, such as Vero, HT1080 and HeLa cells.[36] Antibodies and RNAi directed against Axl potently inhibited infection on some of these lines. However, in Vero cells no inhibition was noted, suggesting the existence of alternate factors for EBOV infection in these cells.[36,37] Furthermore, RNAi knockout of Axl did not alter binding to cells, nor was direct binding between the extracellular domains of Axl and EBOV GP noted, strongly suggesting Axl does not operate directly as a receptor for the virus, but rather acts downstream.[37]

Folate receptor alpha (FRα) has also been implicated as a receptor for filoviruses.[38] Transcripts encoding FRα, that allowed MARV GP-mediated infection of a normally refractory T-cell line, were identified in a cDNA library screen.[38] Furthermore, expression of FRα on T cells reconstituted infection by EBOV GP-bearing pseudotypes, as well as live MARV and EBOV.[38] Surprisingly, the initial cDNA clone of FRα isolated in these studies was not full length, but rather was truncated at the 5' end and hence failed to encode a functional signal peptide for FRα.[38] It is thus unclear how such a protein would be expressed on the cell surface and hence act as a receptor for viral entry. In other studies, transfection of refractory cell lines with plasmids expressing functional FRα was unable to reconstitute EBOV GP-mediated infection.[39] Regardless of the role of FRα in entry, other molecules must be capable of functioning as receptors for filoviruses, as GP can mediate efficient infection of primary and established cell lines that are negative for both FRα mRNA and protein.[38,39] In addition, various ligands to FRα, such as folate and antibodies, were unable to consistently inhibit EBOV GP mediated infection of a range of cell types.[21,39]

A number of other molecules, such as β1 integrins,[40] have been postulated to be involved in filovirus entry, however, none of these potential receptors has passed the acid test of making refractory cells permissive to filoviruses. Thus, these molecules may only play minor, if any, roles in filovirus attachment and entry into target cells. Given the novel mechanisms required for triggering EBOV mediated fusion described below, it may well in fact be that specific cell surface receptors are not required by this virus, but rather any of a multitude of less specific attachment factors or random events leading to internalization are sufficient to lead to entry. Lack of specific receptors for entry has also be hypothesized for highly pH-dependent viruses such as many of the flaviviurses. One could also speculate that with their large, filamentous shape, filoviruses could rely on numerous low-affinity surface interactions for effective initiation of attachment.

ATTACHMENT FACTORS

For many viruses, receptor engagement is a relatively inefficient process.[41] Thus, additional cellular factors that enhance viral attachment to the cell surface can dramatically alter infectivity and cellular tropism without being absolutely required for viral infection (see Chapter 1). These so-called attachment factors often recognize and bind to viruses in a relatively nonspecific manner, for example through carbohydrate modifications on the viral envelope proteins. Calcium-dependent (C-type) lectins represent one of a number of families of molecules termed pattern recognition receptors that are responsible for identifying, and inducing responses to, unique pathogen signatures. Ironically, many of these molecules have been subverted by pathogens in order to either gain entry into antigen-presenting cells such as macrophages, or to subvert immune responses. The prototypical C-type lectin involved in pattern recognition is CD209 (formerly known as DC-SIGN). CD209 maps to chromosome 19p13.3, together with a cluster of related C-type lectins including CD23, CD209L (also called DC-SIGNR or L-SIGN) and LSECtin.[42,43] CD209 and the highly related CD209L (henceforth collectively referred to as CD209(L)) are tetrameric, membrane-anchored lectins reported to act as ligands for the intercellular adhesion molecules (ICAM) -2 and -3, through recognition of N-linked high-mannose carbohydrate moieties.[42,44,45] Transcripts corresponding to CD209 were originally identified in screens of human placental cDNA libraries for molecules capable of binding HIV gp120.[46] Subsequent analysis demonstrated the ability of CD209(L) to interact with glycoproteins from a variety of pathogens, including HIV, hepatitis C virus, dengue, Leishmania and Mycobacterium tuberculosis.[47-52] However, CD209(L) are not universal attachment factors, as glycoproteins from many viruses, including vesicular stomatitis virus, herpes simplex virus and lassa virus, are not enhanced by CD209(L).[53] Some viral glycoproteins that do not interact with CD209(L) efficiently can be engineered to contain high-mannose carbohydrate moieties by treating producer cells with mannosidase I inhibitors.[54] This leads to a marked increase in binding to, and usage of, CD209(L), suggesting that the presence or absence of high mannose is the major determinant of viral glycoprotein interactions with CD209(L).[55] However, it has also been suggested that the spatial arrangement of high-mannose moieties either within a single glycoprotein, or between multiple glycoproteins on the viral surface, is also important in order to allow optimal interactions with these tetrameric calcium-dependent lectins.[56,57]

The composition of the glycans decorating filovirus GP is highly heterologous, consisting of both high-mannose and complex carbohydrate structures.[55,58,59] Thus, it is not surprising that mammalian lectins, such as CD209(L), are able to interact with filovirus GP. The expression of CD209(L) on primary macrophages led to an almost ten-fold enhancement of infection by live, replication-competent EBOV, despite these cells already being a highly competent cell type for EBOV replication.[53] Similarly, transduction of infectible cells by retroviral pseudovirions bearing either EBOV or MARV GP is dramatically enhanced by transient expression of CD209(L), as well as another C-type lectin clustered with CD209(L) on chromosome 19, LSECtin.[53,60-62] Whether expression of CD209(L) on nonpermissive lymphocyte cell lines can make them permissive to filovirus infection is controversial.[53,60] While CD209(L) may inefficiently directly mediate viral infection, this may be a moot point as the majority of CD209(L) positive cells likely also express other receptors for filovirus entry. It is more likely that CD209(L) predominately act to concentrate virus at the cell surface of target cells, and hence increase the likelihood of GP interactions with its cognate receptor. Thus, CD209(L) may act in vivo to target particular cell types for enhanced infection, as well as promoting infection despite very low levels of infectious viral particles, for example during transmission.

Ex-vivo, CD209 is very highly expressed on monocyte-derived dendritic cells but not monocyte-derived macrophages.[45] However, somewhat lower expression is observed on a variety of both dendritic cells and tissue macrophages in vivo, as well as liver sinusoidal endothelium.[45,63-66] CD209L, together with LSECtin, are found predominantly on microvascular endothelial cells in lymph nodes and liver sinusoids.[43,67] Many of these cell types may be important for the establishment and spread of filovirus throughout its target organs.

Hepatic asialoglycoprotein receptor (ASGP-R) was the first attachment factor to be identified for filoviruses.[68] ASGP-R binds galactose on asialylated carbohydrate structures, such as the serum constituent, asialofetuin.[69] Asialofetuin, together with antisera to ASGP-R, were able to block MARV infection of ASGP-R positive HepG2 cells, while transient expression of ASGP-R in mouse cells led to enhanced infection.[68] Similarly, ASGP-R is also able to enhance infection mediated by EBOV GP on already infectible cell lines by over 30-fold.[55] Given that the liver is an important target organ for filovirus infection, ASGP-R may play a role in targeting filovirus infection of liver cells, due to its presence on hepatocytes. Another calcium dependent lectin with specificity for galactose, human macrophage galactose- and N-acetylgalactosamine-specific C-type lectin (hMGL), has also been shown to significantly enhance infection mediated by filovirus glycoproteins.[70] As its name suggests, hMGL is highly expressed on macrophages, as well as dendritic cells, important cell types for filovirus infection in vivo, particularly early during infection. For both galactose-specific lectins, enhancement of infection mediated by EBOV GP lacking the mucin-like domain is greatly reduced[70] (Unpublished observations—G. Simmons). Thus, the determinants for binding likely predominantly lie within this highly glycosylated region.

Given that lectins are expressed on a range of cell types that could function as "doorways" to viral infection within mucosal and epithelial surfaces, as well as target organs, it is tempting to speculate that lectins play an important role both in transmission of filovirus infections, and the dissemination of virus throughout the body.

ROUTES OF ENTRY

A major component of filovirus GP-mediated membrane fusion is an acidic pH-dependent step (see section on mechanisms of membrane fusion below). This implies that the virus requires internalization and trafficking to the low pH environment of an acidified endosome, as demonstrated by a necessity for an intact microtubule network within target cells.[33] Several potential routes of internalization are present within cells and have been hijacked by various viruses for the purpose of entry or trafficking to specific subcellular compartments.[71] These include classical clathrin-mediated endocytosis, lipid raft-associated caveolae, macropinocytosis and less defined nonclathrin, noncaveolae routes of entry. Several groups have determined that cholesterol and lipid rafts are required for EBOV GP-mediated infection, both using pseudovirions, and live virus.[33,72,73] Given that caveolae invaginate from cholesterol-rich lipid rafts, Empig et al looked at the partitioning of GP-bearing pseudovirions following exposure of target cells.[72] Indeed, following internalization, pseudotypes incorporating either EBOV or MARV GP, colocalized with markers of caveolae to a large extent, suggesting a role for caveolae in filovirus entry. However, cells lacking functional components for caveolae formation, such as caveolin-1, remain fully infectious to pseudovirions bearing EBOV GP.[39] Utilizing pseudoparticles, and inhibition of live virus with specific inhibitors, other groups have also implicated clathrin-mediated endocytosis as an efficient mode of entry.[74] However, it appears likely that in this instance pseudoparticles do not adequately mimic authentic virus particles which are typically filamentous and up to 1-2 μm in length. Studies with virus-like particles (VLPs) and replication-competent virus have largely coalesced opinion around the idea that macropinocytosis is the predominant route of entry for filoviruses, with clathrin-dependent endocytosis playing a lesser role.[75-77] Furthermore, it appears that Axl, which had previously been suggested as a receptor, is involved in mediating early events in internalization through macropinocytosis.[78]

MECHANISMS OF MEMBRANE FUSION

Class I viral fusion protein induced membrane fusion, such as that mediated by filovirus GP, occurs through a complex cascade of conformational rearrangements within the glycoprotein. Mature, native class I glycoproteins generally exist as trimers of heterodimers held in a so-called metastable state, primed for fusion.[79] The metastable protein is destabilized during attachment and entry, leading to the exposure of the hydrophobic fusion peptide and subsequently membrane fusion. Three distinct triggers of these conformational rearrangements have been elucidated for the induction of membrane fusion by class I viral glycoproteins; interactions with receptor(s) as seen with HIV, exposure to low pH as is the case with influenza virus, or a two-step process requiring specific interactions with receptor followed by acidic pH treatment.[80]

The exact triggers necessary for filovirus GP conformational change have yet to be elucidated, but it appears that filoviruses present a new paradigm whereby additional non-cell surface cellular factors are required. Low pH clearly plays an important role in filovirus entry as infection by GP-bearing pseudovirions can be inhibited by agents such as bafilomycin A that prevent acidification of endosomes.[25-27] Interestingly, however, in cell-to-cell fusion assays, while acidic pH is required in order to prime the membrane fusion potential of EBOV GP expressing effector cells, treatment of target cells with low pH inhibits fusion.[81] Also, unlike many pH-dependent viruses, acid treatment of virus bound to cells does not induce fusion at the plasma membrane and hence viral entry.[82] Likewise, preincubation at pH 5 does not inactivate EBOV GP, suggesting that low pH does not act as a trigger of irreversible conformational rearrangements within the glycoprotein. These findings suggest, as with other viruses inhibited by Bafilomycin A but not directly sensitive to low pH,[83] that rather than acting purely as a direct trigger, the requirement for low pH indicates the necessity for the action of a cellular factor that is itself sensitive to endosomal pH. Indeed, inhibitors of acid-dependent endosomal cysteine proteases specifically inhibit EBOV GP-mediated entry.[84,85] In particular, a specific inhibitor of the ubiquitous endosomal protease, cathepsin B (CTSB) inhibits both EBOV GP bearing pseudovirions and live EBOV infection.[84] The requirement for CTSB was confirmed by an 80-90% loss of infectivity on both CTSB deficient mouse cells and Vero cells treated with RNAi duplexes capable of reducing CTSB activity by 85%.[84,85] A second cathepsin, cathepsin L (CTSL) was also demonstrated to play a possibly more minor role. Specific inhibitors and loss of function experiments suggest that CTSL has a synergistic effect together with CTSB, but is not sufficient for entry by itself. More recent data suggests that while EBOV, TAFV and BDBV are strongly dependent on CTSB, SUDV, RESTV and MARV have a requirement for as yet unidentified proteases.[86,87]

Cleavage of EBOV GP by CTSB and CTSL can be performed in vitro, however the findings of different laboratories have not been consistent, perhaps due to separate preparations of proteases. Chandran et al demonstrate CTSL can digest GP1 to leave an 18kDa N-terminal fragment associated with GP2.[84] CTSB can also perform this digest, although somewhat less efficiently. Interestingly, the 18kDa form is still infectious further supporting the findings that the receptor binding domain of GP1 lies within its amino terminus.[22] Infection mediated by pseudovirions bearing the 18kDa form remains inhibitible by high concentrations of specific CTSB inhibitors and these viruses do not infect cells from CTSB knockout mice. However, infection by the 18kDa form is significantly enhanced compared to untreated virus on cells undergoing partial inhibition of CTSB (i.e., likely to prevent the inefficient production of the 18kDa form mediated by CTSB) and full inhibition of CTSL. This supports a two-step process whereby CTSL efficiently digests GP1 to leave an 18 kDa fragment, which is then further digested by CTSB. Indeed incubation of the 18 kDa form with CTSB, but not CTSL, leads to a total loss of GP1 from particles. The fact that CTSB can perform both steps, albeit less efficiently, explains why CTSB inhibitors are more effective than those directed against CTSL, but not as potent as both together.

In contrast to the findings of Chandran et al, Schornberg and colleagues[85] report that CTSB treatment reduces GP1 (approximately 130 kDa) to a 50 kDa species—consistent with the loss of the mucin-like domain from GP1. A second minor species at 19 kDa was also noted. CTSL treatment predominantly leads to a 20 kDa form, while combined cleavage gives a doublet of the 20 and 19 kDa fragments.[85] These partially digested pseudovirions demonstrate enhanced infectivity, and while they are no longer sensitive to CTSB inhibition or ablation, they retain a requirement for acidified endosomes.

Cathepsin-mediated proteolysis also plays a role in the entry of several other viruses, including the corona- and reoviruses.[88-90] In the case of the coronavirus, SARS-CoV, temperature-dependent interactions with receptor are required prior to cathepsin-mediated cleavage—an interesting modification of a two-step trigger mechanism for induction of membrane fusion.[89] It appears that almost the reverse situation occurs for filoviruses. Proteolysis proceeds receptor engagement - which uniquely occurs on internal membranes. Following proteolysis, exposure of the receptor binding domain (RBD) allows EBOV GP to interact with Niemann-Pick C1 (NPC1), a cholesterol transporter present in late endosomal membranes.[91,92] NPC1 can be demonstrated to directly bind to EBOV-GP and to allow infection of refractory cell types.[93] Thus, the long search for a true filovirus receptor has finally yielded results in an unexpected place—inside the cell.

CONCLUSION

The recent advances in understanding filovirus entry highlight the requirement by enveloped viruses for a range of host factors other than classical receptors in order to achieve efficient entry. External membrane-bound molecules such as DC-SIGN and glycosaminoglycans are able to concentrate filovirus at the cell surface in a fairly nonspecific manner, while a more specific requirement for proteolysis by individual endosomal proteases is observed post-internalization. The recently identified lysosomal membrane located molecule, NPC-1, can then function as the true receptor, driving membrane fusion and entry.

ACKNOWLEDGMENTS

The author was supported by grant R01AI074986 from the National Institute of Allergy and Infectious Diseases.

REFERENCES

1. Leroy EM, Gonzalez J-P, Baize S. Ebola and Marburg haemorrhagic fever viruses: major scientific advances, but a relatively minor public health threat for Africa. Clin Microbiol Infect 2011; 17:964-976.
2. Leroy EM, Kumulungui B, Pourrut X et al. Fruit bats as reservoirs of Ebola virus. Nature 2005; 438:575-576.
3. Towner JS, Amman BR, Sealy TK et al. Isolation of genetically diverse Marburg viruses from Egyptian fruit bats. PLoS Pathog 2009; 5:e1000536.
4. Barrette RW, Metwally SA, Rowland JM et al. Discovery of swine as a host for the reston Ebola virus. Science 2009; 325:204-206.
5. Volchkov VE, Becker S, Volchkova VA et al. GP mRNA of Ebola virus is edited by the Ebola virus polymerase and by T7 and vaccinia virus polymerases. Virology 1995; 214:421-430.
6. Sanchez A, Ksiazek TG, Rollin PE et al. Detection and molecular characterization of Ebola viruses causing disease in human and nonhuman primates. J Infect Dis 1999; 179(Suppl 1):S164-169.
7. Volchkov VE, Volchkova VA, Muhlberger E et al. Recovery of infectious Ebola virus from complementary DNA: RNA editing of the GP gene and viral cytotoxicity. Science 2001; 291:1965-1969.
8. Yang ZY, Duckers HJ, Sullivan NJ et al. Identification of the Ebola virus glycoprotein as the main viral determinant of vascular cell cytotoxicity and injury. Nat Med 2000; 6:886-889.
9. Simmons G, Wool-Lewis RJ, Baribaud F et al. Ebola virus glycoproteins induce global surface protein down-modulation and loss of cell adherence. J Viro 2002; 76:2518-2528.
10. Marzi A, Wegele A, Pohlmann S. Modulation of virion incorporation of Ebolavirus glycoprotein: Effects on attachment, cellular entry and neutralization. Virology 2006; 352:345-56.
11. Volchkov VE, Feldmann H, Volchkova VA et al. Processing of the Ebola virus glycoprotein by the proprotein convertase furin. Proc Natl Acad Sci USA 1998; 95:5762-5767.
12. Volchkov VE, Volchkova VA, Stroher U et al. Proteolytic processing of Marburg virus glycoprotein. Virology 2000; 268:1-6.
13. Jeffers SA, Sanders DA, Sanchez A. Covalent modifications of the ebola virus glycoprotein. J Virol 2002; 76:12463-12472.
14. Klenk HD, Garten W. Host cell proteases controlling virus pathogenicity. Trends Microbiol 1994; 2:39-43.
15. Wool-Lewis RJ, Bates P. Endoproteolytic processing of the ebola virus envelope glycoprotein: cleavage is not required for function. J Virol 1999; 73:1419-1426.
16. Neumann G, Geisbert TW, Ebihara H et al. Proteolytic processing of the Ebola virus glycoprotein is not critical for Ebola virus replication in nonhuman primates. J Virol 2007; 81:2995-2998.
17. Weissenhorn W, Carfi A, Lee KH et al. Crystal structure of the Ebola virus membrane fusion subunit, GP2, from the envelope glycoprotein ectodomain. Mol Cell 1998; 2:605-616.
18. White JM, Delos SE, Brecher M, Schornberg K. Structures and mechanisms of viral membrane fusion proteins. Crit Rev Biochem Mol Biol 2008; 43(3):189-219.
19. Geyer H, Will C, Feldmann H et al. Carbohydrate structure of Marburg virus glycoprotein. Glycobiology 1992; 2:299-312.
20. Lee JE, Fusco ML, Hessell AJ et al. Structure of the ebola virus glycoprotein bound to an antibody from a human survivor. Nature 2008; 454:177-182.
21. Sinn PL, Hickey MA, Staber PD et al. Lentivirus vectors pseudotyped with filoviral envelope glycoproteins transduce airway epithelia from the apical surface independently of folate receptor alpha. J Virol 2003; 77:5902-5910.
22. Manicassamy B, Wang J, Jiang H et al. Comprehensive analysis of ebola virus GP1 in viral entry. J Virol 2005; 79:4793-4805.
23. Kuhn JH, Radoshitzky SR, Guth AC et al. Conserved receptor-binding domains of Lake Victoria marburgvirus and Zaire ebolavirus bind a common receptor. J Biol Chem 2006; 281:15951-15958.
24. Mpanju OM, Towner JS, Dover JE et al. Identification of two amino acid residues on Ebola virus glycoprotein 1 critical for cell entry. Virus Res 2006; 121:205-14.
25. Takada A, Robison C, Goto H et al. A system for functional analysis of Ebola virus glycoprotein. Proc Natl Acad Sci USA 1997; 94:14764-14769.
26. Wool-Lewis RJ, Bates P. Characterization of Ebola virus entry by using pseudotyped viruses: identification of receptor-deficient cell lines. J Virol 1998; 72:3155-3160.
27. Chan SY, Speck RF, Ma MC et al. Distinct mechanisms of entry by envelope glycoproteins of Marburg and Ebola (Zaire) viruses. J Virol 2000; 74:4933-4937.
28. Ito H, Watanabe S, Takada A et al. Ebola Virus Glycoprotein: Proteolytic Processing, Acylation, Cell Tropism, and Detection of Neutralizing Antibodies. J Virol 2001; 75:1576-1580.
29. Geisbert TW, Hensley LE, Gibb TR et al. Apoptosis induced in vitro and in vivo during infection by Ebola and Marburg viruses. Lab Invest 2000; 80:171-186.
30. Schnittler HJ, Feldmann H. Molecular pathogenesis of filovirus infections: role of macrophages and endothelial cells. Curr Top Microbiol Immunol 1999; 235:175-204.

31. Schnittler HJ, Feldmann H. Marburg and Ebola hemorrhagic fevers: does the primary course of infection depend on the accessibility of organ-specific macrophages? Clin Infect Dis 1998; 27:404-406.
32. Stroher U, West E, Bugany H et al. Infection and activation of monocytes by Marburg and Ebola viruses. J Virol 2001; 75:11025-11033.
33. Yonezawa A, Cavrois M, Greene WC. Studies of ebola virus glycoprotein-mediated entry and fusion by using pseudotyped human immunodeficiency virus type 1 virions: involvement of cytoskeletal proteins and enhancement by tumor necrosis factor alpha. J Virol 2005; 79:918-926.
34. Bosio CM, Aman MJ, Grogan C et al. Ebola and Marburg viruses replicate in monocyte-derived dendritic cells without inducing the production of cytokines and full maturation. J Infect Dis 2003; 188:1630-1638.
35. Kondratowicz AS, Lennemann NJ, Sinn PL et al. T-cell immunoglobulin and mucin domain 1 (tim-1) is a receptor for zaire ebolavirus and lake victoria marburgvirus. Proc Natl Acad Sci USA 2011; 108:8426-8431.
36. Shimojima M, Takada A, Ebihara H et al. Tyro3 family-mediated cell entry of Ebola and Marburg viruses. J Virol 2006; 80:10109-10116.
37. Brindley MA, Hunt CL, Kondratowicz AS et al. Tyrosine kinase receptor axl enhances entry of zaire ebolavirus without direct interactions with the viral glycoprotein. Virology 2011; 415:83-94.
38. Chan SY, Empig CJ, Welte FJ et al. Folate receptor-alpha is a cofactor for cellular entry by Marburg and Ebola viruses. Cell 2001; 106:117-126.
39. Simmons G, Rennekamp AJ, Chai N et al. Folate receptor alpha and caveolae are not required for Ebola virus glycoprotein-mediated viral infection. J Virol 2003; 77:13433-13438.
40. Takada A, Watanabe S, Ito H et al. Downregulation of beta1 integrins by Ebola virus glycoprotein: implication for virus entry. Virology 2000; 278:20-26.
41. O'Doherty U, Swiggard WJ, Malim MH. Human immunodeficiency virus type 1 spinoculation enhances infection through virus binding. J Virol 2000; 74:10074-10080.
42. Soilleux EJ, Barten R, Trowsdale J. DC-SIGN; a related gene, DC-SIGNR; and CD23 form a cluster on 19p13. J Immunol 2000; 165:2937-2942.
43. Liu W, Tang L, Zhang G et al. Characterization of a novel C-type lectin-like gene, LSECtin: demonstration of carbohydrate binding and expression in sinusoidal endothelial cells of liver and lymph node. J Biol Chem 2004; 279:18748-18758.
44. Geijtenbeek TB, Krooshoop DJ, Bleijs DA et al. DC-SIGN-ICAM-2 interaction mediates dendritic cell trafficking. Nat Immunol 2000; 1:353-357.
45. Geijtenbeek TB, Torensma R, van Vliet SJ et al. Identification of DC-SIGN, a novel dendritic cell-specific ICAM-3 receptor that supports primary immune responses. Cell 2000; 100:575-585.
46. Curtis BM, Scharnowske S, Watson AJ. Sequence and expression of a membrane-associated C-type lectin that exhibits CD4-independent binding of human immunodeficiency virus envelope glycoprotein gp120. Proc Natl Acad Sci USA 1992; 89:8356-8360.
47. Geijtenbeek TB, Kwon DS, Torensma R et al. DC-SIGN, a dendritic cell-specific HIV-1-binding protein that enhances trans-infection of T cells. Cell 2000; 100:587-597.
48. Pohlmann S, Zhang J, Baribaud F et al. Hepatitis C virus glycoproteins interact with DC-SIGN and DC-SIGNR. J Virol 2003; 77:4070-4080.
49. Tassaneetrithep B, Burgess TH, Granelli-Piperno A et al. DC-SIGN (CD209) mediates dengue virus infection of human dendritic cells. J Exp Med 2003; 197:823-829.
50. Colmenares M, Puig-Kroger A, Pello OM et al. Dendritic cell (DC)-specific intercellular adhesion molecule 3 (ICAM-3)-grabbing nonintegrin (DC-SIGN, CD209), a C-type surface lectin in human DCs, is a receptor for Leishmania amastigotes. J Biol Chem 2002; 277:36766-36769.
51. Geijtenbeek TB, Van Vliet SJ, Koppel EA et al. Mycobacteria target DC-SIGN to suppress dendritic cell function. J Exp Med 2003; 197:7-17.
52. Tailleux L, Schwartz O, Herrmann JL et al. DC-SIGN is the major Mycobacterium tuberculosis receptor on human dendritic cells. J Exp Med 2003; 197:121-127.
53. Simmons G, Reeves JD, Grogan CC et al. DC-SIGN and DC-SIGNR bind ebola glycoproteins and enhance infection of macrophages and endothelial cells. Virology 2003; 305:115-123.
54. Kaushal GP, Elbein AD. Glycosidase inhibitors in study of glycoconjugates. Methods Enzymol 1994; 230:316-329.
55. Lin G, Simmons G, Pohlmann S et al. Differential N-linked glycosylation of human immunodeficiency virus and Ebola virus envelope glycoproteins modulates interactions with DC-SIGN and DC-SIGNR. J Virol 2003; 77:1337-1346.
56. Mitchell DA, Fadden AJ, Drickamer K. A novel mechanism of carbohydrate recognition by the C-type lectins DC-SIGN and DC-SIGNR. Subunit organization and binding to multivalent ligands. J Biol Chem 2001; 276:28939-28945.
57. Davis CW, Nguyen HY, Hanna SL et al. West Nile virus discriminates between DC-SIGN and DC-SIGNR for cellular attachment and infection. J Virol 2006; 80:1290-1301.

58. Feldmann H, Nichol ST, Klenk HD et al. Characterization of filoviruses based on differences in structure and antigenicity of the virion glycoprotein. Virology 1994; 199:469-473.
59. Powlesland AS, Fisch T, Taylor ME et al. A novel mechanism for LSECtin binding to Ebola virus surface glycoprotein through truncated glycans. J Biol Chem 2008; 283(1):593-602.
60. Alvarez CP, Lasala F, Carrillo J et al. C-type lectins DC-SIGN and L-SIGN mediate cellular entry by Ebola virus in cis and in trans. J Virol 2002; 76:6841-6844.
61. Marzi A, Gramberg T, Simmons G et al. DC-SIGN and DC-SIGNR interact with the glycoprotein of Marburg virus and the S protein of severe acute respiratory syndrome coronavirus. J Virol 2004; 78:12090-12095.
62. Gramberg T, Hofmann H, Moller P et al. LSECtin interacts with filovirus glycoproteins and the spike protein of SARS coronavirus. Virology 2005; 340:224-36.
63. Soilleux EJ, Morris LS, Leslie G et al. Constitutive and induced expression of DC-SIGN on dendritic cell and macrophage subpopulations in situ and in vitro. J Leukoc Biol 2002; 71:445-457.
64. Soilleux EJ, Morris LS, Lee B et al. Placental expression of DC-SIGN may mediate intrauterine vertical transmission of HIV. J Pathol 2001; 195:586-592.
65. McCully ML, Chau TA, Luke P et al. Characterization of human peritoneal dendritic cell precursors and their involvement in peritonitis. Clin Exp Immunol 2005; 139:513-525.
66. Lai WK, Sun PJ, Zhang J et al. Expression of DC-SIGN and DC-SIGNR on human sinusoidal endothelium: a role for capturing hepatitis C virus particles. Am J Pathol 2006; 169:200-208.
67. Bashirova AA, Geijtenbeek TB, van Duijnhoven GC et al. A dendritic cell-specific intercellular adhesion molecule 3-grabbing nonintegrin (DC-SIGN)-related protein is highly expressed on human liver sinusoidal endothelial cells and promotes HIV-1 infection. J Exp Med 2001; 193:671-678.
68. Becker S, Spiess M, Klenk HD. The asialoglycoprotein receptor is a potential liver-specific receptor for Marburg virus. J Gen Virol 1995; 76:393-399.
69. Meier M, Bider MD, Malashkevich VN et al. Crystal structure of the carbohydrate recognition domain of the H1 subunit of the asialoglycoprotein receptor. J Mol Biol 2000; 300:857-865.
70. Takada A, Fujioka K, Tsuiji M et al. Human macrophage C-type lectin specific for galactose and N-acetylgalactosamine promotes filovirus entry. J Virol 2004; 78:2943-2947.
71. Sieczkarski SB, Whittaker GR. Dissecting virus entry via endocytosis. J Gen Virol 2002; 83:1535-1545.
72. Empig CJ, Goldsmith MA. Association of the caveola vesicular system with cellular entry by filoviruses. J Virol 2002; 76:5266-5270.
73. Bavari S, Bosio CM, Wiegand E et al. Lipid raft microdomains: a gateway for compartmentalized trafficking of Ebola and Marburg viruses. J Exp Med 2002; 195:593-602.
74. Bhattacharyya S, Warfield KL, Ruthel G et al. Ebola virus uses clathrin-mediated endocytosis as an entry pathway. Virology 2010; 401:18-28.
75. Saeed MF, Kolokoltsov AA, Albrecht T et al. Cellular entry of ebola virus involves uptake by a macropinocytosis-like mechanism and subsequent trafficking through early and late endosomes. PLoS Pathog 2010; 6:e1001110.
76. Nanbo A, Imai, M, Watanabe S et al. Ebolavirus is internalized into host cells via macropinocytosis in a viral glycoprotein-dependent manner. PLoS Pathog 2010; 6:e1001121.
77. Aleksandrowicz P, Marzi A, Biedenkopf N et al. Ebola virus enters host cells by macropinocytosis and clathrin-mediated endocytosis. J Infect Dis 2011; 204:S957-S967.
78. Hunt CL, Kolokoltsov AA, Davey RA et al. The tyro3 receptor kinase axl enhances macropinocytosis of zaire ebolavirus. J Virol 2011; 85:334-347.
79. Eckert DM, Kim PS. Mechanisms of viral membrane fusion and its inhibition. Annu Rev Biochem. 2001;70:777-810.
80. Mothes W, Boerger AL, Narayan S et al. Retroviral entry mediated by receptor priming and low pH triggering of an envelope glycoprotein. Cell 2000; 103(4):679-689.
81. Bar S, Takada A, Kawaoka Y et al. Detection of cell-cell fusion mediated by Ebola virus glycoproteins. J Virol 2006; 80(6):2815-2822.
82. Ito H, Watanabe S, Sanchez A et al. Mutational analysis of the putative fusion domain of Ebola virus glycoprotein. J Virol 1999; 73(10):8907-8912.
83. Simmons G, Reeves JD, Rennekamp AJ et al. Characterization of severe acute respiratory syndrome-associated coronavirus (SARS-CoV) spike glycoprotein-mediated viral entry. Proc Natl Acad Sci USA 2004; 101(12):4240-4245.
84. Chandran K, Sullivan NJ, Felbor U et al. Endosomal proteolysis of the Ebola virus glycoprotein is necessary for infection. Science 2005; 308(5728):1643-1645.
85. Schornberg K, Matsuyama S, Kabsch K et al. Role of endosomal cathepsins in entry mediated by the Ebola virus glycoprotein. J Virol 2006; 80(8):4174-4178.
86. Misasi J, Chandran K, Yang JY et al. Filoviruses require endosomal cysteine proteases for entry but exhibit distinct protease preferences. J Virol 2012; 86(6):3284-3292.

87. Gnirss K, Kühl A, Karsten C et al. Cathepsins B and L activate Ebola but not Marburg virus glycoproteins for efficient entry into cell lines and macrophages independent of TMPRSS2 expression. Virology 2012; 424(1):3-10.
88. Ebert DH, Deussing J, Peters C et al. Cathepsin L and cathepsin B mediate reovirus disassembly in murine fibroblast cells. J Biol Chem 2002; 277(27):24609-24617.
89. Simmons G, Gosalia DN, Rennekamp AJ et al. Inhibitors of cathepsin L prevent severe acute respiratory syndrome coronavirus entry. Proc Natl Acad Sci USA 2005; 102(33):11876-11881.
90. Qiu Z, Hingley ST, Simmons G et al. Endosomal proteolysis by cathepsins is necessary for murine coronavirus mouse hepatitis virus type 2 spike-mediated entry. J Virol 2006; 80(12):5768-5776.
91. Carette JE, Raaben M, Wong AC et al. Ebola virus entry requires the cholesterol transporter Niemann-Pick C1. Nature 2011; 477(7364):340-343.
92. Cote M, Misasi J, Ren T et al. Small molecule inhibitors reveal Niemann-Pick C1 is essential for Ebola virus infection. Nature 2011; 477(7364):344-348.
93. Miller EH, Obernosterer G, Raaben M et al. Ebola virus entry requires the host-programmed recognition of an intracellular receptor. EMBO J 2012; 31(8):1947-1960.

CHAPTER 6

PARAMYXOVIRUS ENTRY

Katharine N. Bossart,[1,2] Deborah L. Fusco[3] and Christopher C. Broder[3]

[1]Department of Microbiology and [2]National Emerging Infectious Diseases Laboratories Institute, Boston University School of Medicine, Boston, Massachusetts, USA; [3]Department of Microbiology and Immunology, Uniformed Services University, Bethesda, Maryland, USA.
Emails: kbossart@bu.edu; christopher.broder@usuhs.edu

Abstract: The family *Paramyxoviridae* consists of a group of large, enveloped, negative-sense, single-stranded RNA viruses and contains many important human and animal pathogens. Molecular and biochemical characterization over the past decade has revealed an extraordinary breadth of biological diversity among this family of viruses. Like all enveloped viruses, paramyxoviruses must fuse their membrane with that of a receptive host cell as a prerequisite for viral entry and infection. Unlike most other enveloped viruses, the vast majority of paramyxoviruses contain two distinct membrane-anchored glycoproteins to mediate the attachment, membrane fusion and particle entry stages of host cell infection. The attachment glycoprotein is required for virion attachment and the fusion glycoprotein is directly involved in facilitating the merger of the viral and host cell membranes. Here we detail important functional, biochemical and structural features of the attachment and fusion glycoproteins from a variety of family members. Specifically, the three different classes of attachment glycoproteins are discussed, including receptor binding preference, their overall structure and fusion promotion activities. Recently solved atomic structures of certain attachment glycoproteins are summarized, and how they relate to both receptor binding and fusion mechanisms are described. For the fusion glycoprotein, specific structural domains and their proposed role in mediating membrane merger are illustrated, highlighting the important features of protease cleavage and associated tropism and virulence. The crystal structure solutions of both an uncleaved and a cleavage-activated metastable F are also described with emphasis on how small conformational changes can provide the necessary energy to mediate membrane fusion. Finally, the different proposed fusion models are reviewed, featuring recent experimental findings that speculate how the attachment and fusion glycoproteins work in concert to mediate virus entry.

The views expressed in the manuscript are solely those of the authors, and they do not represent official views or opinions of the Department of Defense or the Uniformed Services University of the Health Sciences.

Viral Entry into Host Cells, edited by Stefan Pöhlmann and Graham Simmons.

INTRODUCTION TO PARAMYXOVIRUSES

The family *Paramyxoviridae* is an interesting group of large, enveloped, negative-sense, single-stranded RNA viruses that includes many important human and animal viruses such as measles virus (MeV), mumps virus, respiratory syncytial virus (RSV) and the human parainfluenza viruses (hPIV) in addition to animal viral agents such as Sendai virus (SeV), parainfluenza virus type 5 (PIV5) and canine distemper virus (CDV).[1] Several paramyxoviruses, including Newcastle disease virus (NDV) and rinderpest, pose major economic threats due to their possible impact on poultry and livestock industries.[2,3] In addition, more recently discovered members of the paramyxovirus family, namely Hendra virus (HeV) and Nipah virus (NiV), have been shown to possess a broad host range with the ability to infect and cause disease in a number of animal species as well as humans.[4,5]

Paramyxoviruses were originally classified as "myxoviruses" in the family *Orthomyxoviridae* due to the shared properties of hemagglutination and neuraminidase activity of the envelope glycoproteins of some members. However, paramyxoviruses differ from orthomyxoviruses in a number of critical aspects including genome organization, protein expression and replication strategies and more closely resemble other families in the order *Mononegavirales*, including *Rhabdoviridae*, *Filoviridae* and *Bornaviridae*.[6] The family *Paramyxoviridae* is divided into two subfamilies, *Paramyxovirinae* and *Pneumovirinae*. The classification is based on the organization of the genome, the molecular properties and biological activities of the encoded proteins, and morphological criteria.[1] Existing and proposed genera with examples of family members and unclassified viruses are summarized in Table 1. In 2002, two new genera were added to the *Paramyxovirinae* subfamily such that there are now 5 genera including *Respirovirus, Rubulavirus, Morbillivirus, Henipavirus* and *Avulavirus*.[7] NDV and other avian paramyxoviruses were removed from the genus *Rubulavirus* and reclassified in a new taxon, *Avulavirus*, due to differences in genome organization,[8] RNA editing profiles and phylogenetic comparisons.[9,10] The formation of the genus *Henipavirus*, which presently includes HeV and NiV, was justified in large part because of genome size, unique genome termini, limited homology with other family members and the different biological activities of various encoded proteins.[5,11] The *Pneumovirinae* includes the genera *Pneumovirus* and *Metapneumovirus*, and members include RSV and human metapneumovirus (hMPV), respectively.

Over the past few decades, a number of new paramyxoviruses have been identified, and in spite of the increase in the number of genera in the *Paramyxovirinae*, several members remain unclassified. J-virus (J-V) and the newly discovered Beilong virus (BeV) have been shown to be closely related to one another, yet neither can be placed in any of the existing genera and a new genus, Jeilongvirus, has been proposed.[12] Fer-de-Lance virus (FDLV), a new reptilian paramyxovirus, also has a unique genome, and a further new genus, Ferlavirus, has also been proposed.[13] Most recently, a new paramyxovirus, Cedar virus (CedPV), was isolated from urine samples of flying foxes in Australia and was shown to be genetically and antigenically related to HeV and NiV.[14] CedPV also appeared to utilize the same entry receptor, ephrin-B2, that both HeV and NiV employ, and CedPV is the first new proposed member of the *Henipavirus* genus. Other paramyxoviruses, such as Salem virus, Mossman virus and Nariva virus, have also been described but cannot be placed in any existing genera nor have new genera been proposed; therefore, these viruses remain unclassified.

Until recently, the genomes of paramyxoviruses as a group were generally considered to cluster in the range 15.1-15.9 kb. With the discovery and molecular characterization of HeV, NiV, BeV and CedPV, and the genome sequencing of the previously described Tupaia

Table 1. Existing and proposed genera in the family Paramyxoviridae

Subfamily	Genus	Species
Paramyxovirinae		
	Rubulavirus	Parainfluenza virus type 5*
		Mumps virus
		Human parinfluenza virus types 2, 4a and 4b
		Menangle Virus
	Respirovirus	Sendai virus
		Human parinfluenza virus types 1 and 3
		Bovine parinfluenza virus type 3
	Avulavirus	Newcastle disease virus
		Avian paramyxovirus types 2-9
	Morbillivirus	Measles virus
		Canine distemper virus
		Rinderpest virus
	Henipavirus	Hendra virus
		Nipah Virus
		Cedar virus
Pneumovirinae		
	Pneumovirus	Human respiratory syncytial virus
		Bovine respiratory syncytial virus
		Pneumovirus of mice
	Metapneumovirus	Human metapneumovirus
		Avian pneumovirus**
Proposed		
	TPMV-like viruses***	Tupaia virus
	Jeilongvirus***	J-virus
		Beilong virus
	Ferlavirus***	Fer-de-lance virus
Unclassified		
		Nariva virus
		Mossman virus
		Salem virus

*Formerly known as simian virus 5 (SV5). **Formerly known as turkey rhinotracheitis virus. ***Proposed genus within the subfamily *Paramyxovirinae*

virus and J-V, the genome size range has significantly increased. BeV, with a 19,212-nt genome, now represents the largest genome among all known non-segmented negative-strand RNA viruses, longer than the 19,151-nt genome of Marburg virus.[12] Indeed, aided by discovery and/or sequencing, the genetic diversity within the family *Paramyxoviridae* has rapidly increased within the past decade, and research efforts focused on their molecular and biochemical characterization have revealed an extraordinary breadth of biological diversity among this virus family.

PARAMYXOVIRUS ENTRY

Fusion of enveloped viruses with the plasma membrane of a receptive host cell is a prerequisite for viral entry and infection. As a group, most paramyxoviruses contain two membrane-anchored glycoproteins that are required for the entry process, and these glycoproteins appear as spikes projecting from the envelope membrane of the viral particle when viewed under the electron microscope. Several examples are provided in Figure 1. One glycoprotein is required for virion attachment to the host cell, and depending on the particular virus, has been designated as either the hemagglutinin–neuraminidase glycoprotein (HN), the hemagglutinin glycoprotein (H) or glycoprotein (G), which has neither hemagglutinating nor neuraminidase activities (reviewed in ref. 15). Paramyxovirus attachment glycoproteins are type II membrane proteins where the protein's amino (N)-terminus is oriented towards the cytoplasm and the protein's carboxy (C)-terminus is extracellular. The other glycoprotein is the fusion protein (F), which is directly involved in facilitating the fusion of the viral and host cell membranes (reviewed in ref. 16). The F glycoprotein is a type I integral membrane glycoprotein with an extracellular N-terminus that shares several conserved features with other viral fusion glycoproteins and will be discussed in greater detail in the sections to follow. A cartoon diagram of an attachment and fusion glycoproteins and their important functional domains is depicted in Figure 2.

The attachment and fusion glycoproteins work in concert to mediate membrane fusion and particle entry into susceptible host cells. Following virus attachment to a permissive and receptor-bearing host cell, fusion of the virion and plasma membranes

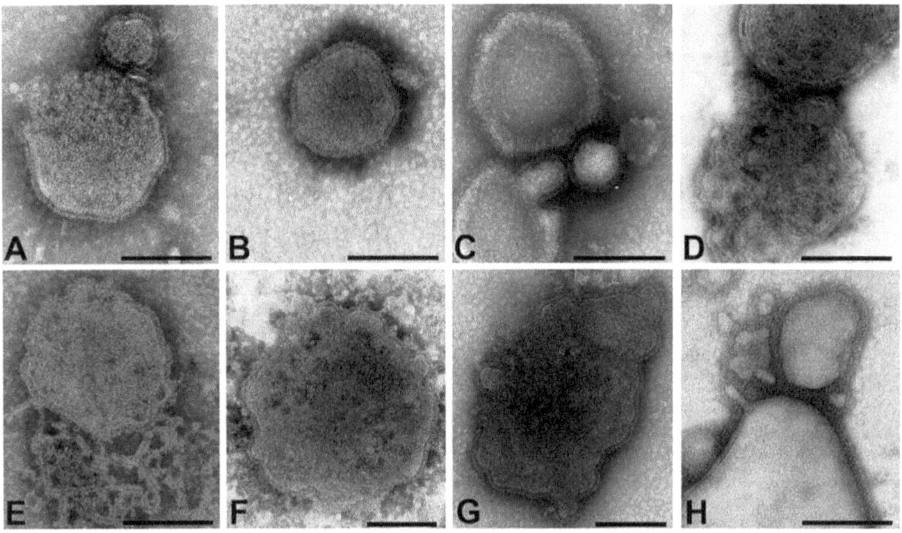

Figure 1. Negatively stained paramyxovirus virions. A) Newcastle disease virus (avulavirus). B) Human parainfluenza virus type 3 (respirovirus). C) Hendra virus (henipavirus). D) Canine distemper virus (morbillivirus). E) Menangle virus (proposed rubulavirus). F) Respiratory syncytial virus (pneumovirus). G) J-virus (proposed jeilongvirus). H) Mossman virus (unclassified). All micrographs were adjusted to the same magnification with exception of panels F and G. For all panels: bar, 200 nm. Images courtesy of the AAHL Biosecurity Microscopy Facility, Australian Animal Health Laboratory (AAHL) Livestock Industries CSIRO, Australia.

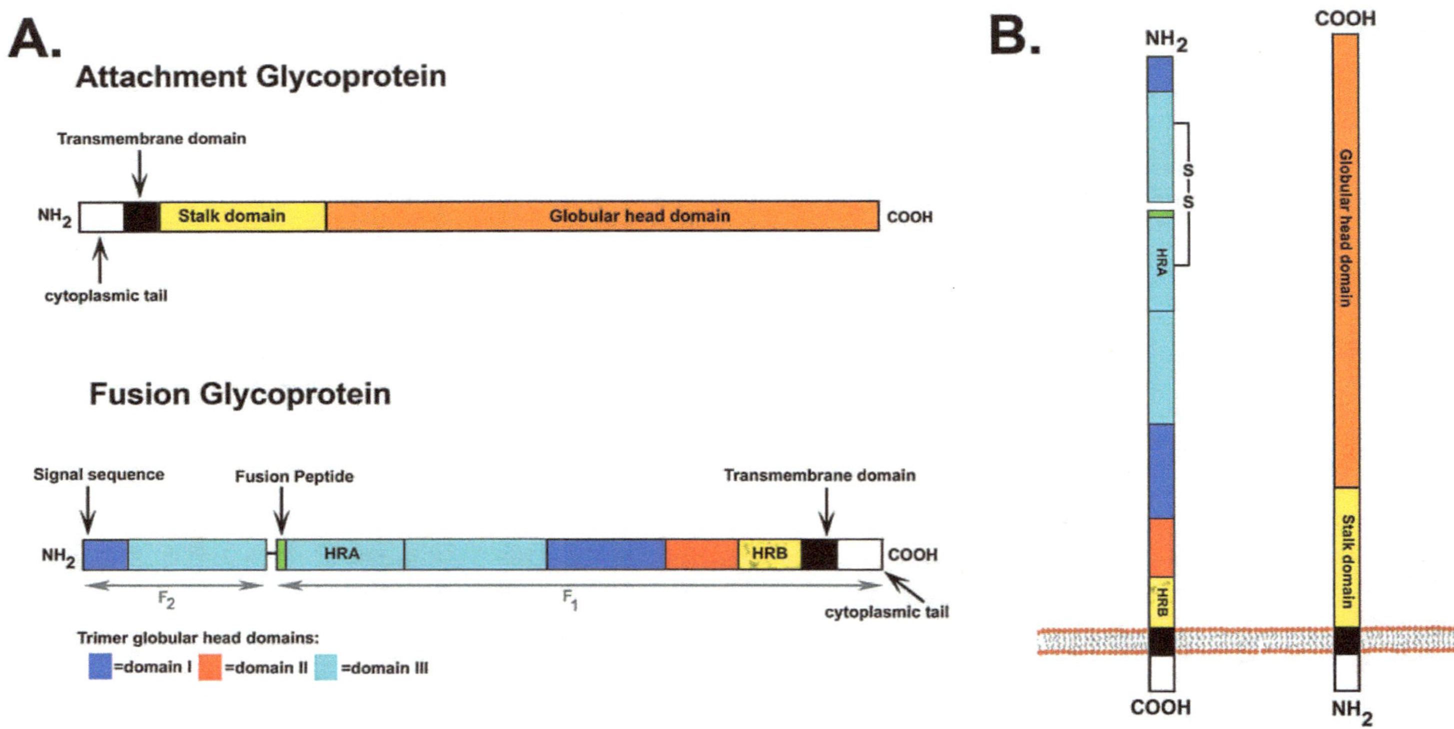

Figure 2. The attachment and fusion envelope glycoproteins. A) Important functional domains. For the attachment glycoprotein the transmembrane domain, cytoplasmic tail, proposed stalk domain and globular head are indicated. For the fusion glycoprotein the F_1 and F_2 subunits are depicted. F_2 contains the signal sequence and F_1 contains the transmembrane domain, cytoplasmic tail, fusion peptide, heptad repeat A (HRA) and heptad repeat B (HRB). The important domains that constitute the globular head of the F trimer, DI, DII and DIII, are represented by different shading. B) Orientation of the attachment and fusion glycoprotein in the virion membrane. F is a typical type I membrane glycoprotein with one membrane spanning domain and an extracellular N-terminus. The disulfide bond that links the F_1 and F_2 subunits is also shown. The attachment glycoproteins are type II membrane proteins where the molecule's amino (N)-terminus is oriented towards the cytoplasm and the protein's carboxy (C)-terminus is extracellular.

occurs, resulting in delivery of the viral nucleocapsid into the cytoplasm. In a related process, cells expressing attachment and fusion glycoproteins on their surface can fuse with receptor-bearing cells under physiological or cell culture conditions, leading to the formation of multinucleated giant cells (syncytia)—a hallmark of many paramyxovirus infections.

ATTACHMENT GLYCOPROTEINS AND THEIR RECEPTORS

General Tertiary Structure of Attachment Glycoproteins

The paramyxovirus attachment glycoproteins consist of a stem (or stalk) and a globular head structure with the latter domain containing both the receptor binding and, if present, enzymatic activities of the molecule.[17-19] The general model for the monomeric structure of HN and H is a globular head comprised of 6-folded antiparallel β-sheets (β1-6) of four strands each (S1-4), with each sheet arranged regularly around and radiating out from the central axis of the molecule.[20] Although the henipavirus G glycoprotein has only limited sequence homology to HN and H glycoproteins it possesses a high structural similarity.[21] The attachment glycoprotein β-propeller shape is maintained by disulfide bonds, which are highly conserved among these three types of attachment glycoproteins (reviewed in ref. 22). The 6-bladed propeller model is similar to both the earlier predicted structures of HN[23] and henipavirus G[21] and the now known HN structures of NDV, hPIV3 and PIV5,[24-27] the structure of MeV H[28,29] and the structures of henipavirus G.[30,31]

Earlier electron micrographs of SeV HN exhibit a box-shaped arrangement consistent with four discrete subunits,[32] similar to influenza virus neuraminidase (NA). The oligomeric forms of HN from different paramyxoviruses have been extensively characterized and depending on the virus consist of pairs of disulfide-linked homodimers that can come together to create noncovalently linked tetramers[32-36] or also disulfide-linked tetramers (dimer of dimers).[37] The disulfide-linked dimers of PIV5 HN are joined through cysteine residue 111 in the stalk domain; the exact residues responsible for tetramer association have yet to be identified although the presence of the stalk domain is critical.[26] PIV5[26] and NDV HN[38] also exist as tetramers in solution; however, significant differences exist in dimer packing.

The oligomeric organization of the MeV H glycoprotein has also been characterized by structural and functional studies identifying disulfide-linked dimers via cysteine residues at positions 139 and 154[39] with a higher order tetramer configuration (dimer of dimers).[40] Likewise, the biochemical characterization of native HeV G, as well as a transmembrane domain/cytoplasmic tail-deleted, soluble version of G has revealed disulfide-linked dimers and both noncovalent and disulfide-linked tetramers of G similar to the oligomeric forms of HN and H glycoproteins.[41] The residues involved in the oligomerization of G are located in the stalk domain with NiV G cysteine residues 158, 162 and 146 having critical roles. Residues 158 and 162 are involved in the covalent dimer formation of NiV G and were found to be absolutely required for its fusion promotion activity, perhaps by maintaining G in a pre-receptor bound conformation, while cysteine residue 146 appeared to stabilize higher-order oligomers (tetramers)[42] (reviewed in ref. 43). Additional details on the structures and functions of the types of attachment glycoproteins will be discussed in the sections below.

Hemagglutinin–Neuraminidase Glycoprotein (HN)

The majority of well-described paramyxoviruses, particularly those in the *Respirovirus*, *Avulavirus* and *Rubulavirus* genera, possess a multifunctional HN glycoprotein that attaches the virion to sialic acid receptors on host cells. In addition to binding, the HN glycoprotein cleaves sialic acid moieties from both host cell molecules and virus particles. Analogous to the role played by influenza NA, the neuraminidase activity of HN prevents re-attachment of the virion to producer cells as well as self-aggregation of progeny virions as they bud from an infected cell into the extracellular environment (reviewed in ref. 44). Because the optimal pH for paramyxovirus neuraminidases is between 4.8 and 5.5 it has been suggested that the removal of sialic acid from carbohydrate chains occurs in the acidic *trans* Golgi network.

Recently, the structure of the NDV HN ectodomain was reported that showed the NA domain dimers flanking the N-terminal stalk domain (Fig. 3). The NDV stalk formed a parallel tetrameric coiled-coil bundle (4HB) that also permitted the classification of existing mutagenesis data, revealing broad insight into the functional roles of the HN stalk and its tetrameric configuration. Many mutations that affected only F-glycoprotein activation mapped to the 4HB surface (Fig. 3A). Two of four NA domains revealed an interaction with the 4HB stalk, and residues at this interface in both the stalk and NA domain have been implicated in HN function.[27] The two independent structures of PIV5 and NDV HN glycoproteins[26,27] have now been referred to as the 'heads-up' versus 'heads-down' conformations, respectively. Alternative models describing how tetrameric HN structures promote fusion have been proposed by both Zaitsev et al[38] and Yuan et al[26,27] and will be discussed in a later section.

HN mediates both binding to sialic acid as a receptor for viral attachment and cleavage of sialic acid via its neuraminidase enzymatic activity. An interesting question remains as to whether this is achieved through one or two separate sialic acid binding sites. Studies have demonstrated two sialic acid binding sites in the NDV HN dimer, one in the globular head domain and the second site at the dimer interface of the molecule.[38] Both of these proposed sialic acid binding sites in NDV HN are shown in Figure 3A. Binding of ligand to the first active site induces a conformational change that then allows formation of the second site to which sialic acid also binds. Structural studies of hPIV3 and PIV5 HN reveal dimers very similar to those of NDV HN and suggest a similar oligomeric arrangement; however, unlike NDV the hPIV3 and PIV5 HN glycoproteins contain only one sialic acid binding site, an enzymatically active site in the globular head domain, while differences in sequence and conformation render the second sialic acid binding site implausible.[25,26] Like hPIV3 HN, PIV5 HN does not undergo conformational change upon ligand binding, highlighting further distinctions from the NDV HN structure.[25,26]

Hemagglutinin Glycoprotein (H)

The morbilliviruses, including MeV and CDV, have a H attachment glycoprotein, which possesses only hemagglutinating activity and does not bind to sialic acid receptors. However, H glycoproteins have significant sequence identity to HN, and similar tertiary structure models analogous to those for the HN glycoproteins of respiroviruses, avulaviruses and rubalviruses have been developed.[20] MeV was also the first paramyxovirus shown to employ a cell-surface protein as a receptor,[45,46] and co-immunoprecipitation experiments

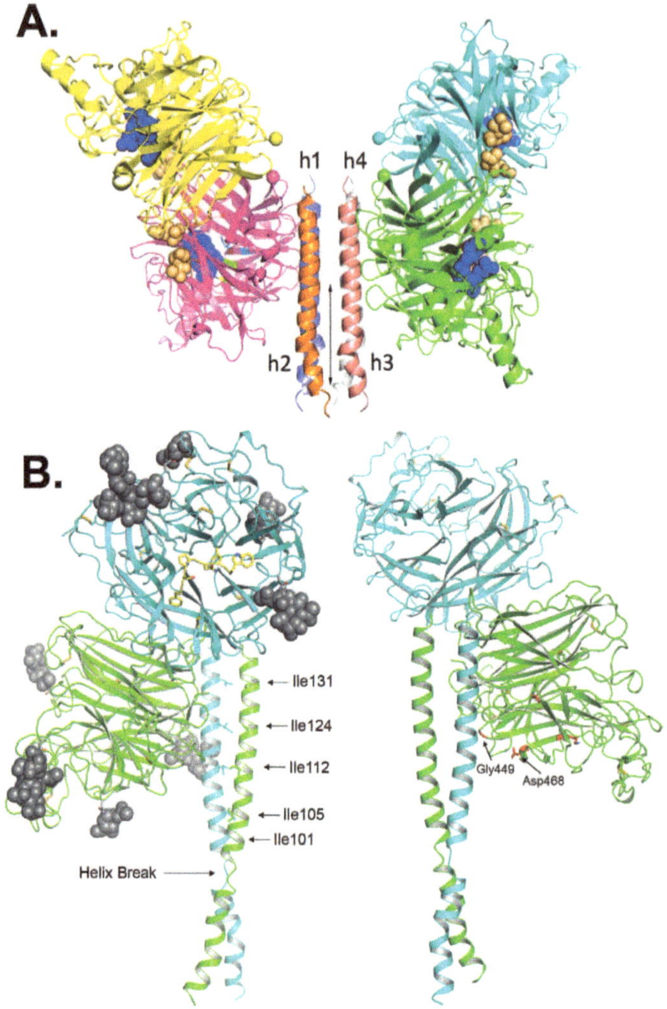

Figure 3. A) Structure of the NDV HN, Australia–Victoria (AV) strain ectodomain. Two dimers of the NDV HN NA domains flank the 4HB (residues 83-114) in the stalk. The four NA active sites are shown as blue spheres. The secondary sialic acid binding sites located at the NA domain dimer interface are shown as orange spheres. Mutations of NDV HN stalk residues R83, A89, L90, L94 and L97 are known to impair F activation specifically and are implicated in forming direct contacts with the F glycoprotein. These mutations reside along the stalk region marked by the arrow in the HN tetramer structure. B) Model of HeV G. Left: The G ectodomain is shown in the dimer conformation with the two globular head domains derived from the crystal structure, colored in green and blue, with predicted N-linked glycosylation sites shown as gray spheres. The G head domain folds as a six-bladed β-propeller with disulfide bonds illustrated as yellow sticks. The residues of the ephrin-B2 G-H loop are also shown in yellow occupying the RBS. Stalk residues 77–136 are modeled for each monomer, and the position of the HeV G head dimer and stalks are oriented based on the alignment with the NDV structure.[27] Ile residues in the HeV G stalk domain that modulate G fusion promotion activity are indicated.[197] Right: Model of the HeV G dimer with globular heads and stalk domains as on right and rotated with residues G449 and D468 highlighted in red showing their proximity to the stalk domain. Mutation of these residues decreases HeV fusion, suggesting they may be involved in interactions between the globular heads and stalk domains that are essential for the fusion process. Figures have been modified from original work with permission from R.A. Lamb and Yuan P et al. Proc Natl Acad Sci USA 2011; 108(36):14920-14925;[27] and Steffen DL et al. Viruses 2012; 4(2):280-30 (the creative commons public license).[43]

demonstrated an interaction between the H glycoprotein of laboratory strains of MeV and CD46.[47] In addition, MeV field isolates as well as vaccine strains can utilize signaling lymphocyte activation molecule (SLAM, CD150) as a receptor[48]—a receptor also employed by CDV and wild-type rinderpest virus.[49,50] Further, CD46 and SLAM expression are down-regulated in MeV-infected cells in a H glycoprotein-dependent manner.[51,52] It was hypothesized that reducing the levels of surface-expressed receptors may circumvent the need for intrinsic neuraminidase activity by MeV, and possibly morbilliviruses in general, which for HN bearing paramyxoviruses cleaves surface-associated sialic acid and prevents virus aggregation during virus budding as discussed earlier. Moreover, the use, and down-modulation, of SLAM by morbilliviruses may play a role in the general immunosuppression seen in infected hosts. Interestingly, the MeV attachment sites for both receptors appeared to overlap on the globular head domain of H,[53] although viruses that preferentially use either CD46 or SLAM could be selected.[54] Yet a third MeV receptor, speculated to exist on epithelial cells,[55,56] was recently discovered. Nectin-4, an adherens junction protein of the immunoglobulin superfamily, is the most recently identified MeV receptor.[57,58] Highly expressed in a variety of tissues including epithelial cells of the human airway, Nectin-4 supports MeV entry and is also down-regulated in infected cells. Of the three MeV receptors, Nectin-4 has the strongest affinity for MeV H[58], and MeV targets Nectin-4 to emerge in tracheobronchial airways. Following the initial infection and spread of MeV within the host, facilitated by macrophages and dendritic cells, infection of epithelial cells occurs later in the disease and is important for aerosol transmission of the virus (reviewed in ref. 59).

In just the past several years a significant amount of new information on the interactions between paramyxovirus H glycoproteins and their binding partners has been obtained and the structures of MeV H alone and in complex with SLAM, CD46 and Nectin-4 have been determined.[28,29,60-62] The most recent structure of MeV H in complex with Nectin-4 has allowed for the first time a detailed comparison of the binding of H with three different receptors, revealing overlapping but distinct binding sites for Nectin-4, CD46 and SLAM.[60] Of particular interest, this latest study revealed a hydrophobic pocket centered in the MeV H β4-β5 groove involved in the binding of all three receptors, suggesting a new potential target for antivirals.

Glycoprotein (G)

There are two distinct and structurally unrelated G lineages within the family *Paramyxoviridae*—those described for the genus *Henipavirus* in the subfamily *Paramyxovirinae* and those described for the genera *Pneumovirus* and *Metapneumovirus* in the subfamily *Pneumovirinae*. Both lineages of G attachment glycoproteins lack hemagglutinin and neuraminidase activities. Only a single amino acid residue of the seven known to be critical for neuraminidase activity is conserved in HeV and NiV G, compared to at least six residues in HN or four residues in H. Additional studies demonstrated that neuraminidase treatment of Vero cells (a cell line used to propagate HeV and NiV stocks) did not inhibit HeV or NiV infection, while such treatment can abrogate their susceptibility to NDV and influenza virus A, which depend on sialic acid structures as receptors.

It was also observed that cell lines from the same species, most notably human cell lines, could be clearly positive or negative for HeV or NiV-mediated membrane fusion and that protease treatment could prevent fusion of an otherwise permissive target cell.[63-65]

Perhaps not unexpected in light of the observed characteristics possessed by HeV and NiV G, it was later discovered that the henipaviruses utilize a host cellular protein as a viral receptor, ephrin-B2 ligand.[66,67] Ephrin-B2 ligand is a widely-expressed and highly conserved cell surface protein across many different species, and its identification as the henipavirus receptor has aided in understanding the broad host range of HeV and NiV as well as their neurotropism.

In addition, ephrin-B3 ligand was identified as a second entry receptor for NiV[68]and HeV.[69] Like MeV H attachment to its protein receptors, the attachment sites for both ephrin receptors in NiV G also revealed overlapping binding sites as binding to ephrin-B2 ligand can inhibit binding to ephrin-B3 ligand.[68,70] More recently, the crystal structures of both NiV and HeV G globular head domains have been determined both alone and in complex with the ephrin-B2 and -B3 receptors, revealing the exact G-receptor interactions and identical receptor binding sites.[30,31,71-73] Also, similar to NDV HN, the henipavirus G stalk domain contains alpha helices with a predicted break from amino acids 95–98, and the stalks with the globular heads of HeV G have been modeled with the resulting structure resembling the heads-down configuration of NDV HN (Fig. 3B).

An obvious difference between the G glycoproteins of henipaviruses and those of the subfamily *Pneumovirinae*, is size, just over 600 amino acids for HeV and NiV whereas the G glycoproteins of pneumoviruses and metapneumoviruses vary between 230 and 300 amino acids.[74] There are also significant sequence differences between the G glycoprotein of pneumoviruses and metapneumonviruses, although both have a similar hydrophobicity profile, a high serine and threonine content (24-34%) and an ectodomain that also contains two mucin-like domains.[75,76] In addition, the G glycoprotein of RSV, a member of the pneumovirus genus, is heavily glycosylated with both N- and O-linked sugars that contribute greater than 50% of the weight of the mature glycoprotein—an unusual feature among viral membrane glycoproteins. The high serine and threonine content facilitates the addition of O-linked carbohydrates, and it is likely that these sugar moieties contribute to the binding of carbohydrate receptors on the cell surface.[75] hMPV G has a serine/threonine content of 34%, slightly higher than RSV,[76] and it is predicted to have a similar profile of O-linked glycosylation and carbohydrate binding as RSV G. In addition to the membrane-bound RSV G, a soluble and secreted G glycoprotein molecule is also observed in infected cell cultures.[77] It has been suggested that this soluble version of RSV G may act as an immunological decoy during infection.

Another notable characteristic of RSV, including bovine (RSV) which is being explored as a human vaccine platform, is that G-deleted viruses are still capable of replication in cell culture or in animals.[77-81] These observations suggested that RSV G may function as an accessory protein that increases the efficiency of virus entry.[82] Indeed, the RSV F glycoprotein has been shown to bind heparin-containing structures and the GTP-binding RhoA protein as well as interact with and subsequently signal through CD14 and toll-like receptor 4.[75,83] Clearly, RSV possesses an alternate mechanism for virus entry in which attachment and fusion can be directly mediated by F and does not strictly require G. More recently, endocytosis has been implicated as a possible route of RSV entry, either via a caveolin[84] or clathrin[85] mediated route(s).

Like RSV, recombinant human metapneumovirus (hMPV), which lacks G, has also been shown to be replication competent in vitro and in vivo,[86,87] and it was recently demonstrated that the first cell surface binding target for hMPV is also heparan sulfate.[88] Although no definitive receptor for the hMPV F protein has been identified, integrin $\alpha v\beta 1$ has been suggested as a host cell factor promoting entry.[89] Altogether, these recent findings suggest that hMPV F glycoprotein has effectively replaced a requirement for an attachment protein with a low pH-induced triggering process,[90,91] a unique feature amongst the paramyxoviruses.

Finally, the newly characterized G genes and their encoded G glycoproteins of J-V and BeV warrant discussion, as they may represent a third lineage of G glycoproteins in the family *Paramyxoviridae*. The J-V and BeV G genes are 4401 and 4527-nt in length, respectively, more than double the size of most family members.[12,92] The significant increase in size is due to the presence of additional open-reading-frames (ORFs) within the G gene. To begin with, for both J-V and BeV, the 5' half of the G mRNA contains an open reading frame (ORF-G) encoding putative proteins 709 or 734 amino acids in length, respectively. Such proteins are 105 and 130 amino acids larger, respectively, than the largest paramyxovirus attachment glycoprotein; HeV G. The putative G glycoproteins of J-V and BeV share many conserved structural elements with other paramyxovirus HN, H and G glycoproteins, and also lack any detectable hemagglutinating or neuraminidase activities.[12,92] In addition, and perhaps of greater interest, both the J-V and BeV G gene contain an additional ORF (ORF-X) within the 3' half of their G mRNAs, encoding putative proteins, 709 and 299 amino acids in length, respectively. These additional coding regions are separated from the ORF-G by only one stop codon. BeV has another ORF downstream from, but overlapping with ORF-X which encodes another putative protein 394 amino acids in length. Probes specific for ORF-G and ORF-X in J-V both identified mRNA transcripts corresponding in size to a monocistronic G gene mRNA. However, no evidence was found for the existence of an mRNA molecule specific to ORF-X alone, nor was the protein encoded by ORF-X or a fusion protein of G-X detected.[93] Although the biological significance of these additional ORFs remains unknown, clearly, the G genes of J-V and BeV are unlike any other within the family *Paramyxoviridae*.

Fusion (F) Glycoprotein

Nearly all paramyxoviruses that have been examined to date require both attachment and F glycoproteins for efficient membrane fusion to occur, although some exceptions have been noted. PIV5 F can mediate moderate levels of membrane fusion in the absence of HN[94], and as described above, RSV and hMPV derivatives that lack the G gene remain fusogenic and infectious. For all paramyxoviruses the F glycoprotein is directly involved in facilitating the fusion between the virus and host cell membranes. F glycoproteins are homotrimeric oligomers with considerable hydrophobicity[36,95-100] and share several conserved features with other viral fusion proteins, including the envelope glycoprotein of retroviruses, such as gp120/gp41 of HIV-1, and the hemagglutinin (HA) of influenza virus.[95,96,101,102] These types of viral fusion glycoproteins have since been categorized as class I viral fusion proteins[103] where from the protein's N- to C-terminus there is a fusion peptide located just C-terminal to the cleavage site of a precursor form of the protein followed by two heptad repeat domains, a transmembrane domain (TMD) and a cytoplasmic tail. These features of the paramyxovirus F glycoprotein will be discussed below. The three prominent classes of viral fusion proteins have been recently reviewed in detail.[104]

Biologically active F consists of two disulfide linked subunits, F_1 and F_2, (Fig. 2) that are generated by the proteolytic cleavage of a precursor known as F_0.[105,106] Likewise, HIV-1 envelope and influenza HA are cleaved by a host cell protease, leading to the generation of a membrane distal subunit analogous to F_2 and a membrane-anchored subunit analogous to F_1. Cleavage of F_0 is thought to play an important role that influences both infectivity and pathogenicity of paramyxoviruses. The various paramyxovirus F glycoproteins fall into two groups—those with multiple basic residues and those with a single basic residue at the cleavage site. Proteolytic activation involves two separate cleavage events mediated by host proteases. The first initially cleaves the carboxyl side of the basic residue, and the next step, mediated by a carboxypeptidase, removes the basic residue. Cleavage of F glycoproteins with multiple basic residues occurs within the cell as they traffic through the *trans* Golgi network of the secretion pathway and is thought to be mediated by furin, a host endoprotease[1] (reviewed in ref. 106). Two different mechanisms exist for cleavage of paramyxoviruses with single basic residues at the cleavage site. The first, which has been more widely studied, has demonstrated extracellular cleavage by an exogenous protease. As an example, SeV replicates poorly in tissue culture; however, after addition of exogenous protease, productive infection significantly increases.[105] When grown in eggs, SeV F is cleaved by an extracellular amniotic endoprotease.[107] For NDV, virulence of the virus is directly correlated to the nature of the cleavage site, where strains with a single basic residue in the cleavage site are avirulent and are restricted to the respiratory tract, while those with multiple basic residues readily disseminate through the host.[108] In general, most viruses that contain a single basic residue at the cleavage site have a more restrictive tropism and do not disseminate. Newly recognized exceptions to these general rules are HeV and NiV. Both henipaviruses have a single basic residue at the cleavage site, however, both viruses readily disseminate within the host upon infection, productively targeting a variety of organ systems. Of particular interest is the discovery that the F_0 glycoprotein of HeV and NiV is cleaved in a novel process that occurs after transportation of the uncleaved molecule to the surface of infected cells. Following re-internalization of F_0, cleavage occurs within the endosomal compartment and is mediated by the endoprotease cathepsin L.[109,110]

Nevertheless, in all cases following F_0 cleavage, the membrane-anchored subunit F_1 remains linked to F_2 by a disulfide bond and contains a new N-terminus, referred to as the fusion peptide (Fig. 2), which is hydrophobic and conserved in its location across virus families.[111,112] The fusion peptides of paramyxoviruses, as well as other viruses including HIV-1 and influenza, are thought to intercalate into target membranes and initiate the fusion process.[113] Although hydrophobic in nature, the absolute conservation of many of the residues within the fusion peptide of paramyxoviruses suggests an as yet unidentified additional function independent of actual membrane insertion.[1]

The paramyxovirus F glycoproteins, like those of retroviruses, contains 2 α-helical domains referred to as heptad repeats that are involved in the formation of a trimer-of-hairpins structure or 6-helix bundle (6-HB) during or immediately following fusion of virus and cell membranes.[111,114-117] For paramyxoviruses, one heptad is located adjacent to the fusion peptide in F_1 and is referred to as the N-terminal heptad or heptad repeat A (HRA). The second heptad is proximal to the transmembrane domain and is referred to as the C-terminal heptad or heptad repeat B (HRB) (Fig. 2). As first noted with the gp41 subunit of HIV-1 envelope glycoprotein,[118,119] peptides corresponding to either of these domains from several paramyxovirus F glycoproteins can inhibit the activity of the fusion glycoprotein when present during the fusion process.[63,64,120-126] It has been generally accepted that significant conformational change occurs during activation of paramyxovirus F fusogenic activity. Differential antibody binding reactivity of precursor and proteolytically processed forms of PIV5 F[127], in conjunction with the structure of the 'postfusion' 6-HB of PIV5 F[111], strongly supported the conformational change model.[102] The postfusion structure of the hPIV3 F core is likely conserved across other paramyxoviruses and has been observed in the F core structures of RSV,[128] MeV,[129] mumps virus[130] and the henipaviruses.[131] A cartoon illustrating how the heptad repeats mediate 6-HB formation and membrane fusion is shown in Figure 4.

More recently, structural studies on the F glycoprotein of NDV reveal a trimer structure that differs from the classic influenza HA structure, principally in the manner in which HRA is oriented. In the NDV F trimer, the HRA segment is located with its C terminus directed towards the head of the molecule; this is the opposite orientation to the observed central coiled coil formed by HRA in the influenza HA trimer.[98,132] Understanding the cascade of conformational changes leading up to 6-HB formation[102,121,127] was hindered by the spontaneous re-arrangement of secreted F glycoproteins to a conformation resembling a postfusion configuration,[98,99,132] recently reviewed in reference 133. However, the first prefusion, metastable structure of a paramyxovirus F glycoprotein has recently been solved by appending a trimerization motif from GCN4 onto the C terminus of secreted PIV5 F, an addition thought to mimic the transmembrane domain of the glycoprotein.[100] The structures of uncleaved and cleaved metastable prefusion PIV5 F glycoprotein and postfusion hPIV3 F conformations are shown in Figure 5. The differences in overall conformations between the pre- and postfusion F glycoproteins merit explanation here.

The first metastable prefusion conformation of a PIV5 F was determined using an uncleaved version of the glycoprotein and was shown to contain a globular head connected to a trimeric coiled-coil stalk formed by HRB.[100] The globular head contains three previously identified domains per subunit, referred to as DI, DII and DIII,[100,132] and in DIII two sets of 6 helices form rings that cover the top of the globular head, while the HRB three-helix bundle seals the bottom. Perhaps the most intriguing aspect of the structure is that DIII undergoes major refolding between the pre- and postfusion conformations of F. In the prefusion conformation HRA is actually divided into four helices, five turn segments and two β–stands and is folded around the core of DIII. Such a conformation suggests that HRA is trapped within DIII as monomeric subunits. Further, in the prefusion conformation, the fusion peptide is wedged between the DII and DIII domains of adjacent F subunits in the trimer. Upon triggering, a total of 11 distinct segments in the prefusion HRA DIII domain refold to generate a single extended α-helical conformation necessary for translocation of the fusion peptide towards the target membrane and pre-hairpin formation. In solution, the prefusion, metastable PIV5 F molecule does not form the 6-HB structure; rather the three fold axis of the HRB three-helix bundle is aligned along the three-fold axis of the globular head with only a slight tilt. The junction of the base of the globular head and the HRB region appear to form an interactive network between trimer subunits. Previously identified residues near HRB that were hypothesized to play a role in conformational switching[134] are now known to reside within this region. The presence of the additional trimeric coiled-coil domain from GCN4[135] in the crystallized metastable PIV5 F is hypothesized to stabilize the trimeric coiled-coil stalk formed by HRB. The transmembrane domain and cytoplasmic tail are hypothesized to perform a similar role in native metastable F and help explain why in their absence secreted, anchorless hPIV3 and NDV F glycoproteins converted to the postfusion F conformation when crystallized.[26]

Very recently, the crystal structure of the cleaved prefusion form of a truncated version of the PIV5 F has been reported.[136] As before; the truncated soluble form of the PIV5 F glycoprotein was appended with a trimeric coiled-coil domain (GCNt) that was able to stabilize the ectodomain of the F glycoprotein in its prefusion conformation. Unlike previous studies, the purified metastable F glycoprotein was cleaved in vitro with trypsin, re-purified and analyzed by electron microscopy, revealing its characteristic prefusion

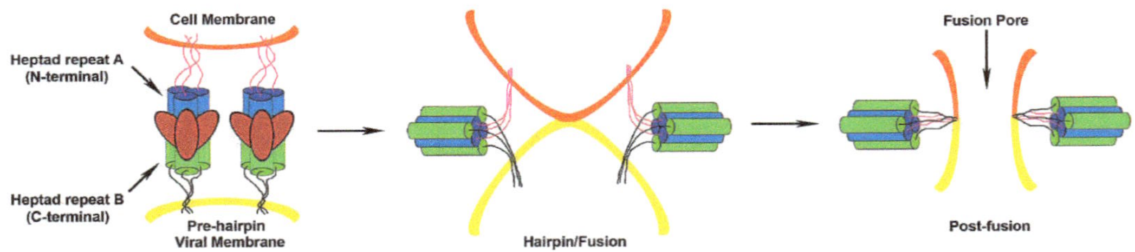

Figure 4. Model showing how paramyxovirus F heptad repeats mediate membrane merger. For simplicity, the attachment glycoprotein and its proposed role in triggering or promoting F activity is not shown. Upon activation, significant conformational changes in F lead to HRA forming a three-stranded α-helical coiled coil and the translocation of the fusion peptide and its intercalation into the target cell membrane. This fusion intermediate conformation of F is referred to as the prehairpin structure. Further conformational changes lead to the packing of α-helical HRB domains into the grooves of the HRA coiled coil in an antiparallel orientation giving rise to six-helix bundle formation. As the six-helix bundle forms, the two membranes are drawn closer together, and the energy released as F transitions to its most stable conformation is thought to drive membrane merger.

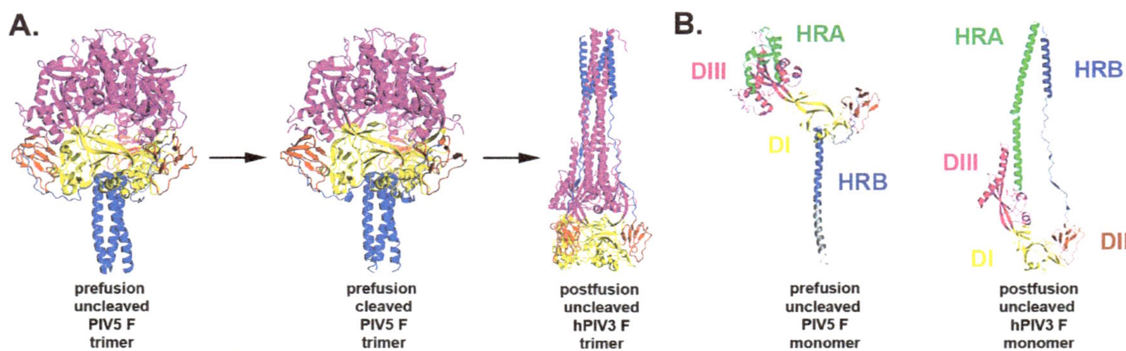

Figure 5. Structural changes between the pre- and postfusion F glycoprotein conformations derived from PIV5 and hPIV3. A) Ribbon diagrams of uncleaved and cleavage-activated prefusion forms and postfusion F trimers are shown side by side. The three domains, DI, DII and DIII, are depicted by different shading (DI yellow, DII red and DIII pink). B) Ribbon diagrams of pre- and postfusion F monomers similarly oriented by DI as shown in Panel A. Monomers are shown side by side to demonstrate more clearly the different domains, DI, DII and DIII, which are depicted by different shading as in A. Adapted with permission from R.A. Lamb and Macmillan Publishers Ltd: Nature ©2006; 439(7072):38-44;[100] and Welch BD et al. Proc Natl Acad Sci USA 2012; 109(41):16672-16677.[136]

form. Interestingly, when heated briefly in vitro, the cleaved metastable F converted to a postfusion conformation.[136] It was notable that other than the newly exposed N-terminus of the trypsin cleaved F_0 precursor, the conformational differences near the cleavage site exhibited no net burying or exposure of hydrophobicity or charge changes[136], and the conformational changes between the uncleaved to cleaved prefusion forms of the PIV5 F were not as dramatic as the prior observations between similar versions of the influenza HA glycoprotein.

Similar soluble forms of the NiV and HeV F glycoproteins have also been recently reported along with the ability to cleave purified prefusion soluble F and generate postfusion forms[137], and a NiV prefusion soluble F crystal structure has been determined (K. Xu, C. Broder and D. Nikolov, unpublished findings). The structure of the cleaved prefusion PIV5 F glycoprotein, together with the previously reported structures of uncleaved prefusion PIV 5 F-GCNt and the postfusion structures of hPIV3 F, NDV F and RSV F, have now provided a detailed high-resolution view of the various static forms of paramyxovirus F glycoproteins (Fig. 5).

Upon appropriate triggering of native F, it is now hypothesized that opening or "melting" of the HRB three-helix bundle stalk triggers the conformational changes in the HRA DIII domain and gives rise to the pre-hairpin fusion intermediate.[100] Indeed, a multi-step process would be consistent with the early inhibition of fusion by HRA-derived peptides but not HRB-derived peptides.[121,138] Accordingly, the model depicting heptad repeat dependent membrane merger (Fig. 4) requires slight adjustment. Specifically, the pre-hairpin conformation of F needs to be altered to portray HRB as three unassociated segments instead of a three-helix bundle. Overall, both prefusion conformations of the cleaved and uncleaved PIV5 F and the postfusion conformation of hPIV3 F suggest how small refolding intermediates can be coupled to the activation of F and ultimately membrane fusion.[139]

It has also long been recognized that truncated or cytoplasmic tail deleted versions of fusion envelope glycoproteins; particularly the human and simian immunodeficiency viruses (HIV-1, HIV-2 and SIV), could often possess an enhanced ability to mediate cell-cell fusion (Aguilar et al[140] and references therein). This feature was also recorded in paramyxovirus SV5 strains where individual isolates possess an F glycoprotein with either a short (20-residue) or long (42-residue) cytoplasmic tail. It was noted earlier that an SV5 strain (W3A) possessing an F glycoprotein with a short tail could mediate syncytium formation in the absence of its HN glycoprotein partner, essentially a hyperfusogenic feature, whereas other strains with the longer tail required HN coexpression for fusion.[141] Further experiments revealed that when the W3A F glycoprotein (short tail) is expressed as the longer tail (42 residues) either by mutation to remove a translational stop codon or by extension using additional sequences, the hyperfusogenic activity of the F glycoprotein is reduced. Additionally, the longer cytoplasmic tail of F modulated the F ectodomain conformation as detected by specific mAb binding, suggesting that the cytoplasmic tail could influence the conformation and function of the protein's ectodomain.[142] Likewise, the cytoplasmic tail of the NiV F glycoprotein has been shown to contain an amino acid motif (KKR) that when mutated can affect the conformation and subsequent fusion activity of the F glycoprotein ectodomain,[140] a feature termed as an inside-out signaling event.

In addition, the potential role(s) of the TMD of a paramyxovirus F glycoprotein in its structure and functional features has also been under investigation. The TMD appears to influence protein folding, prefusion structure stability and the membrane fusion activity of a variety of viral fusion proteins (Smith et al[143] and references therein). However, details on the TMD role have remained poorly understood among the paramyxoviruses. Recent experiments have shown that TMD-TMD interactions within the F glycoprotein trimer of HeV affect protein stability and its fusogenicity, and elements within the C terminus of the HeV F TMD appear to play an important role in the F trimer's TMD-TMD interactions and its membrane fusion activity.[143]

For most paramyxoviruses, the fusion triggering event initiated by the F glycoprotein's attachment glycoprotein partner appears to serve as an effective replacement of the acidification event required by influenza virus HA. However, the conformational changes in the transition from pre- to postfusion PIV5 F are quite different from those observed for the pre- and postfusion influenza virus HA;[144] nevertheless, certain trends do appear. In both viruses, the HRA is prevented from assembling, the fusion peptide is initially buried at the subunit interface, HRA projects the fusion peptide away from the viral membrane and transmembrane domain and finally the transition of HRB to its final state cannot occur due to the absence of an HRA coiled-coil and other structural barriers.

Altogether, understanding the conformational changes that occur in metastable F in its transition to a "fusogenic" and 6-HB structure has now greatly aided our understanding of how the energy required to mediate membrane fusion is captured. However, for paramyxoviruses, the precise trigger that initiates the metastable to "fusogenic" F transition continues to be investigated.

ATTACHMENT AND FUSION GLYCOPROTEINS WORK TOGETHER TO FACILITATE ENTRY

With few exceptions, the fusion activities of most paramyxovirus F glycoproteins are dependent on the activity and availability of their specific partner attachment glycoprotein (reviewed in[22]), and the co-expression of the attachment and fusion glycoproteins is required for virus infectivity for most members of the subfamily *Paramyxovirinae*. For the most part, the attachment glycoprotein-F interaction is also virus type-specific, and fusion mediated by co-expression of the F and attachment glycoproteins of different paramyxoviruses (heterotypic mixing) is rarely seen.[145] Although some examples have been noted, the potency of the fusion process from heterotypic mixing is considerably reduced in comparison to that mediated by the F and attachment glycoproteins from the same virus (homotypic mixing).[47,146] HeV and NiV are closely related henipaviruses and uniquely, heterotypic combinations of the F and G glycoproteins are as potent in mediating fusion as homotypic combinations.[63] Although heterotypic function of the envelope glycoproteins of the morbilliviruses MeV and CDV are not as efficient as the homotypic equivalents, like HeV and NiV, heterotypic activity is bidirectional, and fusion occurs with either heterotypic combination. These bidirectional examples are unlike the heterotypic results observed with the respiroviruses SeV and hPIV1. Here SeV F combined with hPIV1 HN functions efficiently, whereas in the reverse combination, SeV HN is unable to complement hPIV1 F.[147] Given the percent amino acid similarities of F and HN from hPIV1 and SeV, which are greater than those for MeV and CDV, this would not have been expected and may represent not only the need for type-specific interactions but also the possibility that there may be genus-specific factors involved in the interaction between F and attachment glycoproteins. Although the mechanism underlying this process remains obscure, the domains of the attachment glycoprotein necessary to promote fusion have been mapped using functional assays and indicate that regions in the globular head and stalk domain are critical.[148-152] It has now become increasingly clear that following receptor engagement, the attachment glycoprotein somehow signals and/or induces the required conformational changes in F leading to virion/cell fusion.[100,138,153] However, the precise molecular details of how the fusion and attachment glycoproteins function in concert in mediating fusion continue to be gradually elucidated.

Presently there are two widely appreciated models of paramyxovirus glycoprotein-mediated membrane fusion which describe the interactions between the oligomers of an F and of an attachment glycoprotein as they relate to the role of the receptor. Model 1 suggests receptor binding to the attachment glycoprotein induces a subsequent association and triggering of F-mediated fusion. In model 2, however, receptor binds to an attachment glycoprotein that is already in complex with its partner F glycoprotein and induces the dissociation of F and the attachment glycoprotein, initiating F-mediated membrane fusion. These models were first diagrammed by McGinnes et al[154] based on the available data at that time, which focused primarily on HN and F glycoproteins. A comprehensive review of the literature on the paramyxovirus fusion process by Iorio et al[22] has summarized the findings on the interactions between a varied array of paramyxovirus attachment and fusion glycoprotein species and the role of their particular entry receptors. Model 1 is also referred to as the association model and model 2 as the dissociation model.[22,104]

The first model proposes that the F glycoprotein and the attachment glycoprotein are not necessarily physically associated in the membrane and that following receptor binding to the attachment glycoprotein there is some alteration in the receptor-bound complex that facilitates the association of the attachment glycoprotein with its partner F glycoprotein in a manner often referred to as activation of its 'fusion-promotion' activity. This specific association triggers or induces the fusion activity of F, which through its subsequent conformational change drives the membrane fusion process.[153,155] This association model, also recently termed the 'provocateur' model,[156] was recently supported by key data indicating that the prefusion conformation of F (PIV5) was maintained in the absence of HN co-expression. This model is also supported by extensive functional and structural studies on the HN and F glycoproteins from hPIV3, NDV and PIV5[27,94,151,152,156-160] in which the overall theme indicates a positive correlation between HN and F interaction and fusion promotion or triggering activity.

In model 2, F and its partner attachment glycoprotein are pre-associated in some oligomeric complex, and a conformational alteration in the latter following receptor engagement induces some conformational change which facilitates its dissociation or release of F, thereby allowing F to undergo its conformational alterations driving the membrane merger process. Although this model was originally put forth as an alternative possibility based on studies of the HN and F glycoproteins, it was later supported by extensive studies with the MeV-H and -F glycoproteins[161-164] and also with the G and F glycoproteins of the henipaviruses.[69,140,165,166] Here, the overall theme suggests a negative correlation between the attachment and fusion glycoprotein interaction and the membrane fusion activity of the viral species; that is, alterations in H that enhance H-F association adversely affect fusion and those that weaken the H-F interaction yield an enhanced fusion feature. The model 2, or dissociation model, has more recently been referred to as the clamp model, whereby the pre-association of complexes suggests that the attachment glycoprotein maintains the prefusion metastable conformation of its partner F glycoprotein until encountering receptor.[156]

The pre-association of attachment and fusion glycoproteins before receptor binding versus the post-receptor bound inducement of their association has also been addressed by other experiments. For example, as an alternative means to address these models, it has been demonstrated with several attachment glycoproteins and F glycoprotein partners, including those from MeV, NDV and hPIV2, that they interact early during biosynthesis in the endoplasmic reticulum (ER).[161,167-169] However, in other viruses (PIV5 (SV5) and hPIV3) a F and HN interaction prior to fusion was not strong.[94] Additionally, the henipavirus G and F glycoproteins have a more complex biosynthesis and maturation pathway in comparison to other paramyxoviruses, and the G glycoprotein takes longer to traffic through the ER and Golgi. This longer trafficking time of G together with the complex pattern of F maturation suggests that G-F interaction does not occur until both glycoproteins are expressed on the cell membrane.[167,170]

Nevertheless, either model recognizes an interaction between the attachment and fusion glycoproteins that is regulated by receptor binding (recently reviewed refs. 171,172), and the triggering mechanisms between and HN-F pair and H/G-F pair may essentially be the same but with each differing in their general propensity to associate in their pre-receptor bound states. Indeed, recent experiments with NiV employed the addition of N-linked glycosylation modification sites as probes to examine the specific interactions between the G attachment and F glycoproteins.[173] These studies with NiV revealed contrasting findings from two earlier studies with NDV and MeV that demonstrated the NDV HN, with N-glycan additions in the stalk, was defective in both its fusion promotion and F interacting capacity[152] and that N-glycan additions in an implicated F-interactive site of the MeV H blocked H-F complex formation.[174] Rather, the NiV G and its F interaction was not affected by most N-glycan additions and does not appear to be solely mediated by the stalk domain of G.[173] Thus, although the preponderance of data to date have implicated the attachment glycoprotein stalk in the interacting and triggering process with F, these data reveal that there is another level of G-F interaction, which suggests a natural propensity for a specific association between the two glycoproteins. However, this association is not required for maintaining F in a prefusion conformation (clamp) or a trigger for F fusogenic triggering. In either case it also seems likely that upon F glycoprotein triggering, the initiation of its conformational changes leading to 6-HB formation and the driving of the membrane merger process, F would need to be free of any association with its large oligomeric partner (HN, H or G).

The recently characterized soluble forms of trimeric henipavirus F glycoprotein discussed earlier[137] have also been valuable in assessing the themes within the paramyxovirus fusion models. For example, a murine mAb (5B3), which is specific for the prefusion form of henipavirus F glycoprotein, is capable of binding F on the surface of expressing cells and also from cellular lysates containing F glycoprotein in the absence of the co-expression of its partner G glycoprotein, indicating that the clamp model as defined is not accurate.[137] It appears that although receptor-induced G glycoprotein triggering of the F-mediated fusion process likely takes place, the requirement of G association with its partner F glycoprotein in order to maintain F in its prefusion and metastable state is not necessary. Rather, this data is more in line with the provocateur model of fusion discussed earlier, where the prefusion conformation of F (PIV5) is maintained in the absence of the coexpression of HN.[156] Further, the independent trafficking and maturation patterns of HeV and NiV F reviewed above are also in an agreement with such a scenario.

In addition, two MeV F specific mAbs (186A and 19GD) specific for either the prefusion/pretriggered versus fusion-triggered conformation of F, respectively, were recently used as probes to address the MeV F triggering process, and here as well the prefusion MeV F glycoprotein could be recognized strongly by the 186A mAb in absence of H co-expression, revealing that MeV F does not require a physical association with H to maintain its prefusion conformation.[175] Similarly, a second study further explored the morbillivirus F glycoprotein triggering process with similar mAb binding techniques—identifying antibodies to both MeV and CDV F that were specific for either their prefusion, triggered or postfusion conformations.[176] In these studies it was demonstrated that prefusion F-specific mAbs could bind to F in absence of H co-expression, that conversion of F from a prefusion to a triggered conformation by higher temperature revealed loss of binding by those mAbs and that antibodies specific for postfusion forms of F acquired enhanced binding upon heating. Essentially identical antibody binding profiles were observed under physiologic conditions of H triggered F activation.[176] Together these studies support the conclusion recently suggested by Ader et al[176] that the attachment glycoprotein, such as H, serves to lower the F triggering energy barrier rather than to maintain the prefusion conformation of F through a binding mechanism, a function more in line with a provocateur type of model. There is now evidence with at least four different species of F glycoproteins (PIV5, NiV, MeV and CDV), which together span the varied attachment glycoprotein types (HN, H or G), that demonstrate the prefusion metastable F conformation can be maintained in the absence of any ‘clamp’ or physical association with its attachment glycoprotein partner.

In summary, these findings suggest that neither a clamp model nor a provocateur model as presently defined can fully account for all the experimental observations to date on the mechanism of fusion. In fact, a recent report by Porotto et al[177] describes a variation of the hPIV3 fusion mechanism that incorporates features of both the clamp and provocateur models. Here it is suggested that hPIV3 HN must continually engage receptor to activate F as interruption of hPIV3 HN and receptor blocks F-mediated membrane fusion. Although no direct HN-F interactions were assessed in this study, an approach to examine hPIV3 fusion using bimolecular fluorescence complementation to follow the dynamics of HN and F in live cells was conducted.[178] The authors demonstrate that HN and F associate prior to receptor engagement, that HN drives the formation of HN and F interacting clusters at the site of membrane fusion and that the interaction of the HN-F pairs of oligomers modulate the fusion process. In sum, it appears that measurable pre-association of paramyxovirus attachment and fusion glycoproteins, prior to fusion triggering by receptor, is dependent on the particular protein pair of viral species proteins and possibly due to a requirement for maintenance of the prefusion F conformation (clamp model). Nevertheless, the fusion process for most paramyxoviruses is dependent on a receptor mediated binding event by the attachment glycoprotein and this will be discussed in the next section.

RECEPTOR INDUCED CONFORMATIONAL CHANGES IN ATTACHMENT GLYCOPROTEINS

The influence of receptor binding on the fusion-triggering process has been a major focus of research and is a factor that can potentially differentiate the most recently proposed models of the mechanism of paramyxovirus membrane fusion. The large amount of recent structural information, particularly in the many comparisons that have been made between the receptor bound and unbound structures of H and G, has revealed that major conformational changes in the attachment glycoprotein heads are not observed upon receptor binding.[31,60-62,73] Rather, much attention has more recently focused on the role of receptor-induced conformational changes that occur in the higher-order oligomeric structures of the attachment glycoprotein tetramer.

Data derived from NDV HN structural studies suggested that receptor binding leads to conformational changes in HN and a model was suggested in which receptor (sialic acid) engagement would facilitate dimer formation or even tetramer formation in a ligand dependent manner[24] and HN oligomerization would be the trigger for F activation. This model of receptor induced conformational change in HN for triggering F fusion was first proposed by Sergel et al[179] and then diagramed by Lamb[153] which was essentially the first 'association' or 'provocateur' model discussed earlier. The receptor-induced conformational changes in HN were first detailed at the molecular level using mutagenesis analysis.[180,181] These studies revealed differences in the structure of receptor-bound HN and non-receptor-bound HN, although importantly, the structure of HN in complex with its receptor was not then solved. Subsequently, a second sialic acid binding site was identified at the dimer interface of NDV HN,[38] and it was hypothesized to play a role in facilitating membrane fusion. The specific steps of paramyxovirus fusion that were proposed in this revised model[38] are as follows: HN and F may exist in a complex on the cell surface and this complex holds both glycoproteins in the "off states". Upon binding to sialic acid, conformational changes occur in HN that converts the glycoprotein to its "on state," which leads to cleavage and release of the sialic acid from the sialic acid-containing receptor. The release of sialic-acid induces further conformational changes in HN at the dimer interface. These changes were hypothesized to alter the dimeric or tetrameric properties of HN that lead to changes in the stalk domain and in so doing trigger the fusion glycoprotein. Concurrently, the second sialic acid binding site is formed by the conformational change induced by the release of sialic acid from site 1. The existence of a second sialic acid receptor binding site for several different HN glycoproteins has also been functionally identified.[182-184] Recently, using glycan array assays, it was demonstrated that the HN of hPIV1 has a second site for receptor binding masked by an N-linked glycan and that sialic acid receptor binding to the first site triggers the exposure of the second site.[185] The significance of the second sialic acid binding site has been hypothesized to bind cell-surface sialic acid residues and maintain a close proximity of the virion to the cell surface to aid in efficient targeting of the fusion peptide for membrane fusion mediated by the F glycoprotein.[186] This model was recently tested using a series of HN dimer interface mutants, and it was demonstrated that binding of receptor to site 1 triggered HN interaction with F and that site 2 appeared to maintain binding with the target cell membrane during the fusion process.[187] A summary of this model is shown in Figure 6. However, it has also been suggested that receptor binding to the NDV HN site 2 plays an active role in transmitting the fusion activation signal to the stalk region of HN.[188] This function could also be demonstrated with chimeric proteins composed of the globular head of NDV HN and the stalk region of hPIV3 or NiV where receptor binding to site 2 led to the activation of heterotypic F glycoproteins.[188]

A variation of the NDV fusion model described above was later suggested based on the solved structures of PIV5 and hPIV3 HN alone or in complex with receptor.[25,26] Several differences exist between the solved HN structures; however, a key distinction leading to the proposal of this modified model is that for both PIV5 and hPIV3 there are no major conformational changes in HN upon receptor binding. Additionally, neither PIV5 nor hPIV3 HN contain a second sialic acid binding site. Furthermore, dimers and tetramers (PIV5) were evident in the solved structures in the absence of receptor, thus a

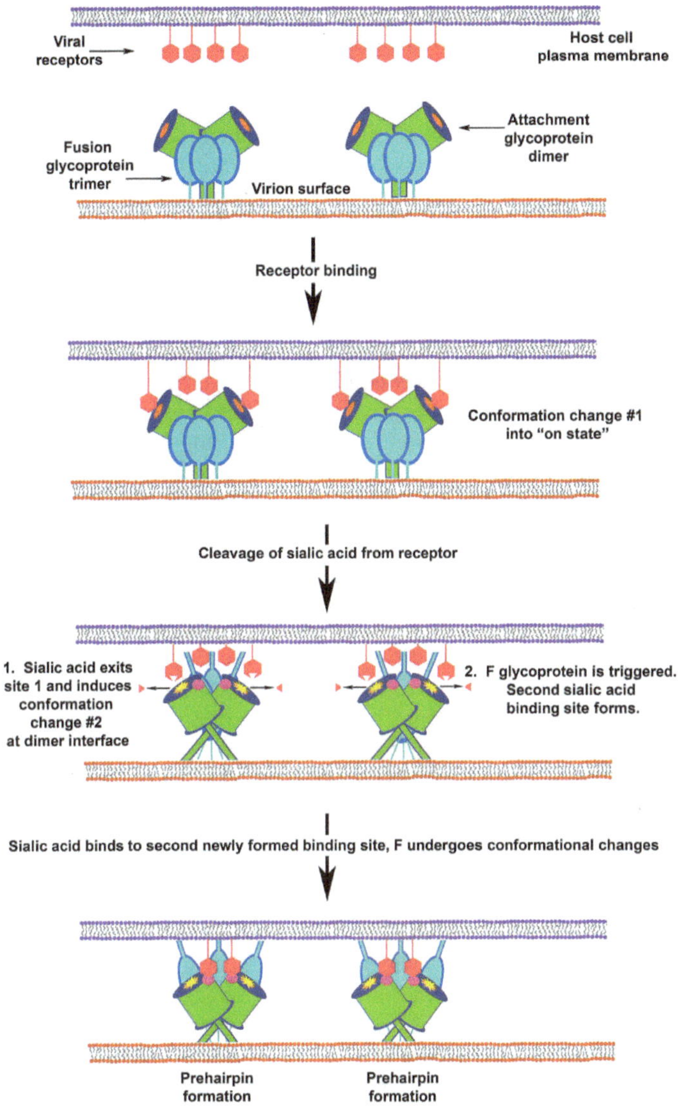

Figure 6. Multiple conformational changes in NDV HN trigger F from metastable to fusogenic. Fusion model as described in Zaitsev et al[38] depicting the receptor-triggered mechanism of NDV fusion. Receptor binding induces conformational changes that convert HN to its "on state". Subsequent cleavage and release of sialic acid leads to conformational changes at the HN dimer interface that not only are critical for triggering F but also generate a second sialic acid binding site. Binding cellular receptors via the newly formed second sialic acid site is hypothesized to keep the virion in close proximity to the target cell membrane. For simplicity F is shown as a trimer with limited conformational changes and membrane merger has been excluded.

ligand induced oligomerization of HN was not at play. NDV, hPIV3 and PIV5 HN all form similar dimer and likely tetrameric conformations. The dimer interactions are formed over a large surface, and it has been hypothesized they most likely are of high affinity, if not disulfide bond linked, requiring significant amounts of energy to destabilize the interaction.[26] The tetramers, that is the dimer of dimers, by comparison appeared to be less conserved and one of weaker association. A revised model proposed that the HN dimer/tetramer is present in the absence of receptor and that receptor binding destabilizes the tetramer and it may partially dissociate. This tetrameric conformational change was suggested to lead to changes in the stalk domain of HN, the proposed site of F interaction, thus providing the necessary trigger for activating F to its fusogenic state.[26]

When the crystal structure of NDV HN alone and in complex with sialic acid (beta-anomer) or a neuraminidase inhibitor was reported, comparisons of the structures also suggested that the catalytic site was activated by a conformational switch in the head, providing roles for both sialic acid binding and hydrolysis activity.[24] It was postulated that significant conformational change in the HN dimer essential for fusion could occur as HN transitions from an initial structure possessing minimal inter-monomeric contacts to a structure containing an extensive dimer interface. This proposed mechanism of oligomeric conformational change and its role in the fusion triggering process was later tested by mutational insertions of inter-monomeric disulfide bonds in the globular head domain of NDV HN. The insertion of disulfide bonds prevented the formation of the minimal interface configuration of HN, however, rather than inhibiting its fusion promotion activity the mutated HN possessed enhanced receptor-binding and fusion promotion activity.[189] This study, using novel disulfide bond engineering to stabilize the HN dimers, showed that neither the minimal interface form of HN nor the proposed conformational changes were required for fusion. In contrast, using the extensive available structural information on H and HN, Navaratnarajah et al[190] modeled and tested the role of conformational changes within the MeV H dimer. Here, the notion of a requirement for a conformational rearrangement of the head domains relative to each other within an individual dimer was also examined by the mutational insertion of strategically placed disulfide bonds. In this instance their placement prevented a required movement of the heads and subsequent F triggering and fusion, and the authors suggested a model in which H and HN transmit the fusion triggering signal in alternative ways, perhaps due to the different locations of the receptor binding site.

As discussed earlier, a tetrameric configuration of the native paramyxovirus attachment glycoprotein has been widely described, and there has recently been a considerable amount of new data on their structure and the role of the tetramer along with the receptor induced conformational changes as they relate to the F triggering fusion process. For a paramyxovirus using a protein receptor, the triggering of MeV F by H is the most extensively explored and understood system. As discussed above, MeV F and H associate intracellularly prior to any role of receptor,[161] and through the use of a bimolecular complementation assay it has been shown that receptor binding to H, as well as the elements in H required for F interaction[174,191] versus F triggering,[163] are distinct, whereby mutants of H in these various domains when expressed together can effectively restore MeV F/H fusion[40] (reviewed in ref. 192)

Structural conformational changes in a central region of either the MeV or CDV H stalk domain were identified by Ader et al[193] as critical in the transfer of the fusion triggering signal. Here, engineered disulfide bonds were introduced within the stalk domain central region spanning residues 91-115, which also contains the F-interacting domain mentioned above.[174,191] Several of these inserted disulfide bridges could block H fusion promotion activity and upon their reduction fusion activity was restored. Altogether, with modeling, the data suggested that the H stalk is a tetramer of subunits which undergo structural rearrangement following receptor binding, promoting an interaction of a specific stalk element of H with the associated F trimer, triggering its fusogenic activity.

Brindley et al[175] later addressed the MeV fusion process by targeting the MeV H stalk through the manipulation of the existing head proximal disulfide bonds in the context of the H tetramer model together with bimolecular complementation and the use of both soluble and membrane associated MeV receptors. These studies found that the H tetramer structure is maintained in that dimers of dimers do not dissociate but that the central stalk region does require flexibility. Furthermore, receptor binding to only one dimer within the context of the tetramer was sufficient to trigger fusion. The triggering of conformational changes in the context of the H tetramer could be accomplished by soluble receptor (SLAM) and also initiate the F triggering and refolding process as detected by pre- and postfusion F specific mAbs described earlier, but cell-cell fusion pore formation required the triggering of H by membrane-anchored receptor.[175]

In another examination of the MeV F/H triggering process using the disulfide bridging approach, Navaratnarajah et al[194] analyzed the H stalk using a comprehensive cysteine residue substitution mutagenesis process. These studies revealed three stalk regions of varying importance in which two of the three stalk segments possessed a tendency for the formation of tetrameric configurations. Some disulfide linked H stalk mutants that were fusion-triggering defective could be chemically reduced with a concomitant restoration of its fusion-promotion activity, whereas another segment of the stalk when covalently linked into H tetramers had no effect on the protein's fusion-promoting activity. A third stalk domain, globular head-proximal, could not be readily disulfide linked and stabilized. In total, this study identified an F-triggering (interacting) domain of the H stalk as residues ~75-127, similar to the CDV stalk residues (91-115). In a companion report, the MeV F glycoprotein was analyzed by modeling to predict surface exposed residues in regions that could be predicted to interact with other glycoproteins, namely H. A large panel of some 50 possible residues were noted and by conducting an iterative mutagenesis and functional analysis, a set of specific mutants were identified that inhibited fusion with four mutants lining a cavity flanked by two monomers of the F trimer model. It was suggested that the stalk region of the H tetramers could be lodged within the sides of their companion F trimers at the site of the modeled cavity with two helices of an H tetramer contacting one side groove of an F trimer, suggesting that one H tetramer could possibly transmit the F-trigging signal to at least two opposing F trimers.[195]

Finally, a recent intriguing study reported on the application of a headless PIV5 HN glycoprotein in triggering PIV5 F-mediated fusion.[196] Here, this study also demonstrated that essentially the entire stalk (PIV5 HN residues 1-117) was required, and it was proposed to fold into its 4-helix bundle or otherwise receptor-bound conformation that could associate with and trigger its F glycoprotein partner. The study also revealed that the F glycoprotein of PIV5 also maintained its metastable prefusion conformation in absence of HN co-expression. Further, and importantly, the 4 helix headless protein structure also displayed and maintained its viral species specificity for triggering F-mediated membrane fusion and could not trigger other F glycoproteins. The globular heads were found clearly dispensable, and the roles of the heads appear to define cell tropism and also perhaps to mask and/or maintain the stalk domains in a pre-receptor bound and non-F triggering form. In hindsight, it seems surprising that this approach was never before tested, and examining this possibility with other paramyxoviruses, particularly the morbilliviruses or henipaviruses, may yield interesting results. Remarkably, in the case of HeV, mutations in the heads that model to the head-stalk interface in a heads-down structure have been identified and shown to block fusion promotion activity but allow competent receptor binding.[69] Perhaps such mutants prevent the receptor-induced movement of the heads that would allow access to the henipavirus G stalk domain required for F triggering.[43]

Transfer of the F-mediated fusion triggering signal appeared to involve an opening and repositioning of the dimeric interface of the H head domains,[190] which is then followed by conformational changes of a central domain of the H stalk.[193] It was suggested that receptor binding and pulling de-stabilizes this H-dimer interface,[190] which would subsequently elicit the conformational change in the central stalk segment of H[191,193] required for the fusion triggering signal. Indeed, such a receptor-pulling process might certainly be envisioned during the cell-cell or virus-cell contact step at the beginning of the process.

FUSION MODELS

In consideration of the large amount of structural and functional data on the paramyxovirus fusion process, with particular attention to the recent reports over just the past few years from leaders in the field, the current modeled scenarios of the receptor binding and fusion triggering steps are shown in a refinement of the originally proposed models 1 and 2 (association and dissociation models) (Fig. 7), and an attempt has been made to include elements of both the non-pre-associated and pre-associated states of a fusion and attachment glycoprotein and the heads-up and heads-down features of the attachment glycoproteins in relation to the fusion triggering process. Initial expression of the tetrameric attachment (HN/H/G) glycoprotein, with each dimer pair distinguished by color, depicts the fusion glycoprotein as not being pre-associated with the attachment glycoprotein, which is in a heads-down configuration (Fig. 7A). This model has most recently been refined and proposed based on the extensive data on the function of F and HN glycoproteins.[196] Here, this model recognizes that a metastable prefusion F can exist without a required association of its attachment glycoprotein partner, the importance of the F-triggering role of the stalk domain and its membrane distal elements, the receptor-mediated movement of the head domains upwards and the provocateur association and activation of F. This model does not necessarily preclude other paramyxovirus species, such as the henipaviruses that possess an F and G glycoprotein pair, because until receptor binding triggers tetrameric conformational changes in G, an F/G oligomeric pair of glycoproteins may simply have an ability to associate in membranes without inducing additional conformational changes. Likewise, the MeV F/H pair could also potentially fit a similar model recognizing that H does not have to maintain F in a prefusion state and that contact between F and H could occur with either a heads-up or heads-down position as discussed by Navaratnarajah et al[194] Alternatively, in Figure 7B, a tetrameric heads-up attachment glycoprotein oligomer is pre-associated with its F glycoprotein partner, a scenario recently refined and modeled with the MeV F and H glycoprotein pairs.[175,193] Here, receptor binding triggers a conformational change in the tetramer, the central features of which are the movement to the heads-down configuration in conjunction with a twist of the stalk region, which all together facilitates a targeted association of elements within the attachment glycoprotein's stalk to its partner F glycoprotein resulting in the disassociation and fusogenic conformational changes in F. However, it is not yet clear whether the heads-up versus heads-down attachment glycoprotein configuration is mechanistically linked to the fusion triggering process.

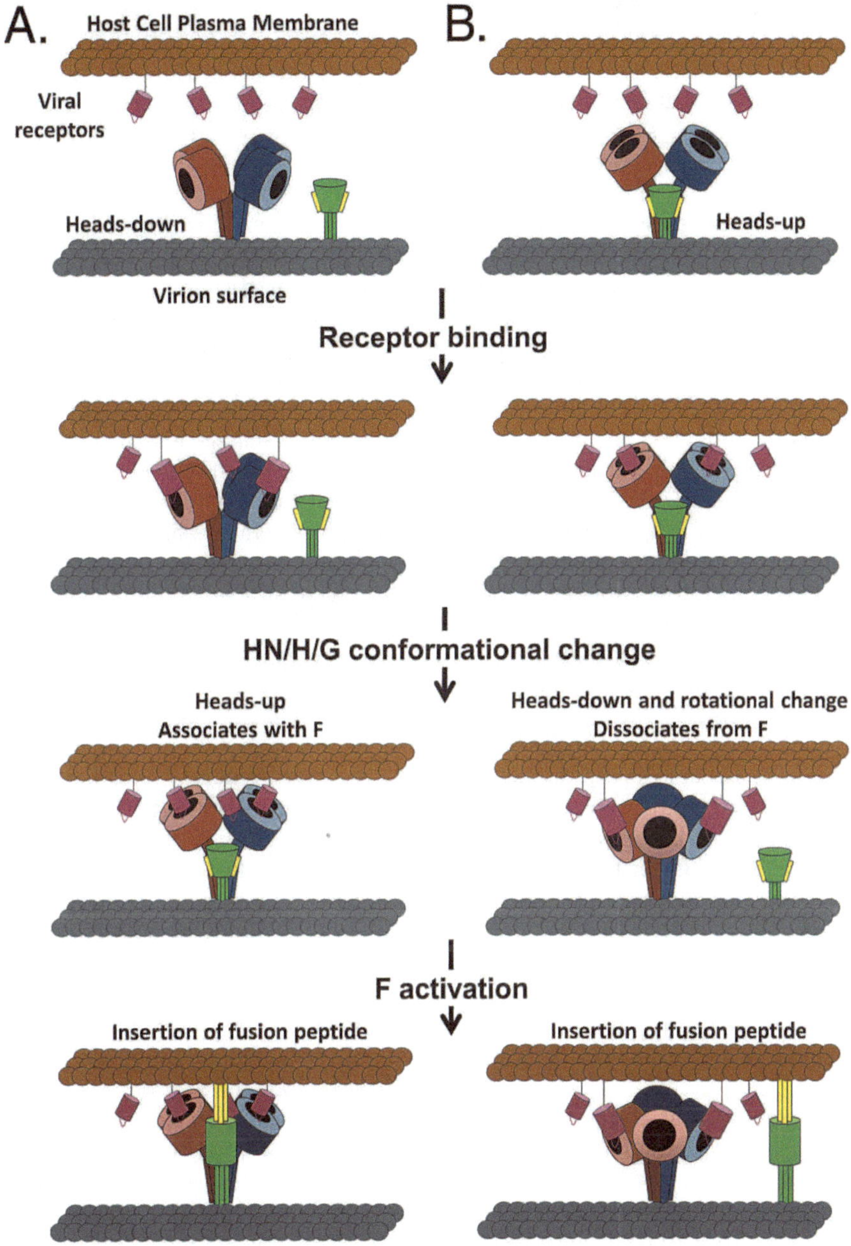

Figure 7. Models of paramyxovirus membrane fusion involving heads-up and heads-down conformations. Initial expression of the tetrameric attachment (HN/H/G) glycoprotein (dimers colored red and blue) and the fusion (F) glycoprotein (green) is depicted in the (A) heads-down, non-F-associated or (B) heads-up, F-associated conformations. In both models, HN/H/G binds receptor (maroon) and undergoes receptor-induced conformational changes, switching from heads-up to heads-down or vice versa. The change in the position of the globular heads allows for (A) association (provocateur model) or (B) dissociation (clamp model) with F, leading to the fusion activation of F and the beginning of membrane fusion by the insertion of the fusion peptide (yellow) into the target cell membrane.

CONCLUSION AND REMAINING QUESTIONS

Paramyxoviruses have evolved a more complex mechanism of attachment and membrane fusion that facilitates delivery of the genome into the host cell, one that for most members requires two independent glycoproteins. The overall structural similarity of the attachment glycoproteins of viruses within the subfamily *Paramyxovirinae*, coupled to the highly conserved functional domains of the fusion glycoproteins in all paramyxoviruses reveal their common ancestry, while their differences in receptor engagement and the events that promote fusion most likely reflect individual adaptations, perhaps as a result of varying hosts and varying tissues within a host. Years of research by many have uncovered important facets of paramyxovirus entry and led to changing hypotheses and the proposal of alternative models of the paramyxovirus entry process. The two current and most favored models that have emerged combine old and new data with the solved structures of the attachment glycoproteins with and without their receptors. Nevertheless, the most recent revised models discussed here that have been derived from extensive data from viruses possessing an HN, H or G, are all in fair agreement and allow for subtle differences that have been experimentally explored. They can also fairly accommodate the differences in whether a sialic acid versus a proteinaceous receptor is employed as well as the locations of the receptor binding site in the attachment glycoprotein. Additionally, the recent structures of both uncleaved and cleaved metastable F glycoprotein forms have provided insights into the significant conformational changes that provide the necessary energy for a paramyxovirus fusion glycoprotein to mediate membrane merger. Although many questions relating to entry have been answered, additional information is still missing, such as the pre- and postfusion F structures from the same viral species. Equally intriguing would be to determine structures of a tetrameric attachment glycoprotein in complex with a receptor(s), particularly a receptor known to trigger fusion promotion activity with one that does not. Also, changes in the biochemical interaction of the fusion and attachment glycoproteins as fusion proceeds have yet to be adequately profiled. Finally, and perhaps difficult to achieve, the structural solution of any F trimer in complex with a stalk containing attachment glycoprotein or perhaps the stalk domain itself would be a significant advance.

ACKNOWLEDGMENTS

We would like to thank Alex Hyatt for generously providing the paramyxovirus electron micrographs and are indebted to Alex, Sandra Crameri and Rosemary Van Driel for their assistance in scanning and scaling all EM negatives. We thank Robert Lamb for generously allowing us to use the NDV HN, uncleaved and cleaved metastable PIV5 F and postfusion hPIV3 F structures in this chapter. Research in the author's laboratories is supported in part by NIH research grants AI077995 and AI054715 (C.C.B.).

REFERENCES

1. Lamb RA, Parks GD. Paramyxoviridae: The viruses and their replication. In: Knipe DM, Howley PM eds. Fields Virology, 5 ed. Philadelphia: Lippincott Williams & Wilkins, 2007:1449-1496.
2. Alexander DJ. Newcastle disease and other avian paramyxoviruses. Rev Sci Tech 2000; 19(2):443-462.
3. Tambi EN, Maina OW, Mukhebi AW et al. Economic impact assessment of rinderpest control in Africa. Rev Sci Tech 1999; 18(2):458-477.
4. Eaton BT, Broder CC, Middleton D et al. Hendra and Nipah viruses: different and dangerous. Nat Rev Microbiol 2006; 4(1):23-35.
5. Eaton BT, Mackenzie JS, Wang L-F. Henipaviruses. In: Knipe DM, Howley PM ed. Fields Virology, 5 ed. Philadelphia: Lippincott Williams & Wilkins, 2007:1587-1600.
6. Lamb RA. Mononegavirales. In: Knipe DM, Howley PM ed. Fields Virology, 5 ed. Philadelphia: Lippincott Williams and Wilkins, 2007:1357-1361.
7. Mayo MA. A summary of taxonomic changes recently approved by ICTV. Arch Virol 2002; 147(8):1655-1663.
8. de Leeuw O, Peeters B. Complete nucleotide sequence of Newcastle disease virus: evidence for the existence of a new genus within the subfamily Paramyxovirinae. J Gen Virol 1999; 80(Pt 1):131-136.
9. Seal BS, Crawford JM, Sellers HS et al. Nucleotide sequence analysis of the Newcastle disease virus nucleocapsid protein gene and phylogenetic relationships among the Paramyxoviridae. Virus Res 2002; 83(1-2):119-129.
10. Seal BS, King DJ, Meinersmann RJ. Molecular evolution of the Newcastle disease virus matrix protein gene and phylogenetic relationships among the paramyxoviridae. Virus Res 2000; 66(1):1-11.
11. Wang LF, Yu M, Hansson E et al. The exceptionally large genome of Hendra virus: support for creation of a new genus within the family Paramyxoviridae. J Virol 2000; 74(21):9972-9979.
12. Li Z, Yu M, Zhang H et al. Beilong virus, a novel paramyxovirus with the largest genome of non-segmented negative-stranded RNA viruses. Virology 2006; 346(1):219-228.
13. Kurath G, Batts WN, Ahne W et al. Complete genome sequence of Fer-de-Lance virus reveals a novel gene in reptilian paramyxoviruses. J Virol 2004; 78(4):2045-2056.
14. Marsh GA, de Jong C, Barr JA et al. Cedar virus: a novel Henipavirus isolated from Australian bats. PLoS Pathog 2012; 8(8):e1002836.
15. Morrison TG. The three faces of paramyxovirus attachment proteins. Trends Microbiol 2001; 9(3):103-105.
16. Lamb RA, Jardetzky TS. Structural basis of viral invasion: lessons from paramyxovirus F. Curr Opin Struct Biol 2007; 17(4):427-436.
17. Parks GD, Lamb RA. Folding and oligomerization properties of a soluble and secreted form of the paramyxovirus hemagglutinin-neuraminidase glycoprotein. Virology 1990; 178(2):498-508.
18. Scheid A, Caliguiri LA, Compans RW et al. Isolation of paramyxovirus glycoproteins. Association of both hemagglutinating and neuraminidase activities with the larger SV5 glycoprotein. Virology 1972; 50(3):640-652.
19. Thompson SD, Portner A. Localization of functional sites on the hemagglutinin-neuraminidase glycoprotein of Sendai virus by sequence analysis of antigenic and temperature-sensitive mutants. Virology 1987; 160(1):1-8.
20. Langedijk JP, Daus FJ, van Oirschot JT. Sequence and structure alignment of Paramyxoviridae attachment proteins and discovery of enzymatic activity for a morbillivirus hemagglutinin. J Virol 1997; 71(8):6155-6167.
21. Yu M, Hansson E, Langedijk JP et al. The attachment protein of Hendra virus has high structural similarity but limited primary sequence homology compared with viruses in the genus Paramyxovirus. Virology 1998; 251(2):227-233.
22. Iorio RM, Melanson VR, Mahon PJ. Glycoprotein interactions in paramyxovirus fusion. Future Virol 2009; 4(4):335-351.
23. Colman PM, Hoyne PA, Lawrence MC. Sequence and structure alignment of paramyxovirus hemagglutinin-neuraminidase with influenza virus neuraminidase. J Virol 1993; 67(6):2972-2980.
24. Crennell S, Takimoto T, Portner A et al. Crystal structure of the multifunctional paramyxovirus hemagglutinin-neuraminidase. Nat Struct Biol 2000; 7(11):1068-1074.
25. Lawrence MC, Borg NA, Streltsov VA et al. Structure of the haemagglutinin-neuraminidase from human parainfluenza virus type III. J Mol Biol 2004; 335(5):1343-1357.
26. Yuan P, Thompson TB, Wurzburg BA et al. Structural studies of the parainfluenza virus 5 hemagglutinin-neuraminidase tetramer in complex with its receptor, sialyllactose. Structure (Camb) 2005; 13(5):803-815.

27. Yuan P, Swanson KA, Leser GP et al. Structure of the Newcastle disease virus hemagglutinin-neuraminidase (HN) ectodomain reveals a four-helix bundle stalk. Proc Natl Acad Sci USA 2011; 108(36):14920-14925.
28. Hashiguchi T, Kajikawa M, Maita N et al. Crystal structure of measles virus hemagglutinin provides insight into effective vaccines. Proc Natl Acad Sci USA 2007; 104(49):19535-19540.
29. Colf LA, Juo ZS, Garcia KC. Structure of the measles virus hemagglutinin. Nat Struct Mol Biol 2007; 14(12):1227-1228.
30. Bowden TA, Aricescu AR, Gilbert RJ et al. Structural basis of Nipah and Hendra virus attachment to their cell-surface receptor ephrin-B2. Nat Struct Mol Biol 2008; 15(6):567-572.
31. Xu K, Rajashankar KR, Chan YP et al. Host cell recognition by the henipaviruses: crystal structures of the Nipah G attachment glycoprotein and its complex with ephrin-B3. Proc Natl Acad Sci USA 2008; 105(29):9953-9958.
32. Thompson SD, Laver WG, Murti KG et al. Isolation of a biologically active soluble form of the hemagglutinin-neuraminidase protein of Sendai virus. J Virol 1988; 62(12):4653-4660.
33. Markwell MA, Fox CF. Protein-protein interactions within paramyxoviruses identified by native disulfide bonding or reversible chemical cross-linking. J Virol 1980; 33(1):152-166.
34. Morrison TG, McQuain C, O'Connell KF et al. Mature, cell-associated HN protein of Newcastle disease virus exists in two forms differentiated by posttranslational modifications. Virus Res 1990; 15(2):113-133.
35. Ng DT, Randall RE, Lamb RA. Intracellular maturation and transport of the SV5 type II glycoprotein hemagglutinin-neuraminidase: specific and transient association with GRP78-BiP in the endoplasmic reticulum and extensive internalization from the cell surface. J Cell Biol 1989; 109(6 Pt 2):3273-3289.
36. Russell R, Paterson RG, Lamb RA. Studies with cross-linking reagents on the oligomeric form of the paramyxovirus fusion protein. Virology 1994; 199(1):160-168.
37. Ng DT, Hiebert SW, Lamb RA. Different roles of individual N-linked oligosaccharide chains in folding, assembly, and transport of the simian virus 5 hemagglutinin-neuraminidase. Mol Cell Biol 1990; 10(5):1989-2001.
38. Zaitsev V, von Itzstein M, Groves D et al. Second sialic acid binding site in Newcastle disease virus hemagglutinin-neuraminidase: implications for fusion. J Virol 2004; 78(7):3733-3741.
39. Plemper RK, Hammond AL, Cattaneo R. Characterization of a region of the measles virus hemagglutinin sufficient for its dimerization. J Virol 2000; 74(14):6485-6493.
40. Brindley MA, Plemper RK. Blue native PAGE and biomolecular complementation reveal a tetrameric or higher-order oligomer organization of the physiological measles virus attachment protein H. J Virol 2010; 84(23):12174-12184.
41. Bossart KN, Crameri G, Dimitrov AS et al. Receptor binding, fusion inhibition and induction of cross-reactive neutralizing antibodies by a soluble G glycoprotein of Hendra virus. J Virol 2005; 79(11):6690-6702.
42. Maar D, Harmon B, Chu D et al. Cysteines in the stalk of the Nipah virus G glycoprotein are located in a distinct subdomain critical for fusion activation. J Virol 2012; 86(12):6632-6642.
43. Steffen DL, Xu K, Nikolov DB et al. Henipavirus Mediated Membrane Fusion, Virus Entry and Targeted Therapeutics. Viruses 2012; 4(2):280-308.
44. Choppin PW, Scheid A. The role of viral glycoproteins in adsorption, penetration, and pathogenicity of viruses. Rev Infect Dis 1980; 2(1):40-61.
45. Dorig RE, Marcil A, Chopra A et al. The human CD46 molecule is a receptor for measles virus (Edmonston strain). Cell 1993; 75(2):295-305.
46. Naniche D, Varior-Krishnan G, Cervoni F et al. Human membrane cofactor protein (CD46) acts as a cellular receptor for measles virus. J Virol 1993; 67(10):6025-6032.
47. Nussbaum O, Broder CC, Moss B et al. Functional and structural interactions between measles virus hemagglutinin and CD46. J Virol 1995; 69(6):3341-3349.
48. Tatsuo H, Ono N, Tanaka K et al. SLAM (CDw150) is a cellular receptor for measles virus. Nature 2000; 406(6798):893-897.
49. Tatsuo H, Ono N, Yanagi Y. Morbilliviruses use signaling lymphocyte activation molecules (CD150) as cellular receptors. J Virol 2001; 75(13):5842-5850.
50. Baron MD. Wild-type Rinderpest virus uses SLAM (CD150) as its receptor. J Gen Virol 2005; 86(Pt 6):1753-1757.
51. Galbraith SE, Tiwari A, Baron MD et al. Morbillivirus downregulation of CD46. J Virol 1998; 72(12):10292-10297.
52. Welstead GG, Hsu EC, Iorio C et al. Mechanism of CD150 (SLAM) down regulation from the host cell surface by measles virus hemagglutinin protein. J Virol 2004; 78(18):9666-9674.
53. Masse N, Ainouze M, Neel B et al. Measles virus (MV) hemagglutinin: evidence that attachment sites for MV receptors SLAM and CD46 overlap on the globular head. J Virol 2004; 78(17):9051-9063.
54. Vongpunsawad S, Oezgun N, Braun W et al. Selectively receptor-blind measles viruses: Identification of residues necessary for SLAM- or CD46-induced fusion and their localization on a new hemagglutinin structural model. J Virol 2004; 78(1):302-313.
55. Leonard VH, Sinn PL, Hodge G et al. Measles virus blind to its epithelial cell receptor remains virulent in rhesus monkeys but cannot cross the airway epithelium and is not shed. J Clin Invest 2008; 118(7):2448-2458.

56. Tahara M, Takeda M, Shirogane Y et al. Measles virus infects both polarized epithelial and immune cells by using distinctive receptor-binding sites on its hemagglutinin. J Virol 2008; 82(9):4630-4637.
57. Noyce RS, Bondre DG, Ha MN et al. Tumor cell marker PVRL4 (nectin 4) is an epithelial cell receptor for measles virus. PLoS Pathog 2011; 7(8):e1002240.
58. Muhlebach MD, Mateo M, Sinn PL et al. Adherens junction protein nectin-4 is the epithelial receptor for measles virus. Nature 2011; 480(7378):530-533.
59. Noyce RS, Richardson CD. Nectin 4 is the epithelial cell receptor for measles virus. Trends Microbiol 2012; 20(9):429-439.
60. Zhang X, Lu G, Qi J et al. Structure of measles virus hemagglutinin bound to its epithelial receptor nectin-4. Nat Struct Mol Biol 2012.
61. Santiago C, Celma ML, Stehle T et al. Structure of the measles virus hemagglutinin bound to the CD46 receptor. Nat Struct Mol Biol 2010; 17(1):124-129.
62. Hashiguchi T, Ose T, Kubota M et al. Structure of the measles virus hemagglutinin bound to its cellular receptor SLAM. Nat Struct Mol Biol 2011; 18(2):135-141.
63. Bossart KN, Wang LF, Flora MN et al. Membrane fusion tropism and heterotypic functional activities of the Nipah virus and Hendra virus envelope glycoproteins. J Virol 2002; 76(22):11186-11198.
64. Bossart KN, Wang LF, Eaton BT et al. Functional expression and membrane fusion tropism of the envelope glycoproteins of Hendra virus. Virology 2001; 290(1):121-135.
65. Eaton BT, Wright PJ, Wang LF et al. Henipaviruses: recent observations on regulation of transcription and the nature of the cell receptor. Arch Virol Suppl 2004(18):122-131.
66. Bonaparte MI, Dimitrov AS, Bossart KN et al. Ephrin-B2 ligand is a functional receptor for Hendra virus and Nipah virus. Proc Natl Acad Sci USA 2005; 102(30):10652-10657.
67. Negrete OA, Levroney EL, Aguilar HC et al. EphrinB2 is the entry receptor for Nipah virus, an emergent deadly paramyxovirus. Nature 2005; 436(7049):401-405.
68. Negrete OA, Wolf MC, Aguilar HC et al. Two key residues in EphrinB3 are critical for its use as an alternative receptor for Nipah virus. PLoS Pathog 2006; 2(2):e7.
69. Bishop KA, Stantchev TS, Hickey AC et al. Identification of Hendra virus G glycoprotein residues that are critical for receptor binding. J Virol 2007; 81(11):5893-5901.
70. Bossart KN, McEachern JA, Hickey AC et al. Neutralization assays for differential henipavirus serology using Bio-Plex Protein Array Systems. J Virol Methods 2007; 142(1-2):29-40.
71. Bowden TA, Crispin M, Harvey DJ et al. Dimeric architecture of the Hendra virus attachment glycoprotein: evidence for a conserved mode of assembly. J Virol 2010; 84(12):6208-6217.
72. Bowden TA, Crispin M, Harvey DJ et al. Crystal structure and carbohydrate analysis of Nipah virus attachment glycoprotein: a template for antiviral and vaccine design. J Virol 2008; 82(23):11628-11636.
73. Xu K, Chan YP, Rajashankar KR et al. New insights into the Hendra virus attachment and entry process from structures of the virus G glycoprotein and its complex with Ephrin-B2. PLoS One 2012; 7(11):e48742.
74. Collins PL, Crowe J, J.E. Respiratory Syncytial Virus and Metapneumovirus. In: Knipe DM, Howley PM ed. Fields Virology, 5 ed. Philadelphia: Lippincott Williams & Wilkins, 2007:1601-1646.
75. Harris J, Werling D. Binding and entry of respiratory syncytial virus into host cells and initiation of the innate immune response. Cell Microbiol 2003; 5(10):671-680.
76. van den Hoogen BG, Bestebroer TM, Osterhaus AD et al. Analysis of the genomic sequence of a human metapneumovirus. Virology 2002; 295(1):119-132.
77. Teng MN, Whitehead SS, Collins PL. Contribution of the respiratory syncytial virus G glycoprotein and its secreted and membrane-bound forms to virus replication in vitro and in vivo. Virology 2001; 289(2):283-296.
78. Karron RA, Buonagurio DA, Georgiu AF et al. Respiratory syncytial virus (RSV) SH and G proteins are not essential for viral replication in vitro: clinical evaluation and molecular characterization of a cold-passaged, attenuated RSV subgroup B mutant. Proc Natl Acad Sci USA 1997; 94(25):13961-13966.
79. Karron RA, Thumar B, Schappell E et al. Evaluation of two chimeric bovine-human parainfluenza virus type 3 vaccines in infants and young children. Vaccine 2012; 30(26):3975-3981.
80. Karger A, Schmidt U, Buchholz UJ. Recombinant bovine respiratory syncytial virus with deletions of the G or SH genes: G and F proteins bind heparin. J Gen Virol 2001; 82(Pt 3):631-640.
81. Techaarpornkul S, Barretto N, Peeples ME. Functional analysis of recombinant respiratory syncytial virus deletion mutants lacking the small hydrophobic and/or attachment glycoprotein gene. J Virol 2001; 75(15):6825-6834.
82. Teng MN, Collins PL. Identification of the respiratory syncytial virus proteins required for formation and passage of helper-dependent infectious particles. J Virol 1998; 72(7):5707-5716.
83. Feldman SA, Hendry RM, Beeler JA. Identification of a linear heparin binding domain for human respiratory syncytial virus attachment glycoprotein G. J Virol 1999; 73(8):6610-6617.
84. Cantin C, Holguera J, Ferreira L et al. Newcastle disease virus may enter cells by caveolae-mediated endocytosis. J Gen Virol 2007; 88(Pt 2):559-569.
85. Kolokoltsov AA, Deniger D, Fleming EH et al. Small interfering RNA profiling reveals key role of clathrin-mediated endocytosis and early endosome formation for infection by respiratory syncytial virus. J Virol 2007; 81(14):7786-7800.

86. Biacchesi S, Skiadopoulos MH, Yang L et al. Recombinant human Metapneumovirus lacking the small hydrophobic SH and/or attachment G glycoprotein: deletion of G yields a promising vaccine candidate. J Virol 2004; 78(23):12877-12887.
87. Biacchesi S, Pham QN, Skiadopoulos MH et al. Infection of nonhuman primates with recombinant human metapneumovirus lacking the SH, G, or M2-2 protein categorizes each as a nonessential accessory protein and identifies vaccine candidates. J Virol 2005; 79(19):12608-12613.
88. Chang A, Masante C, Buchholz UJ et al. Human metapneumovirus (HMPV) binding and infection are mediated by interactions between the HMPV fusion protein and heparan sulfate. J Virol 2012; 86(6):3230-3243.
89. Cseke G, Maginnis MS, Cox RG et al. Integrin alphavbeta1 promotes infection by human metapneumovirus. Proc Natl Acad Sci USA 2009; 106(5):1566-1571.
90. Schowalter RM, Smith SE, Dutch RE. Characterization of human metapneumovirus F protein-promoted membrane fusion: critical roles for proteolytic processing and low pH. J Virol 2006; 80(22):10931-10941.
91. Schowalter RM, Chang A, Robach JG et al. Low-pH triggering of human metapneumovirus fusion: essential residues and importance in entry. J Virol 2009; 83(3):1511-1522.
92. Jack PJM, Boyle DB, Eaton BT et al. The complete genome sequence of J-virus reveals a unique genome structure in the family Paramyxoviridae. J Virol 2005; In Press.
93. Jack PJ, Anderson DE, Bossart KN et al. Expression of novel genes encoded by the paramyxovirus J virus. J Gen Virol 2008; 89(Pt 6):1434-1441.
94. Paterson RG, Johnson ML, Lamb RA. Paramyxovirus fusion (F) protein and hemagglutinin-neuraminidase (HN) protein interactions: intracellular retention of F and HN does not affect transport of the homotypic HN or F protein. Virology 1997; 237(1):1-9.
95. Chan DC, Fass D, Berger JM et al. Core structure of gp41 from the HIV envelope glycoprotein. Cell 1997; 89(2):263-273.
96. Fass D, Harrison SC, Kim PS. Retrovirus envelope domain at 1.7 angstrom resolution. Nat Struct Biol 1996; 3(5):465-469.
97. Wilson IA, Skehel JJ, Wiley DC. Structure of the haemagglutinin membrane glycoprotein of influenza virus at 3 A resolution. Nature 1981; 289(5796):366-373.
98. Chen L, Colman PM, Cosgrove LJ et al. Cloning, expression, and crystallization of the fusion protein of Newcastle disease virus. Virology 2001; 290(2):290-299.
99. Yin HS, Paterson RG, Wen X et al. Structure of the uncleaved ectodomain of the paramyxovirus (hPIV3) fusion protein. Proc Natl Acad Sci USA 2005.
100. Yin HS, Wen X, Paterson RG et al. Structure of the parainfluenza virus 5 F protein in its metastable, prefusion conformation. Nature 2006; 439(7072):38-44.
101. Cao J, Bergeron L, Helseth E et al. Effects of amino acid changes in the extracellular domain of the human immunodeficiency virus type 1 gp41 envelope glycoprotein. J Virol 1993; 67:2747-2755.
102. Hernandez LD, Hoffman LR, Wolfsberg TG et al. Virus-cell and cell-cell fusion. Annu Rev Cell Dev Biol 1996; 12:627-661.
103. Lescar J, Roussel A, Wien MW et al. The Fusion glycoprotein shell of Semliki Forest virus: an icosahedral assembly primed for fusogenic activation at endosomal pH. Cell 2001; 105(1):137-148.
104. White JM, Delos SE, Brecher M et al. Structures and mechanisms of viral membrane fusion proteins: multiple variations on a common theme. Crit Rev Biochem Mol Biol 2008; 43(3):189-219.
105. Scheid A, Choppin PW. Identification of biological activities of paramyxovirus glycoproteins. Activation of cell fusion, hemolysis, and infectivity of proteolytic cleavage of an inactive precursor protein of Sendai virus. Virology 1974; 57(2):475-490.
106. Klenk HD, Garten W. Host cell proteases controlling virus pathogenicity. Trends Microbiol 1994; 2(2):39-43.
107. Gotoh B, Yamauchi F, Ogasawara T et al. Isolation of factor Xa from chick embryo as the amniotic endoprotease responsible for paramyxovirus activation. FEBS Lett 1992; 296(3):274-278.
108. Nagai Y, Klenk HD. Activation of precursors to both glycoporteins of Newcastle disease virus by proteolytic cleavage. Virology 1977; 77(1):125-134.
109. Pager CT, Dutch RE. Cathepsin L is involved in proteolytic processing of the hendra virus fusion protein. J Virol 2005; 79(20):12714-12720.
110. Pager CT, Craft WW, Jr., Patch J et al. A mature and fusogenic form of the Nipah virus fusion protein requires proteolytic processing by cathepsin L. Virology 2006; 346(2):251-257.
111. Baker KA, Dutch RE, Lamb RA et al. Structural basis for paramyxovirus-mediated membrane fusion. Mol Cell 1999; 3(3):309-319.
112. Hunter E. Viral entry and receptors. In: Coffin SH, Hughes SH, and Varmus HE, eds. Retroviruses. New York: Cold Spring Harbor Laboratory Press, 1997:71-119.
113. Novick SL, Hoekstra D. Membrane penetration of Sendai virus glycoproteins during the early stages of fusion with liposomes as determined by hydrophobic photoaffinity labeling. Proc Natl Acad Sci USA 1988; 85(20):7433-7437.

114. Dutch RE, Jardetzky TS, Lamb RA. Virus membrane fusion proteins: biological machines that undergo a metamorphosis. Biosci Rep 2000; 20(6):597-612.
115. Singh M, Berger B, Kim PS. LearnCoil-VMF: computational evidence for coiled-coil-like motifs in many viral membrane-fusion proteins. J Mol Biol 1999; 290(5):1031-1041.
116. Hughson FM. Enveloped viruses: a common mode of membrane fusion? Curr Biol 1997; 7(9):R565-569.
117. Plemper RK, Compans RW. Mutations in the putative HR-C region of the measles virus F2 glycoprotein modulate syncytium formation. J Virol 2003; 77(7):4181-4190.
118. Wild CT, Shugars DC, Greenwell TK et al. Peptides corresponding to a predictive alpha-helical domain of human immunodeficiency virus type 1 gp41 are potent inhibitors of virus infection. Proc Natl Acad Sci USA 1994; 91(21):9770-9774.
119. Jiang S, Lin K, Strick N et al. HIV-1 inhibition by a peptide. Nature 1993; 365:113.
120. Lambert DM, Barney S, Lambert AL et al. Peptides from conserved regions of paramyxovirus fusion (F) proteins are potent inhibitors of viral fusion. Proc Natl Acad Sci USA 1996; 93(5):2186-2191.
121. Joshi SB, Dutch RE, Lamb RA. A core trimer of the paramyxovirus fusion protein: parallels to influenza virus hemagglutinin and HIV-1 gp41. Virology 1998; 248(1):20-34.
122. Wild TF, Buckland R. Inhibition of measles virus infection and fusion with peptides corresponding to the leucine zipper region of the fusion protein. J Gen Virol 1997; 78(Pt 1):107-111.
123. Young JK, Li D, Abramowitz MC et al. Interaction of peptides with sequences from the Newcastle disease virus fusion protein heptad repeat regions. J Virol 1999; 73(7):5945-5956.
124. Young JK, Hicks RP, Wright GE et al. Analysis of a peptide inhibitor of paramyxovirus (NDV) fusion using biological assays, NMR, and molecular modeling. Virology 1997; 238(2):291-304.
125. Rapaport D, Ovadia M, Shai Y. A synthetic peptide corresponding to a conserved heptad repeat domain is a potent inhibitor of Sendai virus-cell fusion: an emerging similarity with functional domains of other viruses. Embo J 1995; 14(22):5524-5531.
126. Bossart KN, Mungall BA, Crameri G et al. Inhibition of Henipavirus fusion and infection by heptad-derived peptides of the Nipah virus fusion glycoprotein. Virol J 2005; 2:57.
127. Dutch RE, Hagglund RN, Nagel MA et al. Paramyxovirus fusion (F) protein: a conformational change on cleavage activation. Virology 2001; 281(1):138-150.
128. Zhao X, Singh M, Malashkevich VN et al. Structural characterization of the human respiratory syncytial virus fusion protein core. Proc Natl Acad Sci USA 2000; 97(26):14172-14177.
129. Zhu J, Zhang CW, Qi Y et al. The fusion protein core of measles virus forms stable coiled-coil trimer. Biochem Biophys Res Commun 2002; 299(5):897-902.
130. Liu Y, Xu Y, Zhu J et al. Crystallization and preliminary X-ray diffraction analysis of central structure domains from mumps virus F protein. Acta Crystallograph Sect F Struct Biol Cryst Commun 2005; 61(Pt 9):855-857.
131. Xu Y, Lou Z, Liu Y et al. Crystallization and preliminary crystallographic analysis of the fusion core from two new zoonotic paramyxoviruses, Nipah virus and Hendra virus. Acta Crystallogr D Biol Crystallogr 2004; 60(Pt 6):1161-1164.
132. Chen L, Gorman JJ, McKimm-Breschkin J et al. The structure of the fusion glycoprotein of Newcastle disease virus suggests a novel paradigm for the molecular mechanism of membrane fusion. Structure (Camb) 2001; 9(3):255-266.
133. Lamb RA, Paterson RG, Jardetzky TS. Paramyxovirus membrane fusion: Lessons from the F and HN atomic structures. Virology 2006; 344(1):30-37.
134. Russell CJ, Kantor KL, Jardetzky TS et al. A dual-functional paramyxovirus F protein regulatory switch segment: activation and membrane fusion. J Cell Biol 2003; 163(2):363-374.
135. Harbury PB, Kim PS, Alber T. Crystal structure of an isoleucine-zipper trimer. Nature 1994; 371(6492):80-83.
136. Welch BD, Liu Y, Kors CA et al. Structure of the cleavage-activated prefusion form of the parainfluenza virus 5 fusion protein. Proc Natl Acad Sci USA 2012; 109(41):16672-16677.
137. Chan YP, Lu M, Dutta S et al. Biochemical, conformational, and immunogenic analysis of soluble trimeric forms of henipavirus fusion glycoproteins. J Virol 2012; 86(21):11457-11471.
138. Russell CJ, Jardetzky TS, Lamb RA. Membrane fusion machines of paramyxoviruses: capture of intermediates of fusion. Embo J 2001; 20(15):4024-4034.
139. Steinhauer DA, Plemper RK. Structure of the primed paramyxovirus fusion protein. Proc Natl Acad Sci USA 2012; 109(41):16404-16405.
140. Aguilar HC, Matreyek KA, Choi DY et al. Polybasic KKR motif in the cytoplasmic tail of Nipah virus fusion protein modulates membrane fusion by inside-out signaling. J Virol 2007; 81(9):4520-4532.
141. Paterson RG, Russell CJ, Lamb RA. Fusion protein of the paramyxovirus SV5: destabilizing and stabilizing mutants of fusion activation. Virology 2000; 270(1):17-30.
142. Waning DL, Russell CJ, Jardetzky TS et al. Activation of a paramyxovirus fusion protein is modulated by inside-out signaling from the cytoplasmic tail. Proc Natl Acad Sci USA 2004; 101(25):9217-9222.
143. Smith EC, Culler MR, Hellman LM et al. Beyond anchoring: the expanding role of the hendra virus fusion protein transmembrane domain in protein folding, stability, and function. J Virol 2012; 86(6):3003-3013.

144. Skehel JJ, Wiley DC. Receptor binding and membrane fusion in virus entry: the influenza hemagglutinin. Annu Rev Biochem 2000; 69:531-569.
145. Hu XL, Ray R, Compans RW. Functional interactions between the fusion protein and hemagglutinin-neuraminidase of human parainfluenza viruses [published erratum appears in J Virol 1992 Aug;66(8):5176]. J Virol 1992; 66(3):1528-1534.
146. Bagai S, Lamb RA. Quantitative measurement of paramyxovirus fusion: differences in requirements of glycoproteins between simian virus 5 and human parainfluenza virus 3 or Newcastle disease virus. J Virol 1995; 69(11):6712-6719.
147. Bousse T, Takimoto T, Gorman WL et al. Regions on the hemagglutinin-neuraminidase proteins of human parainfluenza virus type-1 and Sendai virus important for membrane fusion. Virology 1994; 204(2):506-514.
148. Deng R, Wang Z, Mirza AM et al. Localization of a domain on the paramyxovirus attachment protein required for the promotion of cellular fusion by its homologous fusion protein spike. Virology 1995; 209(2):457-469.
149. Tanabayashi K, Compans RW. Functional interaction of paramyxovirus glycoproteins: identification of a domain in Sendai virus HN which promotes cell fusion. J Virol 1996; 70(9):6112-6118.
150. Tsurudome M, Kawano M, Yuasa T et al. Identification of regions on the hemagglutinin-neuraminidase protein of human parainfluenza virus type 2 important for promoting cell fusion. Virology 1995; 213(1):190-203.
151. Melanson VR, Iorio RM. Amino acid substitutions in the F-specific domain in the stalk of the newcastle disease virus HN protein modulate fusion and interfere with its interaction with the F protein. J Virol 2004; 78(23):13053-13061.
152. Melanson VR, Iorio RM. Addition of N-glycans in the stalk of the Newcastle disease virus HN protein blocks its interaction with the F protein and prevents fusion. J Virol 2006; 80(2):623-633.
153. Lamb RA. Paramyxovirus fusion: A hypothesis for changes. Virology 1993; 197:1-11.
154. McGinnes LW, Gravel K, Morrison TG. Newcastle disease virus HN protein alters the conformation of the F protein at cell surfaces. J Virol 2002; 76(24):12622-12633.
155. Morrison T, McQuain C, McGinnes L. Complementation between avirulent Newcastle disease virus and a fusion protein gene expressed from a retrovirus vector: requirements for membrane fusion. J Virol 1991; 65(2):813-822.
156. Connolly SA, Leser GP, Jardetzky TS et al. Bimolecular complementation of paramyxovirus fusion and hemagglutinin-neuraminidase proteins enhances fusion: implications for the mechanism of fusion triggering. J Virol 2009; 83(21):10857-10868.
157. Yao Q, Hu X, Compans RW. Association of the parainfluenza virus fusion and hemagglutinin-neuraminidase glycoproteins on cell surfaces. J Virol 1997; 71(1):650-656.
158. McGinnes LW, Morrison TG. Inhibition of receptor binding stabilizes newcastle disease virus HN and F protein-containing complexes. J Virol 2006; 80(6):2894-2903.
159. Li J, Quinlan E, Mirza A et al. Mutated form of the Newcastle disease virus hemagglutinin-neuraminidase interacts with the homologous fusion protein despite deficiencies in both receptor recognition and fusion promotion. J Virol 2004; 78(10):5299-5310.
160. Deng R, Wang Z, Mahon PJ et al. Mutations in the Newcastle disease virus hemagglutinin-neuraminidase protein that interfere with its ability to interact with the homologous F protein in the promotion of fusion. Virology 1999; 253(1):43-54.
161. Plemper RK, Hammond AL, Cattaneo R. Measles virus envelope glycoproteins hetero-oligomerize in the endoplasmic reticulum. J Biol Chem 2001; 276(47):44239-44246.
162. Plemper RK, Hammond AL, Gerlier D et al. Strength of envelope protein interaction modulates cytopathicity of measles virus. J Virol 2002; 76(10):5051-5061.
163. Corey EA, Iorio RM. Mutations in the stalk of the measles virus hemagglutinin protein decrease fusion but do not interfere with virus-specific interaction with the homologous fusion protein. J Virol 2007; 81(18):9900-9910.
164. Corey EA, Iorio RM. Measles virus attachment proteins with impaired ability to bind CD46 interact more efficiently with the homologous fusion protein. Virology 2009; 383(1):1-5.
165. Aguilar HC, Matreyek KA, Filone CM et al. N-glycans on Nipah virus fusion protein protect against neutralization but reduce membrane fusion and viral entry. J Virol 2006; 80(10):4878-4889.
166. Aguilar HC, Ataman ZA, Aspericueta V et al. A novel receptor-induced activation site in the Nipah virus attachment glycoprotein (G) involved in triggering the fusion glycoprotein (F). J Biol Chem 2009; 284(3):1628-1635.
167. Whitman SD, Smith EC, Dutch RE. Differential rates of protein folding and cellular trafficking for the Hendra virus F and G proteins: implications for F-G complex formation. J Virol 2009; 83(17):8998-9001.
168. Stone-Hulslander J, Morrison TG. Detection of an interaction between the HN and F proteins in Newcastle disease virus-infected cells. J Virol 1997; 71(9):6287-6295.
169. Tong S, Compans RW. Alternative mechanisms of interaction between homotypic and heterotypic parainfluenza virus HN and F proteins. J Gen Virol 1999; 80(Pt 1):107-115.
170. Whitman SD, Dutch RE. Surface density of the Hendra G protein modulates Hendra F protein-promoted membrane fusion: role for Hendra G protein trafficking and degradation. Virology 2007; 363(2):419-429.

171. Lee B, Ataman ZA. Modes of paramyxovirus fusion: a Henipavirus perspective. Trends Microbiol 2011; 19(8):389-399.
172. Chang A, Dutch RE. Paramyxovirus fusion and entry: multiple paths to a common end. Viruses 2012; 4(4):613-636.
173. Zhu Q, Biering SB, Mirza AM et al. Individual N-glycans added at intervals along the stalk of the Nipah virus G protein prevent fusion, but do not block the interaction with the homologous F protein. J Virol 2013; Jan 2. [Epub ahead of print]
174. Paal T, Brindley MA, St Clair C et al. Probing the spatial organization of measles virus fusion complexes. J Virol 2009; 83(20):10480-10493.
175. Brindley MA, Takeda M, Plattet P et al. Triggering the measles virus membrane fusion machinery. Proc Natl Acad Sci USA 2012; 109(44):E3018-3027.
176. Ader N, Brindley M, Avila M et al. Mechanism for active membrane fusion triggering by morbillivirus attachment protein. J Virol 2013; 87(1):314-326.
177. Porotto M, Devito I, Palmer SG et al. Spring-loaded model revisited: paramyxovirus fusion requires engagement of a receptor binding protein beyond initial triggering of the fusion protein. J Virol 2011; 85(24):12867-12880.
178. Porotto M, Palmer SG, Palermo LM et al. Mechanism of fusion triggering by human parainfluenza virus type III: communication between viral glycoproteins during entry. J Biol Chem 2012; 287(1):778-793.
179. Sergel T, McGinnes LW, Peeples ME et al. The attachment function of the Newcastle disease virus hemagglutinin-neuraminidase protein can be separated from fusion promotion by mutation. Virology 1993; 193(2):717-726.
180. Takimoto T, Taylor GL, Crennell SJ et al. Crystallization of Newcastle disease virus hemagglutinin-neuraminidase glycoprotein. Virology 2000; 270(1):208-214.
181. Takimoto T, Taylor GL, Connaris HC et al. Role of the hemagglutinin-neuraminidase protein in the mechanism of paramyxovirus-cell membrane fusion. J Virol 2002; 76(24):13028-13033.
182. Mishin VP, Watanabe M, Taylor G et al. N-linked glycan at residue 523 of human parainfluenza virus type 3 hemagglutinin-neuraminidase masks a second receptor-binding site. J Virol 2010; 84(6):3094-3100.
183. Bousse T, Takimoto T. Mutation at residue 523 creates a second receptor binding site on human parainfluenza virus type 1 hemagglutinin-neuraminidase protein. J Virol 2006; 80(18):9009-9016.
184. Porotto M, Fornabaio M, Kellogg GE et al. A second receptor binding site on human parainfluenza virus type 3 hemagglutinin-neuraminidase contributes to activation of the fusion mechanism. J Virol 2007; 81(7):3216-3228.
185. Alymova IV, Portner A, Mishin VP et al. Receptor-binding specificity of the human parainfluenza virus type 1 hemagglutinin-neuraminidase glycoprotein. Glycobiology 2012; 22(2):174-180.
186. Bousse TL, Taylor G, Krishnamurthy S et al. Biological significance of the second receptor binding site of Newcastle disease virus hemagglutinin-neuraminidase protein. J Virol 2004; 78(23):13351-13355.
187. Mahon PJ, Mirza AM, Iorio RM. Role of the two sialic acid binding sites on the newcastle disease virus HN protein in triggering the interaction with the F protein required for the promotion of fusion. J Virol 2011; 85(22):12079-12082.
188. Porotto M, Salah Z, DeVito I et al. The second receptor binding site of the globular head of the Newcastle disease virus hemagglutinin-neuraminidase activates the stalk of multiple paramyxovirus receptor binding proteins to trigger fusion. J Virol 2012; 86(10):5730-5741.
189. Mahon PJ, Mirza AM, Musich TA et al. Engineered intermonomeric disulfide bonds in the globular domain of Newcastle disease virus hemagglutinin-neuraminidase protein: implications for the mechanism of fusion promotion. J Virol 2008; 82(21):10386-10396.
190. Navaratnarajah CK, Oezguen N, Rupp L et al. The heads of the measles virus attachment protein move to transmit the fusion-triggering signal. Nat Struct Mol Biol 2011; 18(2):128-134.
191. Lee JK, Prussia A, Paal T et al. Functional interaction between paramyxovirus fusion and attachment proteins. J Biol Chem 2008; 283(24):16561-16572.
192. Plemper RK, Brindley MA, Iorio RM. Structural and mechanistic studies of measles virus illuminate paramyxovirus entry. PLoS Pathog 2011; 7(6):e1002058.
193. Ader N, Brindley MA, Avila M et al. Structural rearrangements of the central region of the morbillivirus attachment protein stalk domain trigger F protein refolding for membrane fusion. J Biol Chem 2012; 287(20):16324-16334.
194. Navaratnarajah CK, Negi S, Braun W et al. Membrane fusion triggering: three modules with different structure and function in the upper half of the measles virus attachment protein stalk. J Biol Chem 2012.
195. Apte-Sengupta S, Negi S, Leonard VH et al. Base of the measles virus fusion trimer head receives the signal that triggers membrane fusion. J Biol Chem 2012; 287(39):33026-33035.
196. Bose S, Zokarkar A, Welch BD et al. Fusion activation by a headless parainfluenza virus 5 hemagglutinin-neuraminidase stalk suggests a modular mechanism for triggering. Proc Natl Acad Sci USA 2012; 109(39):E2625-2634.
197. Bishop KA, Hickey AC, Khetawat D et al. Residues in the stalk domain of the hendra virus g glycoprotein modulate conformational changes associated with receptor binding. J Virol 2008; 82(22):11398-11409.

CHAPTER 7

CELLULAR ENTRY OF RETROVIRUSES

Dirk Lindemann,[1] Imke Steffen[2,3] and Stefan Pöhlmann*[,2,4]

[1]Institute for Virology, Medical Faculty Carl Gustav Carus, Technische Universität Dresden, Dresden, Germany; [2]Institute for Virology, Hannover Medical School, Hannover, Germany; [3]Blood Systems Research Institute, San Francisco, California, USA; [4]German Primate Center, Göttingen, Germany

**Corresponding Author: Stefan Pöhlmann—Email: s.poehlmann@dpz.eu*

Abstract: The retrovirus family contains several important human and animal pathogens, including the human immunodeficiency virus (HIV), the causative agent of acquired immunodeficiency syndrome (AIDS). Studies with retroviruses were instrumental to our present understanding of the cellular entry of enveloped viruses in general. For instance, studies with alpharetroviruses defined receptor engagement, as opposed to low pH, as a trigger for the envelope protein-driven membrane fusion. The insights into the retroviral entry process allowed the generation of a new class of antivirals, entry inhibitors, and these therapeutics are at present used for treatment of HIV/AIDS. In this chapter, we will summarize key concepts established for entry of avian sarcoma and leukosis virus (ASLV), a widely used model system for retroviral entry. We will then review how foamy virus and HIV, primate- and human retroviruses, enter target cells, and how the interaction of the viral and cellular factors involved in the cellular entry of these viruses impacts viral tropism, pathogenesis and approaches to therapy and vaccine development.

INTRODUCTION

The retrovirus family contains important animal and human pathogens, with human immunodeficiency virus (HIV), the causative agent of acquired immunodeficiency syndrome (AIDS), being the most prominent example. Infection by HIV was responsible for 2 million deaths in 2008 and can, at present, neither be cured nor prevented by a vaccine.[1] The failure of several vaccine candidates is due to an incomplete understanding of the structure and function of the viral envelope protein (Env), which mediates entry into target cells.[2] Addressing the question of how HIV and other retroviruses recognize

Viral Entry into Host Cells, edited by Stefan Pöhlmann and Graham Simmons.

and enter host cells can therefore not only provide key insights into the viral cell tropism and pathogenesis, but will be essential for the design of effective vaccines.

The hallmark of retrovirus infection is the retrograde flow of the viral genetic information: The diploid, viral genomic RNA is reverse transcribed into DNA by a viral enzyme, the reverse transcriptase (RT), and subsequently integrated into the host cell genome.[3,4] Integration is essential for gene expression, and the integrated viral DNA, termed provirus, is inherited like a cellular gene. This strategy facilitates establishment of persistent infection, another characteristic of retroviruses. The family *Retroviridae* is large, and members are found in invertebrates and vertebrates.[3,4] Seven retrovirus genera have been defined: alpha-, beta-, gamma-, delta-, epsilon-retroviruses and, lenti-viruses in the *orthoretrovirinae* subfamily and spumaviruses in the *spumaretrovirinae* subfamily. The genera alpha-, beta- and gamma-retroviruses comprise simple retroviruses which only encode for the gag (group specific antigen), pro (protease), pol (polymerase) and env (envelope) genes. The remaining genera comprise complex retroviruses, which apart from gag, pro, pol and env also encode regulatory genes, like tat and rev found in HIV and SIV.[3,4] The Gag proteins make up the viral shell which is located beneath the viral membrane. They are synthesized as polyproteins and are subsequently cleaved into the mature components by the viral Pro proteins. The pol genes encode the remaining viral enzymatic functions, RT and integrase. Finally, the env gene contains the information for the only viral surface protein, the envelope protein (Env, also termed glycoprotein (GP)), which facilitates entry into host cells.

Important concepts underlying viral entry have been established by work with alpharetroviruses, as discussed below. Alpharetroviruses have received particular attention for their ability to induce tumors in avian hosts, and many milestone discoveries like that of RT, integrase and cellular oncogenes are, at least in part, the result of work with these viruses.[3,4] However, retroviral infection and pathogenesis was initially perceived to be confined to non-human hosts and a direct relevance of retroviruses to human health was not obvious, with the notable exception of retroelements and defective retroviruses constituting a substantial portion of the human genome.[5] This has changed dramatically with the discovery of human T-cell leukemia virus[6] (HTLV, a deltaretrovirus) and HIV (a lentivirus),[7] the causative agent of AIDS. The AIDS pandemic has spurred efforts to elucidate the molecular processes underlying retroviral replication and pathogenesis and to establish means for intervention. In the course of these studies key insights into the viral entry process, including structural information on the viral glycoprotein, have been obtained, as described below. In the present chapter, we will introduce common themes in retrovirus entry, which were mainly established by work with alpharetroviruses. Subsequently, we will discuss infectious entry of prototype foamy virus (previously known as human foamy virus; FV, a spumavirus), an in several aspects unusual retrovirus. Finally, we will describe the viral and cellular players involved in entry of HIV.

CELLULAR ENTRY OF AVIAN SARCOMA AND LEUKOSIS VIRUS (ALPHARETROVIRUSES)

The Avian Sarcoma and Leukosis Virus Env is a Class I Fusion Protein

The avian sarcoma and leukosis virus (ASLV), an alpharetrovirus which mainly infects chicken, has been intensively used as a model for analysis of viral entry into target cells. As for other orthoretroviruses including HIV, SIV and HTLV, the ASLV env gene encodes a type I transmembrane protein, which is imported as a precursor into the secretory pathway of the host cell.[8-10] During passage through the endoplasmatic reticulum and Golgi apparatus, Env is heavily N-glycosylated and cleaved by a cellular endoprotease,[11-13] again a recurring theme among retroviral and other viral glycoproteins. Cleavage (and appropriate glycosylation) is essential for function[11,14] and results in the generation of a surface unit (SU) and a transmembrane unit (TM), which remain covalently associated due to a disulfide bond.[8] Trimers of SU and TM heterodimers are finally transported to the cell surface where they are incorporated into nascent progeny particles.[8]

The SU of ASLV Env contains the receptor binding site while the TM harbours the functional elements required for fusion with host cells, a fusion peptide, two heptad repeats (HR) and a transmembrane region (TM). This type of functional organization, in conjunction with the need for Env cleavage, is characteristic for class I fusion proteins (found e.g., in retro-, filo-, orthomyxo-, paramyxoviruses), and is opposed to that of class II (found in e.g., α- and flaviviruses) and class III fusion proteins (found in herpes- and rhabdoviruses).[15,16] In fact, a similar structural organization of the TM of ASLV and Ebola virus, a filovirus, has been noted over a decade ago,[17] and a similar mechanism of membrane fusion has been suggested.

Receptor Interactions of ASLV Env

The SU of ASLV contains variable regions which, similar as for HIV, might protect against antibody-mediated neutralization and are involved in receptor engagement.[18] Based on receptor specificity and interference as well as recognition by neutralizing antibodies ASLVs are divided into ten subgroups (A-J), and entry of group A, B, D and E viruses has been studied in detail.[18,19] Genetic analyses revealed alleles associated with susceptibility to infection by subgroup A viruses (*tva*), subgroup C viruses (*tvc*) and subgroup B, D and E viruses (*tvb*, subdivided into two susceptibility alleles, tvb^{S1} for subgroups B, D and E and tvb^{S3} for subgroups B and D). The identity of the proteins encoded by the *tva, tvb* and *tvc* loci has been identified. Due to alternative splicing, two proteins with an identical ectodomain are produced from the *tva* locus, and both facilitate entry by ASLV-A.[20,21] One is inserted into the membrane due to the presence of a transmembrane domain, while the other is associated with the membrane because of a glycosylphosphatidylinositol (GPI) anchor. TVA is related to low-density lipoprotein (LDL) receptors and a region with homology to the ligand recognition site in LDL receptors, termed LDL-A, has been shown to be sufficient for receptor function.[22] LDL-A interacts with ASLV-A Env and a peptide derived from this sequence blocks ASLV-A infection.[23] The proteins encoded by tvb^{S1} and tvb^{S3} are members of the tumor necrosis factor receptor family, and contain three extracellular cysteine rich domains and an intracellular death domain.[24,25] The presence of the latter might account for the cytopathic effect associated with infection by ASLV, which use this receptor.[18] Notably, $TVBS^{1}$[26] and $TVBS^{3}$[27] differ only in a single amino acid residue, which has been shown to be important for the interaction with ASLV-E,[26] and a 15 amino acid comprising linear sequence in $TVBS^{1}$[26] might be sufficient for binding to ASLV-B Env.[28] Finally, TVC has been identified to belong to the immunoglobulin protein family and is related to butyrophilins.[29,30] Collectively, these studies established that highly specific interactions of the viral Env proteins with cellular receptors are required for ASLV entry into target cells and that receptor specificity constitutes a major determinant of viral cell tropism. In addition, these results showed that the ASLV Env exhibits a functional organization conserved between the envelope proteins of different viruses, which has important implications for the entry process, as discussed below.

The Fusion Activity of the Avian Sarcoma and Leukosis Virus Envelope Protein is Triggered by Receptor Engagement and Low pH

The successful Env-driven fusion of viral and cellular membranes requires that Env receives a trigger, which allows the protein to overcome the energy barrier associated with initiating the membrane fusion reaction. The trigger is usually provided in two forms, either receptor binding or low pH.[15] Thus, it is well established that exposure of influenza virus hemagglutinin (HA) to low pH results in irreversible conformational changes of HA, which facilitate the insertion of the fusion peptide into the target cell membrane and the subsequent membrane merger.[15] Consequently, lysosomotropic agents which interfere with endosomal acidification inhibit influenza entry and exposure of virions to low pH in the absence of target cells abrogates viral infectivity. Alternatively, the energy barrier can be overcome by Env binding to receptor,[15] and this is thought to be the case for most retroviruses, including HIV. For ALSV, however, this concept was challenged by work of Mothes and colleagues, who showed that infectious entry of ASLV-A and ASLV-B depends on both receptor binding and low pH.[31] Specifically, this work demonstrated that lysosomotropic agents and inhibitors of endocytosis block ASLV-A and –B entry and that acid treatment of virions bound to soluble receptor profoundly reduced viral infectivity while a low pH pulse in the absence of receptor or incubation with receptor alone had no pronounced effect (Fig. 1). In addition, cell to cell fusion was shown to occur only upon treatment of cultures with low pH, and structural rearrangement of TM was also dependent on an acidic stimulus.[19,31] Although some of these results are not undisputed,[32-35] they indicate that different, sequential triggers can control the membrane fusion activity of retroviral glycoproteins, as also suggested for a betaretrovirus oncogenic in sheep,[36] and this strategy might be particularly suitable to prevent premature initiation of the fusion reaction.

CELLULAR ENTRY OF FOAMY VIRUS (SPUMAVIRUSES)

Foamy viruses (FVs) occupy a special position among retroviruses, exemplified by the recent reorganization of the retrovirus family in two separate subfamilies, orthoretrovirinae and spumaretrovirinae, with the spumaviruses or FVs as the only genus of the subfamily spumaretrovirinae.[37] This is the result of the identification and characterization of several unique features of the FV replication strategy in the past decade that, interestingly, bear in many aspects strong homology to those of pararetroviruses such as hepatitis B virus (HBV).

The FV Glycoprotein

One of the key structural proteins involved in retroviral entry, the viral glycoprotein (Env), in case of FVs also displays several unique features.[38] At first glance the principal structural organization of the prototype FV (PFV) envelope precursor protein $gp130^{Env}$, composed of an N-terminal signal or leader peptide (LP), a central surface (SU) and a C-terminal transmembrane (TM) domain, resembles that of other orthoretroviral glycoproteins. However, recent reports have demonstrated that it undergoes a highly unusual biosynthesis, maturation, posttranslational modification and intracellular localization.[38]

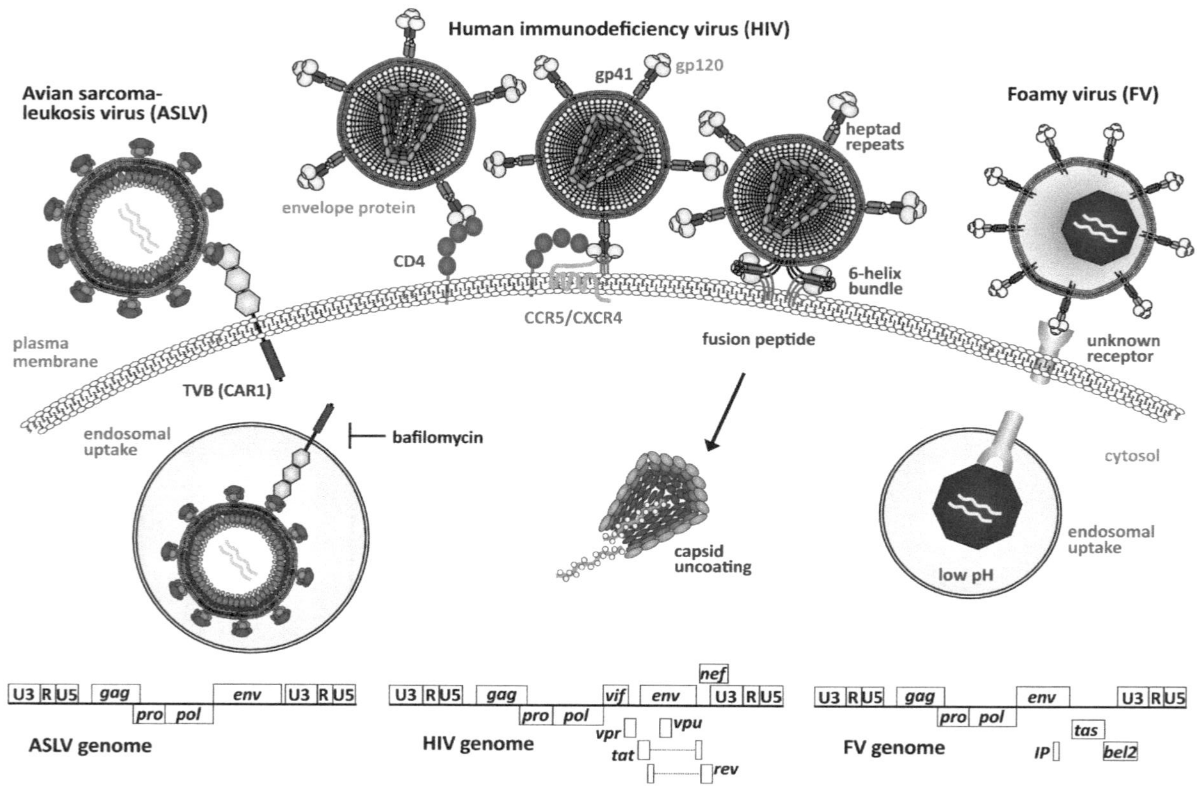

Figure 1. Host cell entry of retroviruses. The glycoproteins of avian sarcoma and leukosis virus (ASLV) require binding to a receptor, for instance TVB, and subsequent exposure to endosomal low pH as trigger to drive fusion of the viral membrane with a host cell membrane. Bafilomycin inhibits endosomal acidification and blocks infectious entry of ASLV. The human immunodeficiency virus (HIV) binds to CD4 and a coreceptor for host cell entry, and binding to these factors is sufficient to trigger the envelope protein-mediated membrane fusion. The receptor of foamy virus (FV) is unknown. Endosomal low pH is essential for cellular entry of FV.

First, gp130^{Env} is translated and inserted into the endoplasmic reticulum as a full-length precursor protein and initially adopts a type III membrane topology with both N- and C-terminus located in the cytoplasm.[39-41] Unlike other retroviral glycoproteins, it is normally not cleaved by the cellular signal peptidase complex at its N-terminus.[42] Instead the 988 aa PFV glycoprotein is processed after aa 126 and 571 only late during its transport to the cell surface by furin or furin-like proteases in the *trans*-Golgi network.[42,43] Precursor processing yields three mature subunits, an N-terminal LP-subunit (gp18^{LP}) with a type II membrane topology, a C-terminal TM-subunit (gp48^{TM}) adopting a type I membrane topology and a central SU-subunit (gp80^{SU}) associating on the luminal side with the LP and SU subunits.[40] SU/TM but not LP/SU cleavage was shown to be required for infectivity of FV particles.[27,42,44] Unique for retroviruses all three glycoprotein subunits, including gp18^{LP}, are integral parts of the secreted foamy viral particle.[40,41] Image reconstruction analysis from electron micrographs of negatively stained virions revealed the characteristic, prominent Env spike structures on FV particles, indicating that the FV Env glycoprotein, similar to other viral glycoproteins, forms trimeric complexes containing three copies of each of the three individual subunits.[45] The unusual particle-association of the FV LP subunit reflects the special function of the Env protein in FV particle morphogenesis. FVs unlike other retroviruses require coexpression of the glycoprotein for particle release.[46,47] An interaction of the N-terminal cytoplasmic domain (CyD) of gp18^{LP}, involving a conserved WXXW motif, with the FV capsid is essential for membrane association and budding to occur.[40,41]

Second, in contrast to orthoretroviruses both the FV Gag and Env protein contain structural information essential for particle release. Therefore, not unsurprisingly, FVs similar to HBV, secrete in addition to viral particles also subviral particles (SVP), capsidless particulate structures containing only the glycoprotein.[48,49] However, in contrast to HBV, FVs naturally release only very low amounts of SVP. Recently, analysis of foamy viral glycoprotein posttranslational modifications revealed attachment of ubiquitin to different lysine residues located in the CyD of the LP subunit.[49] This modification was demonstrated to be involved in regulating the balance of viral vs. subviral particle release.[49,50] Inhibition of glycoprotein ubiquitylation lead to a dramatic increase in SVP release, surprisingly, not at the expense of viral particle release and infectivity. The greater details of this regulatory mechanism of FV particle release are currently unknown. In addition, like the HIV-1 Env, the PFV Env is heavily glycosylated during its intracellular transport to the cell surface. Fourteen out of 15 potential N-glycosylation sites are used of which three were demonstrated to be essential individually for such functions as intracellular transport of the glycoprotein, support of particle release and infectivity.[51] Other potential posttranslational modifications of the viral glycoprotein such as phosphorylation, O-glycosylation or fatty acid addition have so far not been examined.

Third, at the C-terminus of the short PFV TM CyD a di-lysine ER retrieval signal has been identified and characterized that is responsible for the predominant ER localization of PFV Env when expressed by itself.[52,53] This ER retrieval signal is conserved in primate and feline but not bovine and equine FVs.[53,54] Inactivation of the ER retrieval signal by mutagenesis leads to a different intracellular distribution of the PFV glycoprotein although enhanced cell surface expression and fusogenic activity of such mutants are discussed controversially.[44,52] The ER retrieval signal is thought to be responsible, at least in part, for budding of primate and feline FVs into intracellular compartments assumed to be the ER, although the nature of these compartment has been questioned recently.[55]

Cellular Receptor and Viral Entry

One hallmark of FVs is their extremely broad host range. To date only very few cell types or species have been identified that are refractory to FV Env-mediated entry. Susceptible species include, in addition to many mammalian species, reptiles, amphibians and birds.[56] In a recent report, two cell lines, a human erythroid precursor- and a zebrafish-cell line, resistant to FV Env-mediated FV- and HIV-vector transduction were identified.[57] This recently remedied lack of a non-permissive cell line for FVs was one of the reasons why their cellular receptor molecules have so far eluded identification (Fig. 1). However, the receptor seems to be ubiquitously expressed and evolutionary conserved. In addition, superinfection resistance data suggest that different FV species use the same cellular receptor for entry.[56,58]

The receptor-binding-domain (RBD) of the PFV glycoprotein was mapped using recombinant immunoadhesins bearing various extracellular domains of the PFV Env proteins and a mouse IgG2a Fc region.[59] The particle associated LP and TM subunit were shown to be dispensable for specific binding to target cells. Similar to the HIV-1 Env a discontinuous RBD located in the central and C-terminal regions of the PFV SU subunit was characterized. Interestingly, the 8th potential N-glycosylation site, that was previously identified to be the only N-glycosylation site in SU to be important by itself for PFV Env function,[51] was shown to be either directly involved in receptor binding or RBD formation.[59]

FV entry into target cells after receptor recognition is thought to be mediated by endocytosis although this has not been formally demonstrated. However, the fusion activity of the glycoproteins from different FV species was shown to be strongly pH controlled for most FV species examined, highly indicative for an endocytic uptake of FV particles.[60] Electron microscopy and immunofluorescence data suggested that FV capsids travel in a microtubule-dependent fashion to the microtubule-organizing-center (MTOC) after their release into the cytoplasm by fusion of cellular and viral lipid membranes.[61,62] Whereas intact incoming FV capsids were detectable around the MTOC no capsid structures were found near the nuclear membrane or the nuclear pores as reported for Hepatitis B virus or herpes viruses. FV capsid appears to be extremely stable. In G_0 arrested human fibroblasts intact capsids can be detected at the MTOC for weeks after the initial infection event and still proceed to a productive infection upon reentry of the target cell into the cell cycle.[63] This unusual intracellular stability of FV Gag might be associated with a special feature of the FV Gag protein, its extremely low lysine residue content.[64,65] Except for the feline FV (FFV) capsid proteins of all other FV species examined so for contain few or no lysine residues.[64]

Another recent study was unable to reproduce the detection of incoming intact capsids at the MTOC by electron microscopy even when using very high multiplicities of infection (MOI).[55] In contrast the authors report the assembly of newly generated FV capsids at the MTOC upon transfection of proviral expression clones. A sequence motif bearing homology to the cytoplasmic targeting-retention signal (CTRS) of Mason-Pfizer monkey virus is suggested to be responsible for targeting of FV capsid assembly to the MTOC.[55,66] Both studies[55,61] demonstrated the MTOC localization of PFV Gag to be disrupted by nocodazole treatment, a microtubule-network (MT) depolymerizing agent. Taken together these studies clearly indicate that FV intracellular trafficking involves the MT network, however, additional studies are required to clarify whether intact capsids accumulate at the MTOC both after viral entry and during viral egress.

Several structural and functional features of the FV Gag protein are also unusual for a retroviral capsid protein.[67] One of these is the apparently inefficient proteolytic precursor processing not leading to the generation of defined matrix, capsid and nucleocapsid domains as observed for orthoretroviruses. Instead in PFV particles only a single processing event leading to the removal of a C-terminal 3 kD peptide of the $p71^{Gag}$ precursor protein can be detected.[68,69] Both the $p71^{Gag}$ precursor and its $p68^{Gag}$ processing product are found in purified PFV particles at ratios from 1:1 to 1:4 depending on the source of virus examined.[70] The fate of the $p3^{Gag}$ cleavage product is not known. It is undetectable in infected cells or purified viral particles.[69,71] The $p68^{Gag}$ protein by itself is sufficient for PFV capsid morphogenesis and in combination with PFV Env for particle release, although such particles show a 100-fold reduced infectivity.[69,71] However, secondary cleavage sites of the viral protease in PFV Gag have been characterized in vitro.[72] Furthermore, early analysis suggested an important role of these secondary Gag cleavage sites for PFV replication since their mutagenesis led to the generation of non-infectious viral particles without affecting particle egress.[72] A functional role of further PFV Gag protein processing by the viral but also cellular proteases upon entry into target cells for infectivity is supported by the recent characterization of PFV capsids lacking protease activity or bearing mutations at secondary Gag processing sites. These mutant PFV capsids accumulate at the MTOC upon infection of target cells similar to wild type particles but apparently subsequently do not disassemble properly, resulting in the failure of viral genome release and transport into the nucleus.[73]

CELLULAR ENTRY OF HUMAN IMMUNODEFICIENCY VIRUS (RETROVIRUS)

HIV, the Causative Agent of AIDS, Originated from Simian Immunodeficiency Viruses

The human and simian immunodeficiency viruses (HIV, SIV) belong to the lentiviruses, a retrovirus subfamily whose members cause disease after long incubation periods. Human immunodeficiency virus type 1 (HIV-1) evolved upon transmission of SIV from chimpanzees (SIVcpz) to humans in the 1930s in Southern Cameroon and is the cause of the AIDS pandemic.[74,75] Zoonotic transmission of SIV from chimpanzees or gorillas to humans occurred at least four times and the transmitted viruses form the ancestors of group M (Major, Main), N (Non-M, Non-O, New), P and O (Outlier) HIV-1, with group M viruses being responsible for the AIDS pandemic.[74-78] HIV-2 originated from SIV from sooty mangabeys, which crossed the species barrier multiple times, leading to the establishment of the HIV-2 groups A through H.[79-81] Transmission occurred in Western Africa, where HIV-2 is epidemic today.

The continuous destruction of CD4 positive T cells is the hallmark of HIV/AIDS.[82] Depletion of CD4 T cells is due to direct and indirect mechanisms, namely destruction of virus infected CD4+ T cells, virus-induced apoptosis of uninfected, bystander cells and cell death due to persistent T-cell activation.[83] While T cells are continuously replenished over years in HIV infected patients, the restorative capacity of the host and thus the number of CD4 T cells ultimately declines.[82] As a consequence of the compromised CD4 T-cell help infected individuals become susceptible to a variety of ultimately lethal opportunistic infections, which define the clinical presentation of AIDS. In the absence of therapy, AIDS develops about 10 y after infection with HIV-1.[82] HIV-2 also targets CD4+ T cells, but disease progression is slower and AIDS symptoms may take up to 25 y to develop or might not develop at all.[79] The reason for the differential pathogenicity of the two HIV types is likely multifactorial, and pioneer work suggests that differential immune activation, controlled by the viral accessory protein Nef, might play an important role.[84]

Reverse transcription of the viral RNA genome into DNA and processing of viral Gag and Gag-Pol polyproteins by the viral protease can be efficiently blocked by small molecule inhibitors. Cocktails of Reverse Transcriptase (RT), integrase (IN) and protease (PR) inhibitors, termed highly active antiretroviral therapy (HAART), dramatically reduce viral load in patients and allow at least partial immune reconstitution.[85] However, these compounds fail to eradicate the virus in infected individuals and their application is plagued by unwanted side effects as well as the emergence and spread of drug resistant viral variants.[85,86] Moreover, mainly HIV positive individuals in industrialized countries benefit from the availability of HAART. In contrast, patients in resource poor settings, who constitute the vast majority of the HIV/AIDS cases world-wide, frequently do not have adequate access to effective therapy, due to unresolved issues concerning drug distribution. Therefore, new strategies to prevent and treat HIV infection are necessary, and targeting viral entry might be a promising approach, as discussed below.

Synthesis of the HIV Env Protein in the Secretory Pathway of the Host Cell: Adequate Glycosylation and Proteolytic Cleavage are Essential for Env Function

The HIV envelope protein (Env) is incorporated into the viral membrane and mediates viral entry into target cells. Env is a type I transmembrane protein, harbouring an N-terminal signal sequence and a C-terminal membrane anchor. The signal sequence earmarks Env for import into the endoplasmatic reticulum (ER), where calnexin and calreticulin facilitate Env folding,[87] a relatively lengthy process, which involves extensive formation and isomerisation of disulfide bonds.[88] Notably, the Env signal sequence, which is removed posttranslationally, controls the rate of Env folding.[89,90] Concomitantly with folding, Env is extensively modified with N-linked high-mannose glycans,[91] and appropriate glycosylation is essential for function.[92] O-glycosylation of Env was also demonstrated,[93] but target sites and biological relevance are less clear. Adequately folded and glycosylated Env, which forms trimers in the ER membrane,[94,95] is transported into the Golgi apparatus, where the high-mannose glycans are processed. Since the posttranslational glycan modification of Env is the product of the host cell glycosylation machinery, most of the glycans are recognized as "self" by the immune system of the infected host. However, the extremely dense packaging and/or recessed location of glycans often prevents their complete processing, and about 40% of the glycans present on mature HIV Env are of the high-mannose type.[91] As a consequence, some of the glycan epitopes are recognized as "foreign" by the immune system and, in rare cases, elicit the generation of broadly neutralizing antibodies, like 2G12.[91,96,97] The Golgi apparatus, specifically the trans Golgi network, is also the site of Env processing by the cellular proprotein convertases furin and PC6.[98,99] Cleavage of Env results in the generation of the surface unit (SU, gp120) and the transmembrane unit (TM, gp41) and is essential for Env function.[100] Gp120 and gp41 remain non-covalently associated, mainly due to interactions of N- and C-terminal sequences in gp120 with a disulfide bonded loop in gp41 (below). Trimers of gp120/gp41 heterodimers, which have been characterized on the structural level,[101-103] are inserted into the cellular membrane, the site of release of progeny virions.[104,105] Incorporation of Env trimers into budding viral particles is promoted by interactions of the cytoplasmic tail of gp41 and Gag, which are bridged by the cellular factor TIP47.[106,107] While several cellular factors are excluded from the HIV envelope, certain membrane proteins seem to be specifically incorporated[108,109] and their presence can impact attachment and neutralization sensitivity, as discussed below.

Neutralization Resistance is Inherent to the Structural Organization of HIV Env

The two major functions of Env, receptor engagement and membrane fusion, are performed by different Env subunits. Gp120 mediates binding to the CD4 receptor and a chemokine coreceptor while gp41 drives fusion of the viral envelope with the plasma membrane of target cells. Attachment and fusion proceeds through a series of regulated conformational changes in Env, which involve the formation of discrete intermediates.[110,111] The structure of several of these intermediates has been determined on the atomic level. Sequence analysis revealed that gp120 contains 5 constant regions (C1–5) interspersed by 5 variable, disulfide-bonded loops (V1–5), which are organized into an inner and an outer domain in the tertiary structure.[112,113] The outer domain is heavily glycosylated and mainly variable sequences are displayed at its surface.[112,113] The inner domain is more conserved and, in the absence of CD4, composed of several distinct substructures which can move independently. When no CD4 is present, the binding sites for CD4 and coreceptor are not fully formed.[114] Yet, gp120 might drift toward the CD4 bound conformation under these conditions. Attachment to CD4 might then stabilize these conformations and might finally arrest gp120 in the ligand bound state. In the ligand bound conformation, CD4 binds to a recessed area in gp120, which is composed of the interface of the inner domain, the outer domain and a bridging sheet.[113] The bridging sheet connects inner and outer domain and constitutes a highly conserved binding site for a chemokine coreceptor.[114] Engagement of a coreceptor finally activates the fusion machinery in gp41, and fusion of the viral and the cellular membrane allows the viral capsid to enter the host cell lumen as described below. The functional and structural data discussed above highlight why the induction of a broadly neutralizing antibody response against HIV is a formidable challenge: The binding sites for receptor and coreceptor (neutralizing face) are recessed and/or not fully formed in the unliganded protein.[115] Env sequences displayed at the cell surface are highly variable, while conserved sequences are mainly hidden in the interior of the molecule.[115] Glycans protect surface exposed areas from antibody recognition (silent face of gp120). Finally, several epitopes in gp120, which is readily shed from the virus surface, are not accessible in the context of trimers (non-neutralizing face of gp120). It is thus not surprising that vaccine trials with unmodified gp120 yielded disappointing results.[116]

Binding of HIV Env to Lectins on Immune Cells Can Modulate Viral Infectivity

The first step in the entry process is attachment of virus particles to target cells. Attachment can be driven by interactions of Env with CD4, which are highly specific, or by relatively unspecific Env binding to other cellular factors, including heparan sulfate proteoglycans,[117] amyloid fibrils and cellular lectins (see below). Moreover, cellular factors incorporated into the viral membrane can also promote HIV attachment by binding to their receptors on target cells, with ICAM-3 (on the virus surface) mediated HIV-1 binding to LFA-1 on T cells being a prominent example.[108,118] Binding to attachment factors other than CD4 is ultimately dispensable for infectious entry of HIV, but can profoundly augment infection efficiency, at least under certain conditions.[117] Particularly noteworthy is the massive augmentation of HIV infectivity by binding to amyloid fibrils derived from a fragment of prostatic acidic phosphatase, termed SEVI (semen-derived enhancer of virus infection), which are present at high concentrations in human semen.[119] The association of HIV with these fibrils most likely concentrates viruses on the cell surface and boosts infectivity of particles several logs in vitro, suggesting that SEVI fibrils might promote infection of mucosal tissues by sexually transmitted HIV.[119,120] HIV infectivity for T cells can also be profoundly enhanced by interactions with dendritic cells.[121] Enhancement might result from binding of HIV to specific attachment factors on dendritic cells, like the DC-SIGN protein,[122,123] but T-cell activation by dendritic cells could also be involved. DC-SIGN is a calcium-dependent lectin which recognizes mannose-rich glycans on HIV Env.[124-129] Binding of HIV to DC-SIGN on cell lines profoundly augments infection of adjacent target cells (trans-infection) and it has been suggested that this process requires uptake of virions into DC-SIGN expressing cells, escape of HIV from degradation in lysosomes, conservation of viral infectivity in intracellular compartments[130] and finally transmission of virus to T cells within highly conductive microenvironments, termed infectious synapses.[131] Such a scenario would be compatible with the concept that submucosal dendritic cells might capture and transport HIV into lymphoid tissue, where virus loaded dendritic cells could catalyze infection of T cells and thereby contribute to establishment of the primary infection.[126] However, most of these postulates have been challenged and evidence has been presented that virus bound to DC-SIGN on dendritic cells might mainly be degraded and presented via the MHC system.[122,132-136] Trans-infection can occur, but only within hours after virus attachment and the relevance of this process for HIV spread in patients is uncertain.[135,136] Nevertheless, DC-SIGN expression and HIV transmission was also described for B-cells[137,138] and platelets[139,140] and attachment to these cells/cell fragments but not to dendritic cells might modulate HIV dissemination. The observation that polymorphisms in the DC-SIGN promoter modulate the risk of acquiring HIV via the parenteral but not the sexual route is consistent with this concept,[141] since it suggests that DC-SIGN positive cells impact HIV spread once the virus has entered the blood stream. Recent evidence suggests that binding of HIV to DC-SIGN triggers signaling which compromises the immune function of dendritic cells but promotes their ability to transmit virus to T cells.[142-144] Finally, it is noteworthy that HIV can bind to several other cellular lectins, including Langerin,[145] Mannose Receptor (MR)[146-148] and Mannose Binding Lectin (MBL),[149] with diverse consequences for infectivity and neutralization sensitivity.

Determinants of HIV Env Binding to the CD4 Receptor

The CD4 molecule, a member of the immunoglobulin (Ig) superfamily, is the primary receptor for HIV-1, HIV-2 and SIV and its engagement is indispensable for infectious entry (Fig. 1),[150,151] with the exception of some HIV-2 and SIV isolates, as discussed below. CD4 is expressed on T cells, monocytes, macrophages and dendritic cells, all of which are HIV and SIV target cells. The extracellular domain of CD4 contains four Ig-domains and the most membrane distal one is recognized by gp120.[152,153] The CD4 binding site in gp120 is not fully formed in the absence of receptor. Under these conditions, CD4 binds to a region in the outer domain of gp120, which is partially hidden within a cavity formed by α helices in the inner domain, the CD4 binding loop and β sheets 20 and 21.[112,113] CD4 engagement triggers a massive conformational rearrangement of gp120, including an outward rotation and displacement of gp120 monomers,[101,154] which results in the formation of a coreceptor binding site.[114] In addition, CD4 engagement induces conformational changes in gp41, potentially including the exposure and movement of a N-terminal fusion peptide toward the target cell membrane.[101] Analysis of gp120 in the CD4 and coreceptor bound state (a monoclonal antibody was used as a surrogate for coreceptor binding) revealed that under these conditions CD4 binds to a recessed area formed by the interface of inner domain, outer domain and bridging sheet. While the CD4/gp120 interface covers substantial surfaces on CD4 and gp120, the topologies of the interacting regions do not match precisely and large cavities are concealed in the interface. Atomic contacts are formed between amino acids within position 25 to 64 in CD4 and six segments of gp120, with CD4 residues Phe 43 (which inserts into a pocket in gp120) and Arg 59 and the highly conserved gp120 residues Asp 368, Glu 370 and Trp 427 being of particular importance.[112,113] Notably, several HIV-2 and SIV variants have been identified which can infect target cells in the absence of CD4,[155-158] and HIV-1 variants with similar abilities have been selected in cell culture.[159-162] These viruses might harbour partially triggered Env proteins, which can directly interact with a coreceptor. In theory, such viruses might exhibit a broadened tropism, and a link between CD4-independent infection and SIV neurotropism has been suggested.[155] However, CD4 independence comes at the cost of increased neutralization sensitivity,[163,164] making it unlikely that CD4 independent viruses become prevalent in HIV-1 infected patients.

Coreceptor Binding is a Major Determinant of HIV Tropism

Binding to CD4 triggers the formation/exposure of a highly conserved coreceptor binding site, which interacts with a chemokine coreceptor,[114] usually CCR5 and CXCR4 (Fig. 1), both of which are members of the G-protein coupled, seven transmembrane receptor superfamily.[165] The coreceptor binding site is mainly located in the bridging sheet and comprises the V1/V2 stem and sequences in the C4 region.[114] The V3 region, which protrudes from gp120 upon CD4 binding, is the major determinant of coreceptor choice, and also interacts directly with the coreceptor.[166] Mutagenic analysis revealed that the N-terminus (especially sulfated tyrosines in CCR5[167,168]) and the second extracellular loop of CCR5 and CXCR4 are of particular importance for coreceptor function. This is in agreement with a model suggesting that the V3 loop, the structure of which seems to resemble that of chemokines,[169] might contact the second extracellular loop while the coreceptor binding site might bind to N-terminus.[166,170] Viruses transmitted between individuals almost exclusively engage CCR5 for infectious entry,[171] and the preference for CCR5 is independent of the route of HIV acquisition.[172-174] The reason for the preference for CCR5-tropic viruses is at present unclear, but might involve several independent barriers against transmission of CXCR4-using viruses.[175] Homozygous carriers of a ccr5 defect allele are highly resistant against mucosal transmission while heterozygous individuals show delayed diseases progression,[176-180] indicating that CCR5 is of paramount importance to HIV-1 spread. This conclusion is further highlighted by the recent finding that viral spread was fully suppressed in an HIV patient at 20 mo after transplantation of stem cells obtained from an individual homozygous for defective ccr5.[181,182] Homozygosity for the ccr5 defect allele, which is frequent among Caucasians (approximately 1%), is not associated with obvious immune defects or other symptoms, making CCR5 an attractive drug target. Nevertheless, studies in a mouse model indicate that CCR5 might contribute to the control of Cryptococcus neoformans, an AIDS-related pathogen.[183] While CCR5-tropic (R5-tropic) viruses are transmitted between individuals and prevail in the asymptomatic phase of HIV infection, viral variants which use CXCR4 for infectious entry can arise at later stages of the infection and emergence of CXCR4-tropic variants is associated with accelerated disease progression.[171] It is unclear, however, if emergence of X4-tropic viruses is cause or consequence of progression toward AIDS.[184] The observation that many patients develop AIDS in the absence of X4-tropic viruses, argues for the latter. On the other hand, X4-tropic viruses are more adept in depleting T cells in vitro compared with R5-tropic viruses,[185,186] because of their expanded range of target cells: High levels of CXCR4 are expressed on naïve T cells, whereas robust CCR5 expression was detected on memory T cells.[187,188] In any event, CXCR4 is a valid target for antiviral intervention. However, strategies to block CXCR4 usage will be challenging, since CXCR4 and SDF1 (the only CXCR4 ligand) knockout mice are not viable.[189,190] Indeed, clinical evaluation of a small molecule CXCR4 inhibitor for HIV therapy was halted, in part due to unwanted side effects.[191] After the discovery of CCR5 and CXCR4 as HIV coreceptors a variety of structurally related molecules, so called alternative coreceptors, was found to support HIV entry.[165,192-194] Generally, these receptors allow entry upon overexpression in cell lines and are frequently used by HIV-2 and SIV with the same efficiency as CCR5,[195,196] while highly efficient engagement of alternative coreceptors is only observed for some HIV-1 isolates.[197] While several reports indicate that certain alternative coreceptors might sustain spread of some HIV-1 isolates in primary target cells,[198] there is currently no evidence that usage of these receptors contributes to viral spread in patients.[199-202] This might change once potent inhibitors of CCR5 and CXCR4 become available for HIV/AIDS therapy, since adaptation to efficient usage of alternative coreceptors might be one way for HIV to acquire resistance.

The gp41 Subunit Drives Fusion between the Viral and the Target Cell Membrane

Complexation of CD4 and coreceptor by gp120 induces conformational changes in gp41, which bring about fusion of the viral membrane with a host cell membrane, potentially the limiting membrane of endosomes.[203] However, it is incompletely understood how the activation signal is transmitted from gp120 to gp41. The domain organization of gp41 is conserved between class I fusion proteins.[110,204] From N- to C-terminus the following functional elements have been defined: A hydrophobic fusion peptide, which is liberated from the Env precursor gp160 by furin cleavage, two heptad repeats (HR), termed HR1 and HR2, which are connected by a disulphide-bonded loop, followed by a membrane proximal region, a transmembrane domain and a large cytoplasmic tail.[110,204] The rearrangement of these elements in the course of gp41 driven membrane fusion has been studied by employing gp41 specific antibodies or gp41 derived peptides.[110,204] The antibodies 2F5[205] and 4E10[206] are directed against the membrane proximal region while peptides derived from the C-terminal HR (HR2) bind to the N-terminal HR (HR1),[207] and both reagents block viral entry. In addition, a naturally occurring peptide which binds the HIV-1 fusion peptide and thereby inhibits HIV infection has been identified in human hemofiltrate[208] and an optimized version of this peptide was recently shown to potently reduce viral spread in HIV-1 patients.[209] Assessment of the reactivity and inhibitory activity of these reagents at different times after Env engagement combined with structural analyses allowed the establishment of the following model for viral entry: Binding of Env to CD4 is sufficient to induce structural rearrangements in gp41, which expose the HRs and the fusion peptide.[210-212] In contrast, the epitopes of 2F5 and 4E10 are only accessible in the unliganded Env and disappear during membrane fusion. Upon engagement of the coreceptor further conformational changes in gp41 are induced, which result in the insertion of the fusion peptide into the target cell membrane, the formation of a fusion pore and ultimately in the merger of the viral and the host cell membrane. Availability of CD4 and coreceptor, adequate temperature and most likely dissociation of gp120 from gp41 are prerequisites for these processes.[204] A key event in membrane fusion is the transition of gp41 into an energetically stable six-helix bundle structure.[213-215] This gp41 conformation is the result of a back-folding reaction, in which the three HR2 elements are packed, in an antiparallel fashion, onto a central coiled-coil formed by the HR1 elements. As a consequence, the fusion peptide and the transmembrane region of gp41, and therefore the viral and host cell membranes, are pulled into close contact, allowing membrane fusion to proceed.[204] Notably, the final folded six-helix bundle structure is observed only after the opening of a fusion pore.[216] It has thus been suggested that "pre-bundle" conformations are sufficient for pore formation while transition of gp41 into a six-helix bundle seems to be required for stabilization and/or expansion of pores.[204]

CONCLUSION

The Env proteins of retroviruses are class I fusion proteins; the SU engages the receptor while the TM drives fusion of the viral envelope with a cellular membrane. The Env proteins of ASLV and HIV have evolved intricate controls of their fusogenic activity. For ASLV, receptor binding primes the Env-protein for a subsequent stimulus, acidic pH, which then triggers the membrane fusion reaction. In contrast, sequential engagement of a primary receptor, CD4, and a coreceptor, CCR5 and CXCR4, is required for activation of HIV gp41-driven membrane fusion. HIV variants which use CCR5 are transmitted between individuals and absence of functional CCR5 largely protects against sexual acquisition of HIV. In contrast, CXCR4-tropic viruses evolve at later stages of the infection and are associated with a poor clinical prognosis. However, neither the preference for CCR5 usage during establishment of the primary infection nor the reasons for the emergence of CXCR4-tropic viruses in about 40% of the infected individuals are fully understood. Unravelling these questions will provide important insights into HIV transmission and pathogenesis and might help to establish novel strategies for intervention. After the elucidation of the structure of a core HIV Env in complex with CD4 and a chemokine receptor mimicking antibody, novel insights into the Env structure were obtained with trimeric proteins and upon analysis of a V3 loop containing Env. These studies help to understand the sequential rearrangements of Env pivotal to infectious entry and their inhibition by entry inhibitors. Finally, inhibitors targeting gp41 helped to define the intermediates in the membrane fusion reaction, and might ultimately allow the identification of new targets. This speculation is supported by the observation that an endogenous peptide derived from α1 antitrypsin binds to the HIV-1 fusion peptide and thereby prevents viral entry, underlining that structures in Env, which so far have not been considered "druggable," can be successfully targeted.

ACKNOWLEDGMENT

We thank T.F. Schulz for support and the Center for Infection Biology at Hannover Medical School for funding.

REFERENCES

1. Cohen MS, Hellmann N, Levy JA et al. The spread, treatment, and prevention of HIV-1: evolution of a global pandemic. J Clin Invest 2008; 118:1244-54.
2. Karlsson GB, Halloran M, Schenten D et al. The envelope glycoprotein ectodomains determine the efficiency of CD4+ T lymphocyte depletion in simian-human immunodeficiency virus-infected macaques. J Exp Med 1998; 188:1159-71.
3. Coffin JM, Hughes SH, Varmus HE. 1997. Retroviruses, vol. 843. Cold Spring Harbor Laboratory Press, Cold Spring Harbor, N.Y.
4. Goff SP. 2007. Retroviridae: The retroviruses and their replication, p. 1999 - 2070. *In* D. M. Knipe, Howley, P. M. (ed.), Fields Virology, vol. 2. Wolters Kuwer/Lippincott Williams & Wilkins, Philadelphia.
5. Goodier JL, Kazazian HH Jr. Retrotransposons revisited: the restraint and rehabilitation of parasites. Cell 2008; 135:23-35.
6. Poiesz BJ, Ruscetti FW, Gazdar AF et al. Detection and isolation of type C retrovirus particles from fresh and cultured lymphocytes of a patient with cutaneous T-cell lymphoma. Proc Natl Acad Sci USA 1980; 77:7415-9.
7. Barré-Sinoussi F, Chermann JC, Rey F et al. Isolation of a T-lymphotropic retrovirus from a patient at risk for acquired immune deficiency syndrome (AIDS). Science 1983; 220:868-71.
8. Einfeld D, Hunter E. Oligomeric structure of a prototype retrovirus glycoprotein. Proc Natl Acad Sci USA 1988; 85:8688-92.
9. Hunter E, Hill E, Hardwick M et al. Complete sequence of the Rous sarcoma virus env gene: identification of structural and functional regions of its product. J Virol 1983; 46:920-36.
10. Purchio AF, Jovanovich S, Erikson RL. Sites of synthesis of viral proteins in avian sarcoma virus-infected chicken cells. J Virol 1980; 35:629-36.
11. Dong JY, Dubay JW, Perez LG et al. Mutations within the proteolytic cleavage site of the Rous sarcoma virus glycoprotein define a requirement for dibasic residues for intracellular cleavage. J Virol 1992; 66:865-74.
12. Leamnson RN, Halpern MS. Subunit structure of the glycoprotein complex of avian tumor virus. J Virol 1976; 18:956-68.
13. Perez LG, Hunter E. Mutations within the proteolytic cleavage site of the Rous sarcoma virus glycoprotein that block processing to gp85 and gp37. J Virol 1987; 61:1609-14.
14. Delos SE, Burdick MJ, White JM. A single glycosylation site within the receptor-binding domain of the avian sarcoma/leukosis virus glycoprotein is critical for receptor binding. Virology 2002; 294:354-63.
15. Harrison SC. Viral membrane fusion. Nat Struct Mol Biol 2008; 15:690-8.
16. White JM, Delos SE, Brecher M et al. Structures and mechanisms of viral membrane fusion proteins: multiple variations on a common theme. Crit Rev Biochem Mol Biol 2008; 43:189-219.
17. Gallaher WR. Similar structural models of the transmembrane proteins of Ebola and avian sarcoma viruses. Cell 1996; 85:477-8.
18. Barnard RJ, Young JA. Alpharetrovirus envelope-receptor interactions. Curr Top Microbiol Immunol 2003; 281:107-36.
19. Barnard RJ, Elleder D, Young JA. Avian sarcoma and leukosis virus-receptor interactions: from classical genetics to novel insights into virus-cell membrane fusion. Virology 2006; 344:25-9.
20. Bates P, Young JA, Varmus HE. A receptor for subgroup A Rous sarcoma virus is related to the low density lipoprotein receptor. Cell 1993; 74:1043-51.
21. Young JA, Bates P, Varmus HE. Isolation of a chicken gene that confers susceptibility to infection by subgroup A avian leukosis and sarcoma viruses. J Virol 1993; 67:1811-6.
22. Rong L, Bates P. Analysis of the subgroup A avian sarcoma and leukosis virus receptor: the 40-residue, cysteine-rich, low-density lipoprotein receptor repeat motif of Tva is sufficient to mediate viral entry. J Virol 1995; 69:4847-53.
23. Zingler K, Belanger C, Peters R et al. Identification and characterization of the viral interaction determinant of the subgroup A avian leukosis virus receptor. J Virol 1995; 69:4261-6.
24. Brojatsch J, Naughton J, Rolls MM et al. CAR1, a TNFR-related protein, is a cellular receptor for cytopathic avian leukosis-sarcoma viruses and mediates apoptosis. Cell 1996; 87:845-55.
25. Smith EJ, Brojatsch J, Naughton J et al. The CAR1 gene encoding a cellular receptor specific for subgroup B and D avian leukosis viruses maps to the chicken tvb locus. J Virol 1998; 72:3501-3.
26. Adkins HB, Brojatsch J, Naughton J et al. Identification of a cellular receptor for subgroup E avian leukosis virus. Proc Natl Acad Sci USA 1997; 94:11617-22.
27. Bansal A, Shaw KL, Edwards BH et al. Characterization of the R572T point mutant of a putative cleavage site in human foamy virus Env. J Virol 2000; 74:2949-54.
28. Knauss DJ, Young JA. A fifteen-amino-acid TVB peptide serves as a minimal soluble receptor for subgroup B avian leukosis and sarcoma viruses. J Virol 2002; 76:5404-10.

29. Elleder D, Stepanets V, Melder DC et al. The receptor for the subgroup C avian sarcoma and leukosis viruses, Tvc, is related to mammalian butyrophilins, members of the immunoglobulin superfamily. J Virol 2005; 79:10408-19.
30. Munguia A, Federspiel MJ. Efficient subgroup C avian sarcoma and leukosis virus receptor activity requires the IgV domain of the Tvc receptor and proper display on the cell membrane. J Virol 2008; 82:11419-28.
31. Mothes W, Boerger AL, Narayan S et al. Retroviral entry mediated by receptor priming and low pH triggering of an envelope glycoprotein. Cell 2000; 103:679-89.
32. Barnard RJ, Narayan S, Dornadula G et al. Low pH is required for avian sarcoma and leukosis virus Env-dependent viral penetration into the cytosol and not for viral uncoating. J Virol 2004; 78:10433-41.
33. Earp LJ, Delos SE, Netter RC et al. The avian retrovirus avian sarcoma/leukosis virus subtype A reaches the lipid mixing stage of fusion at neutral pH. J Virol 2003; 77:3058-66.
34. Melikyan GB, Barnard RJ, Abrahamyan LG, Mothes W, Young JA. Imaging individual retroviral fusion events: from hemifusion to pore formation and growth. Proc Natl Acad Sci USA 2005; 102:8728-33.
35. Melikyan GB, Barnard RJ, Markosyan RM, Young JA, Cohen FS. Low pH is required for avian sarcoma and leukosis virus Env-induced hemifusion and fusion pore formation but not for pore growth. J Virol 2004; 78:3753-62.
36. Cóte M, Zheng YM, Liu SL. Receptor binding and low pH coactivate oncogenic retrovirus envelope-mediated fusion. J Virol 2009; 83:11447-55.
37. Fauquet C, Mayo M, Maniloff J et al, eds. 2005. Virus Taxonomy: Eighth Report of the International Committee on Taxonomy of Viruses. Academic Press, San Diego.
38. Lindemann D, Goepfert PA. The foamy virus envelope glycoproteins. Curr Top Microbiol Immunol 2003; 277:111-29.
39. Geiselhart V, Schwantes A, Bastone P et al. Features of the Env leader protein and the N-terminal Gag domain of feline foamy virus important for virus morphogenesis. Virology 2003; 310:235-44.
40. Lindemann D, Pietschmann T, Picard-Maureau M et al. A particle-associated glycoprotein signal peptide essential for virus maturation and infectivity. J Virol 2001; 75:5762-71.
41. Wilk T, Geiselhart V, Frech M et al. Specific interaction of a novel foamy virus env leader protein with the n-terminal gag domain. J Virol 2001; 75:7995-8007.
42. Duda A, Stange A, Luftenegger D et al. Prototype foamy virus envelope glycoprotein leader peptide processing is mediated by a furin-like cellular protease, but cleavage is not essential for viral infectivity. J Virol 2004; 78:13865-70.
43. Geiselhart V, Bastone P, Kempf T et al. Furin-mediated cleavage of the feline foamy virus Env leader protein. J Virol 2004; 78:13573-81.
44. Pietschmann T, Zentgraf H, Rethwilm A et al. An evolutionarily conserved positively charged amino acid in the putative membrane-spanning domain of the foamy virus envelope protein controls fusion activity. J Virol 2000; 74:4474-82.
45. Wilk T, de Haas F, Wagner A et al. The intact retroviral Env glycoprotein of human foamy virus is a trimer. J Virol 2000; 74:2885-7.
46. Baldwin DN, Linial ML. The roles of Pol and Env in the assembly pathway of human foamy virus. J Virol 1998; 72:3658-65.
47. Fischer N, Heinkelein M, Lindemann D et al. Foamy virus particle formation. J Virol 1998; 72:1610-5.
48. Shaw KL, Lindemann D, Mulligan MJ et al Foamy virus envelope glycoprotein is sufficient for particle budding and release. J Virol 2003; 77:2338-48.
49. Stanke N, Stange A, Lüftenegger D et al. Ubiquitination of the Prototype Foamy Virus Envelope Glycoprotein Leader Peptide Regulates Subviral Particle Release. J Virol 2005; 79:15074-83.
50. Stange A, Luftenegger D, Reh J et al. Subviral particle release determinants of prototype foamy virus. J Virol 2008; 82:9858-69.
51. Lüftenegger D, Picard-Maureau M, Stanke N et al. Analysis and function of prototype foamy virus envelope N glycosylation. J Virol 2005; 79:7664-72.
52. Goepfert PA, Shaw KL, Ritter GD Jr. et al. A sorting motif localizes the foamy virus glycoprotein to the endoplasmic reticulum. J Virol 1997; 71:778-84.
53. Goepfert PA, Wang G, Mulligan MJ. Identification of an ER retrieval signal in a retroviral glycoprotein. Cell 1995; 82:543-4.
54. Tobaly-Tapiero J, Bittoun P, Neves M et al. Isolation and characterization of an equine foamy virus. J Virol 2000; 74:4064-73.
55. Yu SF, Eastman SW, Linial ML. Foamy virus capsid assembly occurs at a pericentriolar region through a cytoplasmic targeting/retention signal in Gag. Traffic 2006; 7:966-77
56. Hill CL, Bieniasz PD, McClure MO. Properties of human foamy virus relevant to its development as a vector for gene therapy. J Gen Virol 1999; 80:2003-9.
57. Stirnnagel K, Luftenegger D, Stange A et al. Analysis of prototype foamy virus particle-host cell interaction with autofluorescent retroviral particles. Retrovirology 2010; 7:45.

58. Berg A, Pietschmann T, Rethwilm A, Lindemann D. Determinants of foamy virus envelope glycoprotein mediated resistance to superinfection. Virology 2003; 314:243-52.
59. Duda A, Luftenegger D, Pietschmann T et al. Characterization of the prototype foamy virus envelope glycoprotein receptor-binding domain. J Virol 2006; 80:8158-67.
60. Picard-Maureau M, Jarmy G, Berg A et al. Foamy Virus Envelope Glycoprotein-Mediated Entry Involves a pH-Dependent Fusion Process. J Virol 2003; 77:4722-30.
61. Petit C, Giron ML, Tobaly-Tapiero J et al. Targeting of incoming retroviral Gag to the centrosome involves a direct interaction with the dynein light chain 8. J Cell Sci 2003; 116:3433-42.
62. Saib A, Puvion Dutilleul F, Schmid M et al. Nuclear targeting of incoming human foamy virus Gag proteins involves a centriolar step. J Virol 1997; 71:1155-61.
63. Lehmann-Che J, Renault N, Giron ML et al. Centrosomal latency of incoming foamy viruses in resting cells. PLoS Pathog 2007; 3:e74.
64. Matthes D, Wiktorowicz T, Zahn J et al. Basic residues in the foamy virus Gag protein. J Virol 2011; 85:3986-95.
65. Zhadina M, McClure MO, Johnson MC et al. Ubiquitin-dependent virus particle budding without viral protein ubiquitination. Proc Natl Acad Sci USA 2007; 104:20031-6.
66. Eastman SW, Linial ML. Identification of a conserved residue of foamy virus Gag required for intracellular capsid assembly. J Virol 2001; 75:6857-64.
67. Linial ML, Eastman SW. Particle assembly and genome packaging. Curr Top Microbiol Immunol 2003; 277:89-110.
68. Pfrepper KI, Löchelt M, Schnolzer M et al. Expression and molecular characterization of an enzymatically active recombinant human spumaretrovirus protease. Biochem Biophys Res Commun 1997; 237:548-53.
69. Zemba M, Wilk T, Rutten T et al. The carboxy-terminal p3Gag domain of the human foamy virus Gag precursor is required for efficient virus infectivity. Virology 1998; 247:7-13.
70. Cartellieri M, Rudolph W, Herchenröder O et al. Determination of the relative amounts of Gag and Pol proteins in foamy virus particles. Retrovirology 2005; 2:44.
71. Enssle J, Fischer N, Moebes A et al. Carboxy-terminal cleavage of the human foamy virus Gag precursor molecule is an essential step in the viral life cycle. J Virol 1997; 71:7312-7.
72. Pfrepper KI, Löchelt M, Rackwitz HR et al. Molecular characterization of proteolytic processing of the Gag proteins of human spumavirus. J Virol 1999; 73:7907-11.
73. Lehmann-Che J, Giron ML, Delelis O et al. Protease-dependent uncoating of a complex retrovirus. J Virol 2005; 79:9244-53.
74. Keele BF, Van Heuverswyn F, Li Y et al. Chimpanzee reservoirs of pandemic and nonpandemic HIV-1. Science 2006; 313:523-6.
75. Van Heuverswyn F, Li Y, Neel C et al. Human immunodeficiency viruses: SIV infection in wild gorillas. Nature 2006; 444:164.
76. Gao F, Bailes E, Robertson DL et al. Origin of HIV-1 in the chimpanzee Pan troglodytes troglodytes. Nature 1999; 397:436-41.
77. Hahn BH, Shaw GM, De Cock KM et al. AIDS as a zoonosis: scientific and public health implications. Science 2000; 287:607-14.
78. Plantier JC, Leoz M, Dickerson JE et al. A new human immunodeficiency virus derived from gorillas. Nat Med 2009; 15:871-2.
79. de Silva TI, Cotten M, Rowland-Jones SL. HIV-2: the forgotten AIDS virus. Trends Microbiol 2008; 16:588-95.
80. Gao F, Yue L, White AT et al. Human infection by genetically diverse SIVSM-related HIV-2 in west Africa. Nature 1992; 358:495-9.
81. Hirsch VM, Olmsted RA, Murphey-Corb M et al. An African primate lentivirus (SIVsm) closely related to HIV-2. Nature 1989; 339:389-92.
82. Levy JA. HIV pathogenesis: 25 years of progress and persistent challenges. AIDS 2009; 23:147-60.
83. Douek DC, Roederer M, Koup RA. Emerging concepts in the immunopathogenesis of AIDS. Annu Rev Med 2009; 60:471-84.
84. Schindler M, Munch J, Kutsch O et al. Nef-mediated suppression of T cell activation was lost in a lentiviral lineage that gave rise to HIV-1. Cell 2006; 125:1055-67.
85. Richman DD, Margolis DM, Delaney M et al. The challenge of finding a cure for HIV infection. Science 2009; 323:1304-7.
86. Trono D, Van Lint C, Rouzioux C et al. HIV persistence and the prospect of long-term drug-free remissions for HIV-infected individuals. Science 2010; 329:174-80.
87. Otteken A, Moss B. Calreticulin interacts with newly synthesized human immunodeficiency virus type 1 envelope glycoprotein, suggesting a chaperone function similar to that of calnexin. J Biol Chem 1996; 271:97-103.

88. Otteken A, Earl PL, Moss B. Folding, assembly, and intracellular trafficking of the human immunodeficiency virus type 1 envelope glycoprotein analyzed with monoclonal antibodies recognizing maturational intermediates. J Virol 1996; 70:3407-15.
89. Li Y, Bergeron JJ, Luo L, Ou WJ et al. Effects of inefficient cleavage of the signal sequence of HIV-1 gp120 on its association with calnexin, folding, and intracellular transport. Proc Natl Acad Sci USA 1996; 93:9606-11.
90. Li Y, Luo L, Thomas DY, Kang CY. The HIV-1 Env protein signal sequence retards its cleavage and down-regulates the glycoprotein folding. Virology 2000; 272:417-28.
91. Scanlan CN, Offer J, Zitzmann N, Dwek RA. Exploiting the defensive sugars of HIV-1 for drug and vaccine design. Nature 2007; 446:1038-45.
92. Li Y, Luo L, Rasool N, Kang CY. Glycosylation is necessary for the correct folding of human immunodeficiency virus gp120 in CD4 binding. J Virol 1993; 67:584-8.
93. Bernstein HB, Tucker SP, Hunter E et al. Human immunodeficiency virus type 1 envelope glycoprotein is modified by O-linked oligosaccharides. J Virol 1994; 68:463-8.
94. Lu M, Blacklow SC, Kim PS. A trimeric structural domain of the HIV-1 transmembrane glycoprotein. Nat Struct Biol 1995; 2:1075-82.
95. Thomas DJ, Wall JS, Hainfeld JF et al. gp160, the envelope glycoprotein of human immunodeficiency virus type 1, is a dimer of 125-kilodalton subunits stabilized through interactions between their gp41 domains. J Virol 1991; 65:3797-803.
96. Sanders RW, Venturi M, Schiffner L et al. The mannose-dependent epitope for neutralizing antibody 2G12 on human immunodeficiency virus type 1 glycoprotein gp120. J Virol 2002; 76:7293-305.
97. Trkola A, Purtscher M, Muster T et al. Human monoclonal antibody 2G12 defines a distinctive neutralization epitope on the gp120 glycoprotein of human immunodeficiency virus type 1. J Virol 1996; 70:1100-8.
98. Hallenberger S, Bosch V, Angliker H, Shaw E, Klenk HD, Garten W. Inhibition of furin-mediated cleavage activation of HIV-1 glycoprotein gp160. Nature 1992; 360:358-61.
99. Miranda L, Wolf J, Pichuantes S, Duke R et al. Isolation of the human PC6 gene encoding the putative host protease for HIV-1 gp160 processing in CD4+ T lymphocytes. Proc Natl Acad Sci USA 1996; 93:7695-700.
100. McCune JM, Rabin LB, Feinberg MB et al. Endoproteolytic cleavage of gp160 is required for the activation of human immunodeficiency virus. Cell 1988; 53:55-67.
101. Liu J, Bartesaghi A, Borgnia MJ et al. Molecular architecture of native HIV-1 gp120 trimers. Nature 2008; 455:109-13.
102. Zanetti G, Briggs JA, Grunewald K et al. Cryo-electron tomographic structure of an immunodeficiency virus envelope complex in situ. PLoS Pathog 2006; 2:e83.
103. Zhu P, Liu J, Bess J Jr. et al. Distribution and three-dimensional structure of AIDS virus envelope spikes. Nature 2006; 441:847-52.
104. Ganser-Pornillos BK, Yeager M, Sundquist WI. The structural biology of HIV assembly. Curr Opin Struct Biol 2008; 18:203-17.
105. Sundquist WI, Hill CP. How to assemble a capsid. Cell 2007; 131:17-9.
106. Bauby H, Lopez-Verges S, Hoeffel G et al. TIP47 is required for the production of infectious HIV-1 particles from primary macrophages. Traffic 2010; 11:455-67.
107. Lopez-Verges S, Camus G, Blot G et al. Tail-interacting protein TIP47 is a connector between Gag and Env and is required for Env incorporation into HIV-1 virions. Proc Natl Acad Sci USA 2006; 103:14947-52.
108. Cantin R, Methot S, Tremblay MJ. Plunder and stowaways: incorporation of cellular proteins by enveloped viruses. J Virol 2005; 79:6577-87.
109. Ott DE. Cellular proteins in HIV virions. Rev Med Virol 1997; 7:167-80.
110. Steffen I, Pöhlmann S. Peptide-based inhibitors of the HIV envelope protein and other class I viral fusion proteins. Curr Pharm Des 2010; 16:1143-58.
111. Tilton JC, Doms RW. Entry inhibitors in the treatment of HIV-1 infection. Antiviral Res 2010; 85:91-100.
112. Kwong PD, Wyatt R, Robinson J et al. Structure of an HIV gp120 envelope glycoprotein in complex with the CD4 receptor and a neutralizing human antibody. Nature 1998; 393:648-59.
113. Wyatt R, Kwong PD, Desjardins E et al. The antigenic structure of the HIV gp120 envelope glycoprotein. Nature 1998; 393:705-11.
114. Rizzuto CD, Wyatt R, Hernandez-Ramos N et al. A conserved HIV gp120 glycoprotein structure involved in chemokine receptor binding. Science 1998; 280:1949-53.
115. Wyatt R, Sodroski J. The HIV-1 envelope glycoproteins: fusogens, antigens, and immunogens. Science 1998; 280:1884-8.
116. Karlsson Hedestam GB, Fouchier RA, Phogat S et al. The challenges of eliciting neutralizing antibodies to HIV-1 and to influenza virus. Nat Rev Microbiol 2008; 6:143-55.
117. Ugolini S, Mondor I, Sattentau QJ. HIV-1 attachment: another look. Trends Microbiol 1999; 7:144-9.
118. Fortin JF, Cantin R, Lamontagne G et al. Host-derived ICAM-1 glycoproteins incorporated on human immunodeficiency virus type 1 are biologically active and enhance viral infectivity. J Virol 1997; 71:3588-96.

119. Münch J, Rucker E, Standker L et al. Semen-derived amyloid fibrils drastically enhance HIV infection. Cell 2007; 131:1059-71.
120. Kim KA, Yolamanova M, Zirafi O et al. Semen-mediated enhancement of HIV infection is donor-dependent and correlates with the levels of SEVI. Retrovirology 2010; 7:55.
121. Cameron PU, Freudenthal PS, Barker JM et al. Dendritic cells exposed to human immunodeficiency virus type-1 transmit a vigorous cytopathic infection to CD4+ T cells. Science 1992; 257:383-7.
122. Tsegaye TS, Pöhlmann S. The multiple facets of HIV attachment to dendritic cell lectins. Cell Microbiol 2010; 12:1553-61.
123. Wu L, KewalRamani VN. Dendritic-cell interactions with HIV: infection and viral dissemination. Nat Rev Immunol 2006; 6:859-68.
124. Curtis BM, Scharnowske S, Watson AJ. Sequence and expression of a membrane-associated C-type lectin that exhibits CD4-independent binding of human immunodeficiency virus envelope glycoprotein gp120. Proc Natl Acad Sci USA 1992; 89:8356-60.
125. Feinberg H, Mitchell DA, Drickamer K et al. Structural basis for selective recognition of oligosaccharides by DC-SIGN and DC-SIGNR. Science 2001; 294:2163-6.
126. Geijtenbeek TB, Kwon DS, Torensma R et al. DC-SIGN, a dendritic cell-specific HIV-1-binding protein that enhances trans-infection of T cells. Cell 2000; 100:587-97.
127. Geijtenbeek TB, Torensma R, van Vliet SJ et al. Identification of DC-SIGN, a novel dendritic cell-specific ICAM-3 receptor that supports primary immune responses. Cell 2000; 100:575-85.
128. Guo Y, Feinberg H, Conroy E et al. Structural basis for distinct ligand-binding and targeting properties of the receptors DC-SIGN and DC-SIGNR. Nat Struct Mol Biol 2004; 11:591-8.
129. Lin G, Simmons G, Pöhlmann S et al. Differential N-linked glycosylation of human immunodeficiency virus and Ebola virus envelope glycoproteins modulates interactions with DC-SIGN and DC-SIGNR. J Virol 2003; 77:1337-46.
130. Kwon DS, Gregorio G, Bitton N et al. DC-SIGN-mediated internalization of HIV is required for trans-enhancement of T cell infection. Immunity 2002; 16:135-44.
131. McDonald D, Wu L, Bohks SM et al. Recruitment of HIV and its receptors to dendritic cell-T cell junctions. Science 2003; 300:1295-7.
132. Boggiano C, Manel N, Littman DR. Dendritic cell-mediated trans-enhancement of human immunodeficiency virus type 1 infectivity is independent of DC-SIGN. J Virol 2007; 81:2519-23.
133. Moris A, Nobile C, Buseyne F et al. DC-SIGN promotes exogenous MHC-I-restricted HIV-1 antigen presentation. Blood 2004; 103:2648-54.
134. Moris A, Pajot A, Blanchet F et al. Dendritic cells and HIV-specific CD4+ T cells: HIV antigen presentation, T-cell activation, and viral transfer. Blood 2006; 108:1643-51.
135. Nobile C, Petit C, Moris A et al. Covert human immunodeficiency virus replication in dendritic cells and in DC-SIGN-expressing cells promotes long-term transmission to lymphocytes. J Virol 2005; 79:5386-99.
136. Turville SG, Santos JJ, Frank I et al. Immunodeficiency virus uptake, turnover, and 2-phase transfer in human dendritic cells. Blood 2004; 103:2170-9.
137. He B, Qiao X, Klasse PJ et al. HIV-1 envelope triggers polyclonal Ig class switch recombination through a CD40-independent mechanism involving BAFF and C-type lectin receptors. J Immunol 2006; 176:3931-41.
138. Rappocciolo G, Piazza P, Fuller CL et al. DC-SIGN on B lymphocytes is required for transmission of HIV-1 to T lymphocytes. PLoS Pathog 2006; 2:e70.
139. Boukour S, Masse JM, Benit L et al. Lentivirus degradation and DC-SIGN expression by human platelets and megakaryocytes. J Thromb Haemost 2006; 4:426-35.
140. Chaipan C, Soilleux EJ, Simpson P et al. DC-SIGN and CLEC-2 mediate human immunodeficiency virus type 1 capture by platelets. J Virol 2006; 80:8951-60.
141. Martin MP, Lederman MM, Hutcheson HB et al. Association of DC-SIGN promoter polymorphism with increased risk for parenteral, but not mucosal, acquisition of human immunodeficiency virus type 1 infection. J Virol 2004; 78:14053-6.
142. Gringhuis SI, den Dunnen J, Litjens M et al. Carbohydrate-specific signaling through the DC-SIGN signalosome tailors immunity to Mycobacterium tuberculosis, HIV-1 and Helicobacter pylori. Nat Immunol 2009; 10:1081-8.
143. Gringhuis SI, den Dunnen J, Litjens M et al. C-type lectin DC-SIGN modulates Toll-like receptor signaling via Raf-1 kinase-dependent acetylation of transcription factor NF-kappaB. Immunity 2007; 26:605-16.
144. Gringhuis SI, van der Vlist M, van den Berg LM et al. HIV-1 exploits innate signaling by TLR8 and DC-SIGN for productive infection of dendritic cells. Nat Immunol 2010; 11:419-26.
145. de Witte L, Nabatov A, Pion M et al. Langerin is a natural barrier to HIV-1 transmission by Langerhans cells. Nat Med 2007; 13:367-71.
146. Nguyen DG, Hildreth JE. Involvement of macrophage mannose receptor in the binding and transmission of HIV by macrophages. Eur J Immunol 2003; 33:483-93.

147. Trujillo JR, Rogers R, Molina RM et al. Noninfectious entry of HIV-1 into peripheral and brain macrophages mediated by the mannose receptor. Proc Natl Acad Sci USA 2007; 104:5097-102.
148. Turville SG, Cameron PU, Handley A et al. Diversity of receptors binding HIV on dendritic cell subsets. Nat Immunol 2002; 3:975-83.
149. Saifuddin M, Hart ML, Gewurz H et al. Interaction of mannose-binding lectin with primary isolates of human immunodeficiency virus type 1. J Gen Virol 2000; 81:949-55.
150. Klatzmann D, Champagne E, Chamaret S et al. T-lymphocyte T4 molecule behaves as the receptor for human retrovirus LAV. Nature 1984; 312:767-8.
151. Maddon PJ, Dalgleish AG, McDougal JS et al. The T4 gene encodes the AIDS virus receptor and is expressed in the immune system and the brain. Cell 1986; 47:333-48.
152. Bour S, Geleziunas R, Wainberg MA. The human immunodeficiency virus type 1 (HIV-1) CD4 receptor and its central role in promotion of HIV-1 infection. Microbiol Rev 1995; 59:63-93.
153. Jameson BA, Rao PE, Kong LI et al. Location and chemical synthesis of a binding site for HIV-1 on the CD4 protein. Science 1988; 240:1335-9.
154. Chen B, Vogan EM, Gong H et al. Determining the structure of an unliganded and fully glycosylated SIV gp120 envelope glycoprotein. Structure 2005; 13:197-211.
155. Edinger AL, Mankowski JL, Doranz BJ et al. CD4-independent, CCR5-dependent infection of brain capillary endothelial cells by a neurovirulent simian immunodeficiency virus strain. Proc Natl Acad Sci USA 1997; 94:14742-7.
156. Endres MJ, Clapham PR, Marsh M et al. CD4-independent infection by HIV-2 is mediated by fusin/CXCR4. Cell 1996; 87:745-56.
157. Martin KA, Wyatt R, Farzan M et al. CD4-independent binding of SIV gp120 to rhesus CCR5. Science 1997; 278:1470-3.
158. Reeves JD, Schulz TF. The CD4-independent tropism of human immunodeficiency virus type 2 involves several regions of the envelope protein and correlates with a reduced activation threshold for envelope-mediated fusion. J Virol 1997; 71:1453-65.
159. Dumonceaux J, Nisole S, Chanel C et al. Spontaneous mutations in the env gene of the human immunodeficiency virus type 1 NDK isolate are associated with a CD4-independent entry phenotype. J Virol 1998; 72:512-9.
160. Hoffman TL, LaBranche CC, Zhang W et al. Stable exposure of the coreceptor-binding site in a CD4-independent HIV-1 envelope protein. Proc Natl Acad Sci USA 1999; 96:6359-64.
161. Kolchinsky P, Mirzabekov T, Farzan M et al. Adaptation of a CCR5-using, primary human immunodeficiency virus type 1 isolate for CD4-independent replication. J Virol 1999; 73:8120-6.
162. Taylor BM, Foulke JS, Flinko R et al. An alteration of human immunodeficiency virus gp41 leads to reduced CCR5 dependence and CD4 independence. J Virol 2008; 82:5460-71.
163. Edwards TG, Hoffman TL, Baribaud F et al. Relationships between CD4 independence, neutralization sensitivity, and exposure of a CD4-induced epitope in a human immunodeficiency virus type 1 envelope protein. J Virol 2001; 75:5230-9.
164. Kolchinsky P, Kiprilov E, Sodroski J. Increased neutralization sensitivity of CD4-independent human immunodeficiency virus variants. J Virol 2001; 75:2041-50.
165. Lusso P. HIV and the chemokine system: 10 years later. EMBO J 2006; 25:447-56.
166. Huang CC, Tang M, Zhang MY et al. Structure of a V3-containing HIV-1 gp120 core. Science 2005; 310:1025-8.
167. Farzan M, Mirzabekov T, Kolchinsky P et al. Tyrosine sulfation of the amino terminus of CCR5 facilitates HIV-1 entry. Cell 1999; 96:667-76.
168. Huang CC, Lam SN, Acharya P et al. Structures of the CCR5 N terminus and of a tyrosine-sulfated antibody with HIV-1 gp120 and CD4. Science 2007; 317:1930-4.
169. Sharon M, Kessler N, Levy R et al. Alternative conformations of HIV-1 V3 loops mimic beta hairpins in chemokines, suggesting a mechanism for coreceptor selectivity. Structure 2003; 11:225-36.
170. Cormier EG, Dragic T. The crown and stem of the V3 loop play distinct roles in human immunodeficiency virus type 1 envelope glycoprotein interactions with the CCR5 coreceptor. J Virol 2002; 76:8953-7.
171. Connor RI, Sheridan KE, Ceradini D et al. Change in coreceptor use correlates with disease progression in HIV-1–infected individuals. J Exp Med 1997; 185:621-8.
172. Philpott S, Burger H, Charbonneau T et al. CCR5 genotype and resistance to vertical transmission of HIV-1. J Acquir Immune Defic Syndr 1999; 21:189-93.
173. van't Wout AB, Kootstra NA, Mulder-Kampinga GA et al. Macrophage-tropic variants initiate human immunodeficiency virus type 1 infection after sexual, parenteral, and vertical transmission. J Clin Invest 1994; 94:2060-7.
174. Zhu T, Mo H, Wang N et al. Genotypic and phenotypic characterization of HIV-1 patients with primary infection. Science 1993; 261:1179-81.

175. Margolis L, Shattock R. Selective transmission of CCR5-utilizing HIV-1: the 'gatekeeper' problem resolved? Nat Rev Microbiol 2006; 4:312-7.
176. Dean M, Carrington M, Winkler C et al. Genetic restriction of HIV-1 infection and progression to AIDS by a deletion allele of the CKR5 structural gene. Hemophilia Growth and Development Study, Multicenter AIDS Cohort Study, Multicenter Hemophilia Cohort Study, San Francisco City Cohort, ALIVE Study. Science 1996; 273:1856-62.
177. Garred P, Eugen-Olsen J, Iversen AK et al. Dual effect of CCR5 delta 32 gene deletion in HIV-1-infected patients. Copenhagen AIDS Study Group. Lancet 1997; 349:1884.
178. Huang Y, Paxton WA, Wolinsky SM et al. The role of a mutant CCR5 allele in HIV-1 transmission and disease progression. Nat Med 1996; 2:1240-3.
179. Michael NL, Chang G, Louie LG et al. The role of viral phenotype and CCR-5 gene defects in HIV-1 transmission and disease progression. Nat Med 1997; 3:338-40.
180. Samson M, Libert F, Doranz BJ et al. Resistance to HIV-1 infection in caucasian individuals bearing mutant alleles of the CCR-5 chemokine receptor gene. Nature 1996; 382:722-5.
181. Hütter G, Nowak D, Mossner M et al. Long-term control of HIV by CCR5 Delta32/Delta32 stem-cell transplantation. N Engl J Med 2009; 360:692-8.
182. Hütter G, Thiel E. Allogeneic transplantation of CCR5-deficient progenitor cells in a patient with HIV infection: an update after 3 years and the search for patient no. 2. AIDS 2011; 25:273-4.
183. Huffnagle GB, McNeil LK, McDonald RA et al. Cutting edge: Role of C-C chemokine receptor 5 in organ-specific and innate immunity to Cryptococcus neoformans. J Immunol 1999; 163:4642-6.
184. Regoes RR, Bonhoeffer S. The HIV coreceptor switch: a population dynamical perspective. Trends Microbiol 2005; 13:269-77.
185. Grivel JC, Margolis LB. CCR5- and CXCR4-tropic HIV-1 are equally cytopathic for their T-cell targets in human lymphoid tissue. Nat Med 1999; 5:344-6.
186. Penn ML, Grivel JC, Schramm B et al. CXCR4 utilization is sufficient to trigger CD4+ T cell depletion in HIV-1-infected human lymphoid tissue. Proc Natl Acad Sci USA 1999; 96:663-8.
187. Bleul CC, Wu L, Hoxie JA et al. The HIV coreceptors CXCR4 and CCR5 are differentially expressed and regulated on human T lymphocytes. Proc Natl Acad Sci USA 1997; 94:1925-30.
188. Lee B, Sharron M, Montaner LJ et al. Quantification of CD4, CCR5, and CXCR4 levels on lymphocyte subsets, dendritic cells, and differentially conditioned monocyte-derived macrophages. Proc Natl Acad Sci USA 1999; 96:5215-20.
189. Nagasawa T, Hirota S, Tachibana K et al. Defects of B-cell lymphopoiesis and bone-marrow myelopoiesis in mice lacking the CXC chemokine PBSF/SDF-1. Nature 1996; 382:635-8.
190. Zou YR, Kottmann AH, Kuroda M et al. Function of the chemokine receptor CXCR4 in haematopoiesis and in cerebellar development. Nature 1998; 393:595-9.
191. Hendrix CW, Collier AC, Lederman MM et al. Safety, pharmacokinetics, and antiviral activity of AMD3100, a selective CXCR4 receptor inhibitor, in HIV-1 infection. J Acquir Immune Defic Syndr 2004; 37:1253-62.
192. Calado M, Matoso P, Santos-Costa Q et al. Coreceptor usage by HIV-1 and HIV-2 primary isolates: the relevance of CCR8 chemokine receptor as an alternative coreceptor. Virology 2010; 408:174-82.
193. Deng, HK, Unutmaz D, KewalRamani VN et al. Expression cloning of new receptors used by simian and human immunodeficiency viruses. Nature 1997; 388:296-300.
194. Farzan M, Choe H, Martin K et al. Two orphan seven-transmembrane segment receptors which are expressed in CD4-positive cells support simian immunodeficiency virus infection. J Exp Med 1997; 186:405-11.
195. Blaak H, Boers PH, Gruters RA et al. CCR5, GPR15, and CXCR6 are major coreceptors of human immunodeficiency virus type 2 variants isolated from individuals with and without plasma viremia. J Virol 2005; 79:1686-700.
196. Neil SJ, Aasa-Chapman MM, Clapham PR et al. The promiscuous CC chemokine receptor D6 is a functional coreceptor for primary isolates of human immunodeficiency virus type 1 (HIV-1) and HIV-2 on astrocytes. J Virol 2005; 79:9618-24.
197. Pöhlmann S, Krumbiegel M, Kirchhoff F. Coreceptor usage of BOB/GPR15 and Bonzo/STRL33 by primary isolates of human immunodeficiency virus type 1. J Gen Virol 1999; 80:1241-51.
198. Cilliers T, Willey S, Sullivan WM et al. Use of alternate coreceptors on primary cells by two HIV-1 isolates. Virology 2005; 339:136-44.
199. Pöhlmann S, Stolte N, Munch J et al. Co-receptor usage of BOB/GPR15 in addition to CCR5 has no significant effect on replication of simian immunodeficiency virus in vivo. J Infect Dis 1999; 180:1494-502.
200. Zhang Y, Lou B, Lal RB et al. Use of inhibitors to evaluate coreceptor usage by simian and simian/human immunodeficiency viruses and human immunodeficiency virus type 2 in primary cells. J Virol 2000; 74:6893-910.
201. Zhang YJ, Dragic T, Cao Y et al. Use of coreceptors other than CCR5 by non-syncytium-inducing adult and pediatric isolates of human immunodeficiency virus type 1 is rare in vitro. J Virol 1998; 72:9337-44.

202. Zhang YJ, Moore JP. Will multiple coreceptors need to be targeted by inhibitors of human immunodeficiency virus type 1 entry? J Virol 1999; 73:3443-8.
203. Miyauchi K, Kim Y, Latinovic O et al. HIV enters cells via endocytosis and dynamin-dependent fusion with endosomes. Cell 2009; 137:433-44.
204. Melikyan GB. Common principles and intermediates of viral protein-mediated fusion: the HIV-1 paradigm. Retrovirology 2008; 5:111.
205. Conley AJ, Kessler JA 2nd, Boots LJ et al. Neutralization of divergent human immunodeficiency virus type 1 variants and primary isolates by IAM-41-2F5, an anti-gp41 human monoclonal antibody. Proc Natl Acad Sci USA 1994; 91:3348-52.
206. Stiegler G, Kunert R, Purtscher M et al. A potent cross-clade neutralizing human monoclonal antibody against a novel epitope on gp41 of human immunodeficiency virus type 1. AIDS Res Hum Retroviruses 2001; 17:1757-65.
207. Eckert DM, Kim PS. Mechanisms of viral membrane fusion and its inhibition. Annu Rev Biochem 2001; 70:777-810.
208. Munch J, Standker L, Adermann K et al. Discovery and optimization of a natural HIV-1 entry inhibitor targeting the gp41 fusion peptide. Cell 2007; 129:263-75.
209. Forssmann WG, The YH, Stoll M et al. Short-term monotherapy in HIV-infected patients with a virus entry inhibitor against the gp41 fusion peptide. Sci Transl Med 2010; 2:re63.
210. Furuta RA, Wild CT, Weng Y et al. Capture of an early fusion-active conformation of HIV-1 gp41. Nat Struct Biol 1998; 5:276-9.
211. Gallo SA, Clore GM, Louis JM et al. Temperature-dependent intermediates in HIV-1 envelope glycoprotein-mediated fusion revealed by inhibitors that target N- and C-terminal helical regions of HIV-1 gp41. Biochemistry 2004; 43:8230-3.
212. Kilgore NR, Salzwedel K, Reddick M et al. Direct evidence that C-peptide inhibitors of human immunodeficiency virus type 1 entry bind to the gp41 N-helical domain in receptor-activated viral envelope. J Virol 2003; 77:7669-72.
213. Chan DC, Fass D, Berger JM et al. Core structure of gp41 from the HIV envelope glycoprotein. Cell 1997; 89:263-73.
214. Melikyan GB, Markosyan RM, Hemmati H et al. Evidence that the transition of HIV-1 gp41 into a six-helix bundle, not the bundle configuration, induces membrane fusion. J Cell Biol 2000; 151:413-23.
215. Weissenhorn W, Dessen A, Harrison SC et al. Atomic structure of the ectodomain from HIV-1 gp41. Nature 1997; 387:426-30.
216. Markosyan RM, Cohen FS, Melikyan GB. HIV-1 envelope proteins complete their folding into six-helix bundles immediately after fusion pore formation. Mol Biol Cell 2003; 14:926-38.

CHAPTER 8

CLASS II FUSION PROTEINS

Yorgo Modis

Department of Molecular Biophysics and Biochemistry, Yale University, New Haven, Connecticut, USA
Email: yorgo.modis@yale.edu

Abstract: Enveloped viruses rely on fusion proteins in their envelope to fuse the viral membrane to the host-cell membrane. This key step in viral entry delivers the viral genome into the cytoplasm for replication. Although class II fusion proteins are genetically and structurally unrelated to class I fusion proteins, they use the same physical principles and topology as other fusion proteins to drive membrane fusion. Exposure of a fusion loop first allows it to insert into the host-cell membrane. Conserved hydrophobic residues in the fusion loop act as an anchor, which penetrates only partway into the outer bilayer leaflet of the host-cell membrane. Subsequent folding back of the fusion protein on itself directs the C-terminal viral transmembrane anchor towards the fusion loop. This fold-back forces the host-cell membrane (held by the fusion loop) and the viral membrane (held by the C-terminal transmembrane anchor) against each other, resulting in membrane fusion. In class II fusion proteins, the fold-back is triggered by the reduced pH of an endosome, and is accompanied by the assembly of fusion protein monomers into trimers. The fold-back occurs by domain rearrangement rather than by an extensive refolding of secondary structure, but this domain rearrangement and the assembly of monomers into trimers together bury a large surface area. The energy that is thus released exerts a bending force on the apposed viral and cellular membranes, causing them to bend towards each other and, eventually, to fuse.

INTRODUCTION

Enveloped viruses acquire a lipid bilayer membrane when they bud across the plasma membrane or the membrane of the endoplasmic reticulum (ER) during assembly of the virion.[1,2] During infection, the viral membrane must be fused to the host-cell membrane to deliver the viral genome into the cytoplasm for replication (Fig. 1). The fusion of the viral and host-cell membranes is therefore the central molecular event during the entry of

Viral Entry into Host Cells, edited by Stefan Pöhlmann and Graham Simmons.

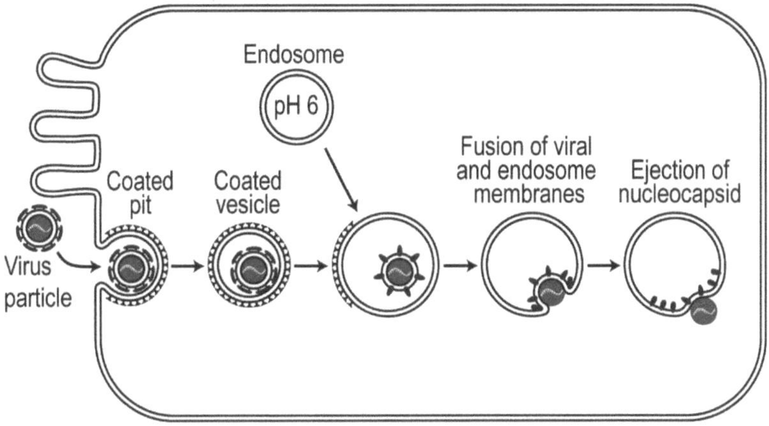

Figure 1. Cell entry of class II enveloped viruses. Virus particles bind target cells through a surface receptor, which is linked to the clathrin-dependent endocytic pathway. Internalized vesicles fuse with endosomal compartments. The acidic pH of these compartments causes conformational rearrangements in the viral envelope proteins that catalyze the fusion of the viral and host-cell membranes. Upon membrane fusion, the viral genome enters the cytoplasm.

enveloped viruses into cells. Adjacent membranes do not fuse spontaneously; membrane fusion requires considerable energy (on the order of 100 kJ mol^{-1} or 40 *kT*).[3,4] Envelope proteins anchored in the viral membrane provide this energy in the form of a conformational rearrangement that bends the apposed membranes towards each other, inducing them to fuse.[5-7] Most 'fusion proteins' (or their cleavage products) also effect cellular attachment of the virus prior to the membrane fusion event by binding to a receptor on the cell surface, except the paramyxo- and alphaviruses, in which a second envelope protein binds the receptor.

Fusion proteins of enveloped viruses fall into two structural classes. The influenza virus haemagglutinin (HA) is the prototype of class I fusion proteins,[8] which encompass those of other orthomyxo- and paramyxoviruses such as measles virus, retroviruses such as human immunodeficiency virus (HIV), filoviruses such as Ebola virus, and coronaviruses such as SARS (see Chapters 4-6). Class II fusion proteins are a structurally and evolutionarily distinct class of proteins found in *Flaviviridae*, such as dengue, yellow fever, and West Nile viruses, and on alphaviruses, such as Semliki Forest and Sindbis viruses. Hepatitis C has a similar genomic organization to the flaviviruses, and therefore most likely relies on a Class II fusion protein as well. Crystal structures of several class I and class II fusion proteins before[9-15] and after[5,16-29] their fusogenic conformational rearrangements have provided us with a detailed molecular understanding of the fusion mechanism (Table 1). The structures show that, despite the absence of similarities in the protein folds of the two classes, fusion proteins from both classes use the same physical principles and general topology to drive membrane fusion. First the fusion protein inserts a hydrophobic fusion anchor partway into the outer bilayer leaflet of the host-cell membrane. The fusion anchor is either an N-terminal peptide,[30] as in influenza and HIV,[31] or an internal loop, as in SARS coronavirus,[32] avian sarcoma leucosis virus[33] and all class II enveloped viruses.[34] Second, the fusion protein folds back on itself, directing the (C-terminal) viral transmembrane anchor towards the fusion anchor. This fold-back forces the host-cell membrane (held by the fusion anchor) and the viral membrane (held by the C-terminal transmembrane

Table 1. Class II fusion protein crystal structures and corresponding electron cryomicroscopy structures

Virus	Fusion Protein	Fusion State	Quaternary Structure in Solution	References
Tick-borne encephalitis	E	Prefusion	Dimer	10
Semliki Forest virus	E1	Prefusion	Monomer	12, 46, 50
Dengue virus type 2	E	Prefusion	Dimer	13, 35, 45, 48
Dengue virus type 2	E	Postfusion	Trimer	29
Semliki Forest virus	E1	Postfusion	Trimer	55
Dengue virus type 3	E	Prefusion	Dimer	14
Tick-borne encephalitis	E	Postfusion	Trimer	56
West Nile virus	E	Prefusion	Monomer	93
Dengue virus type 1	E	Postfusion	Trimer	94
Chikungunya virus	E1	Prefusion	Monomer	95
Sindbis virus	E1	Intermediate	Trimer	96
Dengue virus type 4	E	Prefusion	Dimer	97
Japanese encephalitis virus	E	Prefusion	Monomer	98

anchor) against each other, resulting in fusion of the two membranes. In this chapter, I describe our current picture of how class II fusion proteins drive viral membrane fusion, based on the structural and biochemical data available to date.

OVERALL ARCHITECTURE

Three-dimensional structures of eight class II fusion proteins in their native, or prefusion states,[10,12-14,35,93-95,97] that is, the conformation that they adopt on the surface of a mature virus particle, have been determined at near atomic resolution. Figure 2 shows the three-domain structures of E[13] and E1,[12] the fusion proteins of dengue virus (a representative flavivirus) and of Semliki Forest Virus (an alphavirus), respectively. The two proteins share a common molecular architecture, despite a lack of significant sequence similarity. Domain I, an eight-stranded β-barrel, organizes the structure. Two long insertions between pairs of consecutive β-strands in domain I form the elongated domain II, which bears the fusion anchor, a fusion loop in class II proteins, at its tip (Figs. 2, 4). Domain II contains twelve β-strands and two α-helices. Domain III is an IgC-like module, with ten β-strands. Domain III contains most of the antigenic sites on E, as well as most of the structural determinants of virulence and tropism.[10] This observation, and the widespread occurrence of immunoglobulin modules in cell-adhesion proteins, suggest that domain III participates in attachment to a cellular receptor.[10] Indeed, positively charged patches on the surface of domain III in dengue virus have been suggested to promote attachment by binding heparan sulfate on the cell surface.[36] Both E1 and E have one or more glycosylation sites. These glycans can aid viral attachment to the cell surface, in

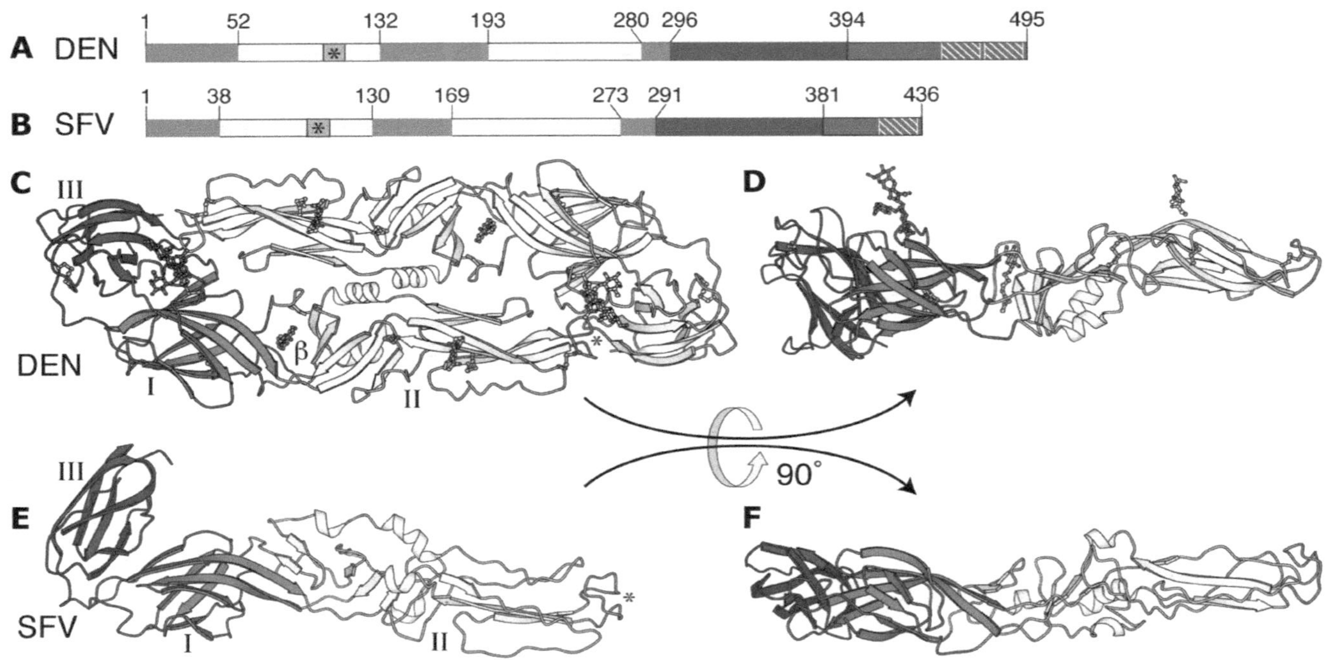

Figure 2. Representative class II fusion protein structures. A) The three domains of the flaviviral fusion protein of dengue virus (DEN), E: Domain I (residues 1-52, 133-193, 281-296), domain II (residues 53-132, 194-280), domain III (residues 297-394). A 53-residue 'stem' links the ectodomains to a two-helix C-terminal transmembrane anchor. B) The domains of an alphaviral fusion protein, Semliki Forest virus (SFV) E1: Domain I (residues 1-38, 131-169, 274-291), domain II (residues 39-130, 170-273), domain III (residues 292-381). A 32-residue 'stem' links the ectodomains to a single-helix C-terminal transmembrane anchor. C) Crystal structure of DEN E in the prefusion conformation,[13] as found in the mature virus particle.[48] The fusion loop in A-C is marked with an asterisk. A second subunit of E, forming the dimer found on the viral surface and in solution, is shown in light gray. D) View rotated 90° relative to C, with the second subunit omitted for clarity. E) Crystal structure of SFV E1 in the prefusion conformation,[12] as found in the mature virus particle.[50] The fusion loop is marked with an asterisk. F) View rotated 90° relative to E.

the case of dengue virus by binding to the lectin DC-SIGN.[37,38] As expected from their sequence identities (≥37%), flaviviral E proteins have very similar overall structures, and differ only in the length and structure of surface-exposed loops, some of which have been implicated in receptor binding.[10,39,40] Despite these hints on the basis of cellular attachment, however, a cellular receptor that specifically recognizes an envelope protein on a class II enveloped virus has yet to be conclusively identified, although candidate receptors for dengue virus type 1[41] and West Nile virus[42] were recently suggested.

It is important to note that all the crystal structures of fusion proteins determined so far, from both classes and regardless of their conformational state, lack the C-terminal viral membrane anchor. This anchor consists of one or two transmembrane helices, and has been intentionally omitted in constructs targeted for crystallization to facilitate expression and handling, and to promote crystallization. The crystallized species are therefore referred to as soluble fragments of the ectodomains of the full-length fusion protein. Furthermore, all available crystal structures of class II fusion proteins lack the 'stem' region,[43] a 30-55 amino acid linker between Domain III and the C-terminal transmembrane anchor (Figs. 2A-B, 3). As I will discuss below, the stem region plays a key role in the final stages of membrane fusion. Its function is analogous to that of the 'outer helix' in class I fusion proteins.[8]

MATURATION AND PRIMING

Both class I and class II fusion proteins rely on a proteolytic cleavage event to become primed to respond to the environmental conditions appropriate for fusion. These conditions are usually the acidic pH of an endosome (Fig. 1), but for some class I enveloped viruses, such as HIV, coreceptor binding is required instead. In contrast to class I fusion proteins, however, class II fusion proteins rely on a priming proteolytic cleavage that does not cleave the fusion protein itself. Instead, class II proteins associate with a second, 'protector' protein, called M (for membrane protein) in flaviviruses or E2 in alphaviruses. The protector protein is cleaved by furin when immature virus particles assembled in the ER reach the trans-Golgi network.[44] The cleavage produces mature virus particles, which are then released from the host cell by exocytosis. The cleavage of the protector protein releases a conformational constraint on the fusion protein, which allows it to adopt its mature conformation (described above) in a large rearrangement on the viral surface. In the mature conformation, the fusion protein is primed to respond to acidic pH and induce membrane fusion with a further conformational rearrangement (described below).

Structures from electron cryomicroscopy of both immature[45,46] and mature[47-50] flavivirus and alphavirus particles, provide a detailed picture of the rearrangement that accompanies maturation in these viruses. Alphaviruses retain the T = 4 icosahedral packing of their envelope proteins, but domains that form spikes on the immature virion swing in towards the threefold symmetry axis, during maturation.[46,50] The rearrangement is more dramatic in flaviviruses, in which the fusion protein E breaks the T = 3 icosahedral symmetry of the immature virion[45] to adopt an unusual icosahedral herringbone pattern in the mature virion.[47,51] In both alphaviruses and flaviviruses, the fusion proteins form dimers in the mature virion albeit in different configurations.[10,12] The key feature of the maturation process in both genera, however, is that cleavage of the protector protein allows the fusion loop to reposition itself so that it is poised to insert into the host-cell membrane in response to acidification of the solute in the endosome. Mature virus particles are therefore infectious,[44,52] unlike immature virions,[53,54] which are insensitive to pH. The fusion loop is shielded from the viral surface in mature virions by E-E dimer contacts in flaviviruses, or by protein E2 in alphaviruses (Figs. 3A,B, 5A).

THE FUSOGENIC CONFORMATIONAL REARRANGEMENT

The three-dimensional structures of four class II fusion proteins in their postfusion states[29,55,56,94] reveal striking differences from the prefusion forms (Fig. 3), and suggest a

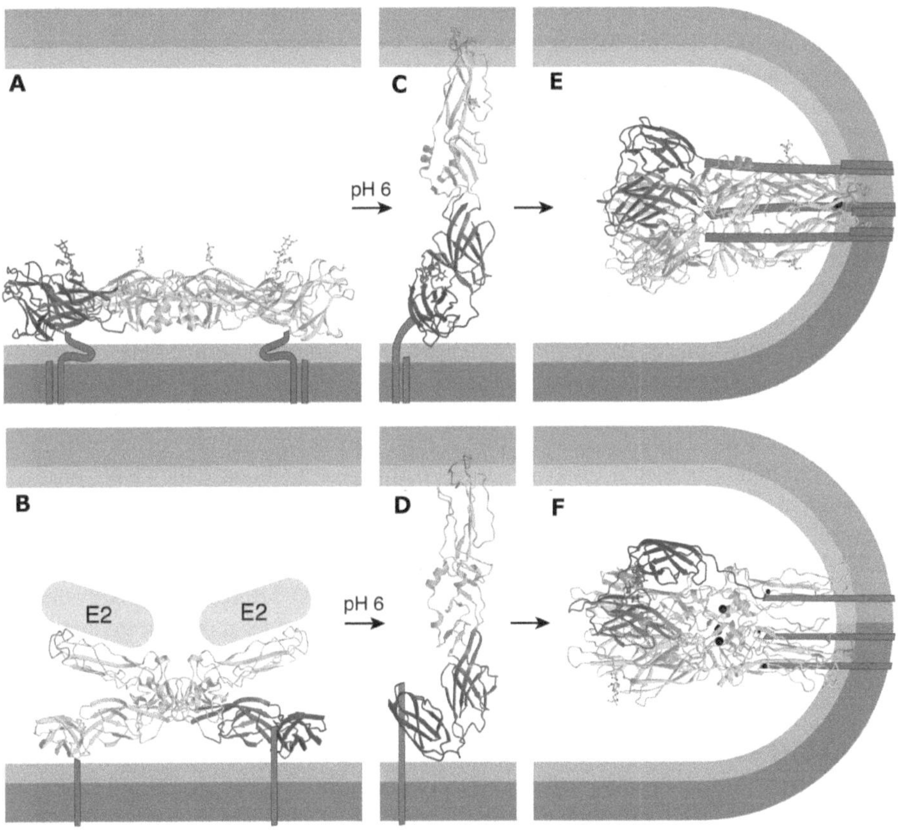

Figure 3. Pre- and postfusion structures of class II fusion proteins, and proposed intermediates. A) A dimer of DEN E,[13] and B) two SFV E1[12] molecules in the prefusion conformation as found on the viral surface, viewed perpendicular to the viral membrane. The fusion loop is buried, either in the dimer interface (A), or under E2 (B). The outer (proximal) bilayer leaflets of the cellular and viral membranes are shown to scale as solid rectangles. The thin outer layer within each leaflet represents the polar headgroup layer, and the thicker inner layer represents the hydrocarbon layer. The stem-anchor segments are absent from the crystal structure, but are represented here schematically as rods in the viral membrane. C,D) Upon acidification of the solute in the endosome, the domain II rotates 15-30° about a hinge in the domain II-domain I interface. This exposes the fusion loop, which then inserts into the host cell membrane. The postfusion, trimeric structures of DEN E[29] (E) and SFV E1 (F).[55] After insertion of their fusion loops into the target membrane, the fusion proteins form trimers and fold back on themselves, bringing the fusion loops close to the C-terminal transmembrane anchors.

molecular mechanism for membrane fusion (see below and Fig. 5). Like class I fusion proteins, flaviviral E proteins and alphaviral E1 proteins are both homotrimers in their postfusion conformations. Class II proteins form trimers from monomers on the viral surface, while class I proteins are trimeric in their prefusion state.[8] However, a comparison of the pre- and postfusion states of influenza HA-the only example in its class where both structures are known for the same protein-shows that, as in class II fusion proteins, nearly all of the trimer contacts in the postfusion state are formed during the fusogenic conformational rearrangement.

Unlike influenza HA, which undergoes extensive refolding during membrane fusion, the three domains of class II fusion proteins retain most of their folded structures (Fig. 3). Instead, the domains undergo major rearrangements in their relative orientations, through flexion of the interdomain linkers. Domain III undergoes the most significant displacement in the fusion transition. It rotates by about 70°, and its center of mass shifts by 30-40 Å towards domain II. This folding-over brings the C-terminus of domain III about 40 Å closer to the fusion loop, at tip of domain II (Fig. 3). Domain II rotates 15-30° with respect to domain I about a hinge region[13] in which mutations affect the pH threshold of fusion in various flaviviruses.[57-62] These conformational rearrangements position the end of domain III-and the beginning of the stem region that links domain III to the C-terminal viral transmembrane anchor-towards the fusion loop (Fig. 3E-F).[29,55] A deep channel extends from the C-terminus of the crystallized fragment along the intersubunit contact between domains II to the fusion loops, in both the dengue and Semliki Forest virus postfusion trimer structures. In the full-length fusion proteins, it is thought that the stem binds in this channel in an extended, but mainly α-helical conformation.[29,48] This proposed stem conformation places the viral transmembrane anchor in the immediate vicinity of the fusion loop, just as in the postfusion conformation of class I viral fusion proteins.

The fusion transition in class II viral proteins is irreversible. The refoldings just described may impart irreversibility by contributing a high barrier to initiation of trimerization and an even higher barrier to dissociation of postfusion trimers once they have formed. Moreover, many new polar and nonpolar contacts are formed during the fusion transition, in several different areas along the threefold axis of the trimer. The total surface buried is 13,000-15,000 Å[2,29,55] nearly four times more than is buried in the prefusion dimer. The stem, which is missing from currently available crystal structures, most likely forms additional contacts with the core trimer structure. The stem does indeed promote trimer assembly even in the absence of liposomes.[43]

THE FUSION LOOP

The process of viral membrane fusion in both class I and class II enveloped viruses begins with the exposure of a fusion anchor, and its subsequent insertion into the host-cell membrane. Fusion anchors from both viral classes vary in length but are in general rich in glycines and hydrophobic residues, particularly aromatic residues such as Trp or Phe. Sequence conservation is poor between fusion proteins of both classes. The fusion anchor in class I fusion proteins-the 'fusion peptide'-is a region of approximately 20 residues at or near the N-terminus of the envelope protein. The crystal structure of the parainfluenza virus 5 fusion (F) protein in its prefusion form reveals the fusion peptide wedged between two subunits of the protein, in a partly extended, partly β-sheet and partly α-helical conformation.[15] Structural studies on influenza HA in its postfusion conformation using NMR and other spectroscopic techniques show that the fusion peptide is mostly α-helical

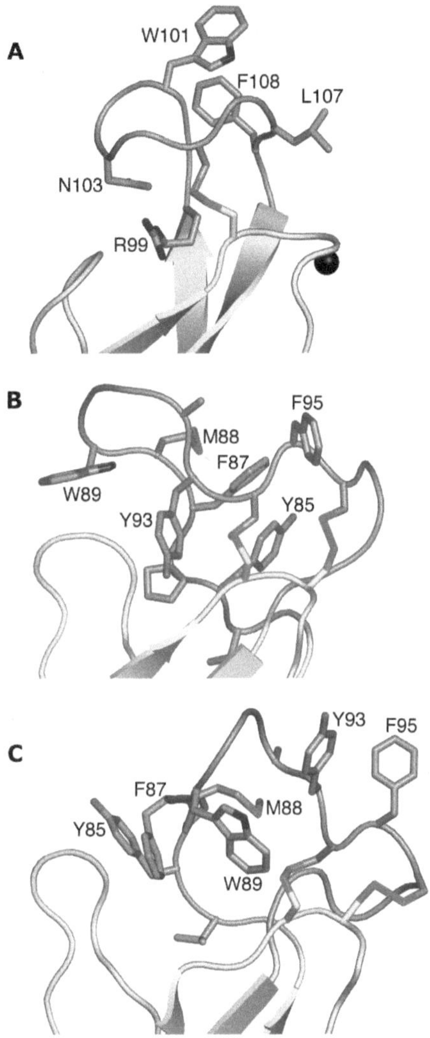

Figure 4. Close-up of the aromatic anchor formed by the fusion loop in: A) dengue virus E, and B,C) SFV E1. In flaviviruses (A), three clustered fusion loops form a nonpolar, bowl-shaped apex, with three residues (Trp, Phe and Leu) protruding at the tip of domain II. These three residues insert into the hydrocarbon layer of the target cell membrane. The fusion loop has a rigid structure. In alphaviruses, the fusion loop is flexible and can adopt very different conformations. Two conformations of the fusion loop from different subunits of the SFV E1 postfusion crystal structure are shown in B and C. The alphaviral fusion loops do not appear to cluster around the threefold axis of the trimer.

in character and that its structure changes only subtly as it inserts partway into the outer leaflet of the host-cell lipid bilayer.[63,64] None of the currently available postfusion class I protein crystal structures contain information on the fusion peptide.

The recently determined crystal structures of class II fusion proteins in pre[10,12-14] and postfusion[29,55,56] conformations offer the first direct views of fusion anchors-in this case, the fusion loops-as they insert into a target membrane (Fig. 4). Like the class I fusion peptide, the class II fusion loop penetrates only partway into the hydrocarbon layer of the target membrane. Exposed carbonyls and charged residues prevent the fusion loop from penetrating further than 6 Å.[29,56] In flaviviruses, the fusion loop adopts a tightly folded conformation, which is stabilized by a disulfide bond (Fig. 4A). The structure of the fusion loop is essentially identical in the pre- and postfusion conformations of the protein, suggesting that membrane insertion has no effect on the structure of the fusion loop. During the fusion transition, three hydrophobic residues in the fusion loop (Trp, Leu, and Phe) become exposed on the molecular surface. Three fusion loops end up in close proximity at the tip of the trimer in the postfusion conformation, where they form a crater-like surface with a hydrophobic rim (Fig. 3E). Electron cryomicroscopy[29] and mutagenesis studies[34] confirm that these hydrophobic, mostly aromatic residues on the crater rim insert into the host-cell membrane, acting as an 'aromatic anchor' for the fusion protein. The concave shape of the crater is thought to be important in generating distortions or perturbations in the host-cell membrane,[29] which are required for fusion.[65]

In alphaviruses, the fusion loop is also rich in aromatic and other hydrophobic residues. Unlike flaviviral fusion loops, however, alphaviviral fusion loops do not form trimer contacts (Fig. 3F). Indeed, in the postfusion structure of the Semliki Forest virus E1 trimer, the fusion loops have high temperature factors and exhibit a high degree of flexibility despite the presence of two disulfide bonds. Thus, the structures of the fusion loops are poorly defined, but each fusion loop seems to adopt a very different conformation (Fig. 4B,C).[55] The fusion loops in the postfusion Semliki Forest virus E1 structure form quite polar surfaces, with many mainchain carbonyls and some polar or charged sidechains exposed on the surface. This suggests that, in contrast to flaviviral fusion loops, alphaviral fusion loops either change their conformation upon membrane insertion to shield polar groups from the membrane, or the fusion loops only interact with the polar headgroups of the lipids, and do not penetrate into the hydrocarbon layer.

Semliki Forest virus E1 trimers form irregular clusters, or 'rosettes' of about 40-60 trimers through contacts between fusion loops in adjacent trimers.[55] This is reminiscent of influenza virus HA, which aggregates into rosettes through interactions between the fusion peptide, at low pH and after proteolytic activation.[66] This fusion loop/peptide clustering may provide a mechanism for the direct coupling of several E1/HA trimers to work in concert around a single fusion site (see below).

MECHANISM OF MEMBRANE FUSION

Combined with previous knowledge, the structures of the fusion proteins from class II viruses in their postfusion states[29,55,56] have led to a much better understanding of how conformational changes in the proteins drive membrane fusion. The structures confirm two major principles of membrane fusion machineries: (1) the fusion protein must insert an anchor into each of the two membranes to be fused, and (2) the protein folds back on itself in a thermodynamically favorable conformational rearrangement that drives membrane fusion by forcing the two anchors into close proximity.

In the current model, viral membrane fusion proceeds as follows (Fig. 5). First, receptor binding by an envelope protein, which in flaviviruses is also the fusion protein, leads to clathrin-mediated endocytosis of the virus (Figs. 1, 5A). When the virus reaches endosomal compartments the low pH of the lumen (pH 6) causes an initial conformational rearrangement that leads to the exposure of the previously buried fusion loop[48,50] at the

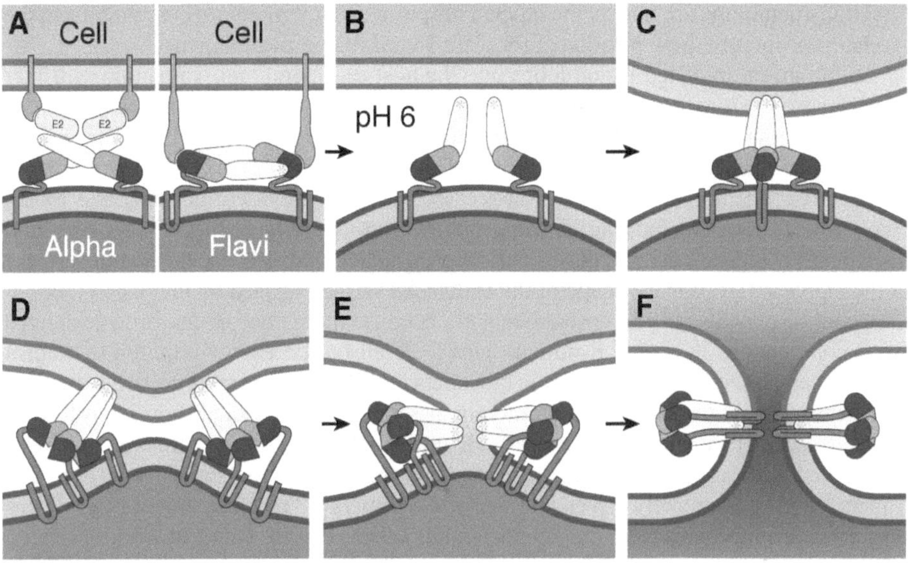

Figure 5. Proposed fusion mechanism for fusion mediated by class II fusion proteins. A) The virus binds to a receptor on the cell surface. In flaviviruses, the fusion protein E binds the receptor, while in alphaviruses, the 'protector' protein E2 binds the receptor. Following attachment, the virus is internalized to an endosome. B) Acidic pH in the endosome causes domain II to hinge outward from the virion surface, exposing the fusion loop, and allowing E monomers to rearrange laterally in the plane of the membrane. C) The fusion loop inserts into the hydrocarbon layer of the host-cell membrane, promoting trimer formation. D) Formation of trimer contacts spreads from the fusion loop at the tip of the trimer, to the base of the trimer. The protein folds back on itself, directing the fusion loop towards the C-terminal transmembrane anchor. Energy release by this refolding bends the apposed membranes. E) Creation of additional trimer contacts between the stem-anchor and domain II leads first to hemifusion and then (F) to formation of a lipidic fusion pore.

tip of domain II. In flaviviruses, domains I and II flex relative to each other by 30°.[29] This hinge motion causes domain II, and therefore the fusion loop, to swing away from the viral surface and towards the host-cell membrane (Fig. 5B). Indeed, mutations at the domain I-domain II interface in various flaviviruses alter the pH threshold of fusion.[13,57-62] As domain II swings away from the viral surface, constraints imposed by the tight packing of E on the viral surface are released, allowing E to diffuse freely in the plane of the viral membrane. The stem may also be able to extend away from the membrane at this stage. In alphaviruses, constraints are released in response to low pH by the dissociation of the protector (and receptor-binding) protein E2. This exposes the fusion loop and allows domain II of E1 to swing towards the nearest threefold symmetry axis in the virus particle in a 15° hinge motion relative to domain I, leading to the formation of trimer contacts with adjacent E1 molecules.[46]

The second key step in the fusion process is insertion of the exposed fusion loop into the host-cell membrane (Fig. 5C). Alphaviral E1 has already formed some trimer contacts at this stage, but flaviviral E proteins probably insert their fusion loops as monomers. Membrane insertion probably catalyzes trimerization of the fusion loops,[67] by lateral rearrangement of E monomers. This trimeric prefusion intermediate (Fig. 5C) bridges host-cell and viral membranes, anchored by its fusion loops in the former and by the viral transmembrane anchors in the latter. This proposed intermediate is analogous to the 'prehairpin' intermediate postulated for class I viral fusion mechanisms.[68]

Upon insertion of the fusion loops into the host-cell membrane, formation of trimer contacts spreads from the fusion loops at the trimer tip to domain I at the trimer base. Domain II shifts and rotates, folding the stem and C-terminal anchor back towards the fusion loop (Fig. 5D), and burying additional protein surfaces. Free energy released by this refolding drives the two membranes to bend towards each other,[5-7] as the C-terminal anchor is forced closer to the fusion loop, forming apposing nipples in the membranes (Fig. 5D).[3] Fusion-loop insertion may induce positive bilayer curvature, which would stabilize the lateral surfaces of the nipples. The concave shape of the crater-like surface formed by the fusion loops at the trimer tip may also have a destabilizing effect on the membrane, as has been postulated for fusion peptides in class I fusion proteins.[65] Based on the energy required to deform lipid bilayers, it seems likely that a ring of trimers refolding in concert is needed to provide sufficient energy to form nipples in the membranes.[3,4] It is unclear exactly how many trimers are needed to drive membrane fusion in class II viruses, nor how their conformational changes are coupled. In the case of influenza, fusion requires the concerted action of at least three HA trimers,[69] and is more likely driven by rings of 6-8 trimers.[70] The clustering of fusion loops may provide a mechanism for the direct coupling of several E1 trimers to work in concert around a single fusion site in alphaviruses, but such clustering has not been observed in flaviviruses. It is possible that coupling occurs via the membrane: only when several trimers fold back in concert can they overcome the resistance of the membrane to deformation and reach their final, most stable postfusion conformation (Figs. 5D-F).

As the fusion transition proceeds, the stem zippers up onto the core of the trimer, along a channel that spans domain II, at the intersubunit contact regions (Figs. 3, 5D-F). The zippering up of the stem onto the domain II forces the fusion loop and the viral transmembrane anchor closer and closer, until the proximal leaflets of the two membranes fuse to form a 'hemifusion stalk' (Fig. 5E). Hemifusion is thought to be an essential intermediate of membrane fusion.[3,4,71] (Fig. 5E) illustrates the need for shallow penetration of the viral fusion anchor into the host-cell membrane: assuming several trimers do in fact act in concert around a single fusion site, fusion anchors from different trimers would collide if they inserted beyond the outer (proximal) lipid bilayer leaflet. This constraint on the length of the fusion anchor holds true for both class I fusion peptides and class II fusion loops.

Hemifusion stalks can 'flicker' open into narrow fusion pores.[71] In order to prevent the transient fusion pores from closing, the stem must complete its zippering up onto the core of the trimer, and the C-terminal transmembrane anchor must migrate into the pore (Fig. 5F). Indeed, the transition from hemifusion stalk to full fusion pore appears to require that the viral transmembrane anchor span the membrane completely, in all biological membrane fusion systems. Thus, the replacement of the C-terminal transmembrane anchor of influenza HA with a glycosylphophatidylinositol (GPI) lipid anchor,[72-74] or with a half-length protein α-helical anchor,[75] stalls the fusion reaction at the stage of hemifusion. Other viral fusion proteins and cellular SNARE fusion proteins also require at least one transmembrane anchor.[76-83] Upon completion of fusion, the trimer has reached the conformation seen in the postfusion crystal structures.[29,55,56] The stems (not present in the structures) are docked along the surface of domains II, and the fusion loops and transmembrane anchors lie next to each other in the fused membrane (Fig. 5F).

Some class II fusion proteins, including those of alphaviruses, can only fuse membranes containing cholesterol and sphingolipids.[84] The structural basis for this requirement is still not well understood. Several mutations in different regions of the Semliki Forest virus fusion protein E1 lower its dependence on cholesterol and/or sphingolipids for membrane fusion.[85,86] It is unclear, however, whether the lower dependence on cholesterol of these mutants is due to an apparent destabilization of the E1 homotrimer,[87] or to the different physical properties of membranes lacking cholesterol and sphingolipids. In flaviviruses, cholesterol facilitates fusion, but neither cholesterol nor sphingolipids are essential for fusion.[88]

STRATEGIES FOR FUSION INHIBITION

Many class II viruses, especially the flaviviruses, represent important human pathogens such as dengue, hepatitis C, yellow fever, West Nile, Japanese encephalitis and tick-borne encephalitis viruses.[89] For most of these viruses, there are no specific treatments for infection, their control by vaccination has proved elusive,[89] and the number of infections is on the rise. Recently determined three-dimensional structures of class II fusion proteins suggest new strategies for inhibiting viral entry by blocking membrane fusion. One such strategy stems for the discovery in dengue virus E of a long, tapering channel lined with hydrophobic side chains.[13] In the crystal structure, the channel is occupied by a molecule of the detergent n-octyl-β-D-glucoside. In the absence of detergent, a β-hairpin covering the channel swings towards the protein, and closes up the channel.[13] The location of this 'ligand-binding pocket' at the domain I-domain II interface coincides with that of mutations affecting the pH threshold of fusion in various flaviviruses.[57-62] Most of these mutations involve side chains lining the ligand-binding pocket. The postfusion structure of dengue virus E shows that this region acts as a hinge between domains I and II during the fusogenic conformational rearrangement (see above).[29] The opening up of a ligand-binding pocket just at the locus of a hinge suggests that compounds tightly inserted at this position might hinder the conformational changes required for membrane fusion (Fig. 6A). The mechanism of action of such compounds might resemble that of some of the well-studied antipicornaviral compounds, which block a concerted structural transition in the icosahedral assembly.[90] Alternatively, small molecules that pry open the β-hairpin on binding in the pocket may inhibit infection by facilitating the low-pH conformational change, causing premature triggering. Knowledge of the structure of the binding pocket with a bound ligand will guide efforts to design derivative ligands with higher affinities for use as inhibitors of flaviviral membrane fusion.

The postfusion structures of dengue[29] and Semliki Forest[55] viruses suggest a second possible strategy for fusion inhibition, related to an approach successful in developing an HIV antiviral compound.[91] Peptides corresponding to the stem region of the gp41 fusion protein inhibit HIV entry by binding to the trimeric, N-terminal 'inner core' of the protein and interfering with the folding back against it of the stem and C-terminal viral transmembrane anchor. The way in which the stem is likely to fold back in class II viral fusion proteins (Figs. 3, 5D-F) suggests that an analogous strategy may be successful with class II viruses. Peptides derived from stem sequences could block completion of the fusogenic conformational change, by competing with the stem for interaction with surfaces on domain II, at the trimer interface (Fig. 6B). Stem-like peptides or peptidomimetic compounds could thus inhibit viral membrane fusion in class II enveloped viruses by preventing the final folding back of the fusion protein that is required to drive the viral and host-cell membranes together to fuse.

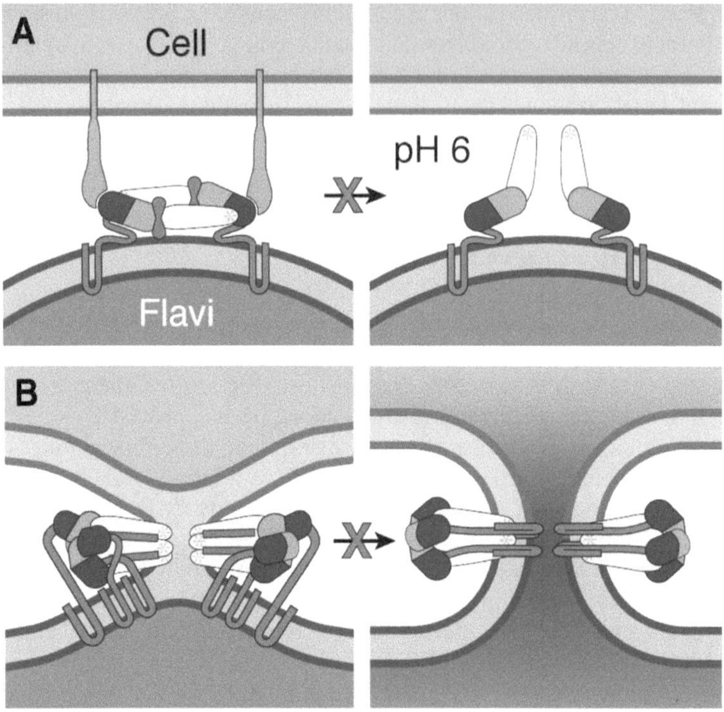

Figure 6. Fusion inhibition strategies. A) The discovery of a ligand-binding pocket at the interface between domains I and II in dengue virus E,[13] just at the locus of a hinge motion required for fusion, suggests that compounds inserted in the pocket might hinder the hinge motion and hence inhibit the fusion transition. This approach would block the first step in the fusion mechanism (Fig. 5A-B). B) Peptides corresponding to the stem region of the fusion protein may inhibit viral entry by binding to the trimeric core of the protein in its postfusion conformation,[29,55] and interfering with the folding back against it of the fusion protein's own stem. An analogous strategy has been successful with HIV gp41.[91,92] This approach would block the last step in the fusion mechanism (Fig. 5E-F).

CONCLUSION

All viral membrane fusion proteins use the same physical principles and topology to drive membrane fusion. Class II fusion proteins are structurally and evolutionarily distinct class of proteins found in Flaviviridae, such as dengue, yellow fever, and West Nile viruses, and on alphaviruses, such as Semliki Forest and Sindbis viruses. Unlike class I fusion proteins such as influenza HA, which undergoes extensive refolding during membrane fusion, the three domains of class II fusion proteins retain most of their folded structures. Instead, the domains undergo major rearrangements in their relative orientations, through flexion of the interdomain linkers. Class II fusion proteins rely on a hydrophobic fusion loop to anchor themselves in the target cellular membrane. Like the class I fusion peptide, the class II fusion loop penetrates only partway into the hydrocarbon layer of the target membrane. Class II fusion proteins drive membrane fusion in a foldback rearrangement of a trimeric protein assembly. Crystal structures of class II envelope proteins have suggested two specific strategies for fusion inhibition, with hydrophobic small molecules and "stem"-like peptides or peptidomimetics, respectively.

NOTE ADDED AFTER PROOFS

This chapter was originally written in 2006 and was updated in 2012 to add recent relevant advances.

REFERENCES

1. Lindenbach BD, Rice CM. Flaviviridae: The viruses and their replication. In: Knipe DM, Howley PM, eds. Fields Virology. 4th ed. Philadelphia: Lippincott Williams and Wilkins, 2001:991-1041.
2. Schlesinger S, Schlesinger MJ. Togaviridae: The viruses and their replication. In: Knipe DM, Howley PM, eds. Fields Virology. 4th ed. Philadelphia: Lippincott Williams and Wilkins, 2001:895-916.
3. Kuzmin PI, Zimmerberg J, Chizmadzhev YA et al. A quantitative model for membrane fusion based on low-energy intermediates. Proc Natl Acad Sci USA 2001; 98(13):7235-7240.
4. Kozlov MM, Chernomordik LV. A mechanism of protein-mediated fusion: Coupling between refolding of the influenza hemagglutinin and lipid rearrangements. Biophys J 1998; 75(3):1384-1396.
5. Baker KA, Dutch RE, Lamb RA et al. Structural basis for paramyxovirus-mediated membrane fusion. Mol Cell 1999; 3(3):309-319.
6. Melikyan GB, Markosyan RM, Hemmati H et al. Evidence that the transition of HIV-1 gp41 into a six-helix bundle, not the bundle configuration, induces membrane fusion. J Cell Biol 2000; 151(2):413-423.
7. Russell CJ, Jardetzky TS, Lamb RA. Membrane fusion machines of paramyxoviruses: Capture of intermediates of fusion. EMBO J 2001; 20(15):4024-4034.
8. Skehel JJ, Wiley DC. Receptor binding and membrane fusion in virus entry: The influenza hemagglutinin. Annu Rev Biochem 2000; 69:531-569.
9. Wilson IA, Skehel JJ, Wiley DC. Structure of the haemagglutinin membrane glycoprotein of influenza virus at 3 A resolution. Nature 1981; 289(5796):366-373.
10. Rey FA, Heinz FX, Mandl C et al. The envelope glycoprotein from tick-borne encephalitis virus at 2 A resolution. Nature 1995; 375(6529):291-298.
11. Rosenthal PB, Zhang X, Formanowski F et al. Structure of the haemagglutinin-esterase-fusion glycoprotein of influenza C virus. Nature 1998; 396(6706):92-96.
12. Lescar J, Roussel A, Wien MW et al. The Fusion glycoprotein shell of Semliki Forest virus: An icosahedral assembly primed for fusogenic activation at endosomal pH. Cell 2001; 105(1):137-148.
13. Modis Y, Harrison SC. A ligand-binding pocket in the dengue virus envelope glycoprotein. Proc Natl Acad Sci USA 2003; 100:6986-6991.
14. Modis Y, Ogata S, Clements D et al. Variable surface epitopes in the crystal structure of dengue virus type 3 envelope glycoprotein. J Virol 2005; 79(2):1223-1231.
15. Yin HS, Wen X, Paterson RG et al. Structure of the parainfluenza virus 5 F protein in its metastable, prefusion conformation. Nature 2006; 439(7072):38-44.
16. Bullough PA, Hughson FM, Skehel JJ et al. Structure of influenza haemagglutinin at the pH of membrane fusion. Nature 1994; 371(6492):37-43.
17. Fass D, Harrison SC, Kim PS. Retrovirus envelope domain at 1.7 angstrom resolution. Nat Struct Biol 1996; 3(5):465-469.
18. Chan DC, Fass D, Berger JM et al. Core structure of gp41 from the HIV envelope glycoprotein. Cell 1997; 89(2):263-273.
19. Tan K, Liu J, Wang J et al. Atomic structure of a thermostable subdomain of HIV-1 gp41. Proc Natl Acad Sci USA 1997; 94(23):12303-12308.
20. Weissenhorn W, Dessen A, Harrison SC et al. Atomic structure of the ectodomain from HIV-1 gp41. Nature 1997; 387(6631):426-430.
21. Malashkevich VN, Chan DC, Chutkowski CT et al. Crystal structure of the simian immunodeficiency virus (SIV) gp41 core: Conserved helical interactions underlie the broad inhibitory activity of gp41 peptides. Proc Natl Acad Sci USA 1998; 95(16):9134-9139.
22. Caffrey M, Cai M, Kaufman J et al. Three-dimensional solution structure of the 44 kDa ectodomain of SIV gp41. EMBO J 1998; 17(16):4572-4584.
23. Weissenhorn W, Carfi A, Lee KH et al. Crystal structure of the Ebola virus membrane fusion subunit, GP2, from the envelope glycoprotein ectodomain. Mol Cell 1998; 2(5):605-616.
24. Kobe B, Center RJ, Kemp BE et al. Crystal structure of human T cell leukemia virus type 1 gp21 ectodomain crystallized as a maltose-binding protein chimera reveals structural evolution of retroviral transmembrane proteins. Proc Natl Acad Sci USA 1999; 96(8):4319-4324.

25. Chen J, Skehel JJ, Wiley DC. N- and C-terminal residues combine in the fusion-pH influenza hemagglutinin HA(2) subunit to form an N cap that terminates the triple-stranded coiled coil. Proc Natl Acad Sci USA 1999; 96(16):8967-8972.
26. Zhao X, Singh M, Malashkevich VN et al. Structural characterization of the human respiratory syncytial virus fusion protein core. Proc Natl Acad Sci USA 2000; 97(26):14172-14177.
27. Xu Y, Lou Z, Liu Y et al. Crystal structure of severe acute respiratory syndrome coronavirus spike protein fusion core. J Biol Chem 2004; 279(47):49414-49419.
28. Xu Y, Liu Y, Lou Z et al. Structural basis for coronavirus-mediated membrane fusion. Crystal structure of mouse hepatitis virus spike protein fusion core. J Biol Chem 2004; 279(29):30514-30522.
29. Modis Y, Ogata S, Clements D et al. Structure of the dengue virus envelope protein after membrane fusion. Nature 2004; 427(6972):313-319.
30. Gething MJ, White JM, Waterfield MD. Purification of the fusion protein of Sendai virus: Analysis of the NH2-terminal sequence generated during precursor activation. Proc Natl Acad Sci USA 1978; 75(6):2737-2740.
31. Gallaher WR. Detection of a fusion peptide sequence in the transmembrane protein of human immunodeficiency virus. Cell 1987; 50(3):327-328.
32. Supekar VM, Bruckmann C, Ingallinella P et al. Structure of a proteolytically resistant core from the severe acute respiratory syndrome coronavirus S2 fusion protein. Proc Natl Acad Sci USA 2004; 101(52):17958-17963.
33. Cheng SF, Wu CW, Kantchev EA et al. Structure and membrane interaction of the internal fusion peptide of avian sarcoma leukosis virus. Eur J Biochem 2004; 271(23-24):4725-4736.
34. Allison SL, Schalich J, Stiasny K et al. Mutational evidence for an internal fusion peptide in flavivirus envelope protein E. J Virol 2001; 75(9):4268-4275.
35. Zhang Y, Zhang W, Ogata S et al. Conformational changes of the flavivirus E glycoprotein. Structure (Camb) 2004; 12(9):1607-1618.
36. Chen Y, Maguire T, Hileman RE et al. Dengue virus infectivity depends on envelope protein binding to target cell heparan sulfate. Nat Med 1997; 3(8):866-871.
37. Navarro-Sanchez E, Altmeyer R, Amara A et al. Dendritic-cell-specific ICAM3-grabbing nonintegrin is essential for the productive infection of human dendritic cells by mosquito-cell-derived dengue viruses. EMBO Rep 2003; 4(7 Suppl):1-6.
38. Tassaneetrithep B, Burgess TH, Granelli-Piperno A et al. DC-SIGN (CD209) mediates dengue virus infection of human dendritic cells. J Exp Med 2003; 197(7):823-829.
39. Crill WD, Roehrig JT. Monoclonal antibodies that bind to domain III of dengue virus E glycoprotein are the most efficient blockers of virus adsorption to Vero cells. J Virol 2001; 75(16):7769-7773.
40. Hung JJ, Hsieh MT, Young MJ et al. An external loop region of domain III of dengue virus type 2 envelope protein is involved in serotype-specific binding to mosquito but not mammalian cells. J Virol 2004; 78(1):378-388.
41. Thepparit C, Smith DR. Serotype-specific entry of dengue virus into liver cells: Identification of the 37-kilodalton/67-kilodalton high-affinity laminin receptor as a dengue virus serotype 1 receptor. J Virol 2004; 78(22):12647-12656.
42. Chu JJ, Ng ML. Interaction of West Nile virus with alpha v beta 3 integrin mediates virus entry into cells. J Biol Chem 2004; 279(52):54533-54541.
43. Allison SL, Stiasny K, Stadler K et al. Mapping of functional elements in the stem-anchor region of tick-borne encephalitis virus envelope protein E. J Virol 1999; 73(7):5605-5612.
44. Stadler K, Allison SL, Schalich J et al. Proteolytic activation of tick-borne encephalitis virus by furin. J Virol 1997; 71(11):8475-8481.
45. Zhang Y, Corver J, Chipman PR et al. Structures of immature flavivirus particles. EMBO J 2003; 22(11):2604-2613.
46. Ferlenghi I, Gowen B, de Haas F et al. The first step: Activation of the Semliki Forest virus spike protein precursor causes a localized conformational change in the trimeric spike. J Mol Biol 1998; 283(1):71-81.
47. Kuhn RJ, Zhang W, Rossmann MG et al. Structure of dengue virus: Implications for flavivirus organization, maturation, and fusion. Cell 2002; 108(5):717-725.
48. Zhang W, Chipman PR, Corver J et al. Visualization of membrane protein domains by cryo-electron microscopy of dengue virus. Nat Struct Biol 2003; 10(11):907-912.
49. Mukhopadhyay S, Kim BS, Chipman PR et al. Structure of West Nile virus. Science 2003; 302(5643):248.
50. Mancini EJ, Clarke M, Gowen BE et al. Cryo-electron microscopy reveals the functional organization of an enveloped virus, Semliki Forest virus. Mol Cell 2000; 5(2):255-266.
51. Zhang W, Mukhopadhyay S, Pletnev SV et al. Placement of the structural proteins in Sindbis virus. J Virol 2002; 76(22):11645-11658.
52. Elshuber S, Allison SL, Heinz FX et al. Cleavage of protein prM is necessary for infection of BHK-21 cells by tick-borne encephalitis virus. J Gen Virol 2003; 84(Pt 1):183-191.

53. Guirakhoo F, Heinz FX, Mandl CW et al. Fusion activity of flaviviruses: Comparison of mature and immature (prM-containing) tick-borne encephalitis virions. J Gen Virol 1991; 72(Pt 6):1323-1329.
54. Guirakhoo F, Bolin RA, Roehrig JT. The Murray Valley encephalitis virus prM protein confers acid resistance to virus particles and alters the expression of epitopes within the R2 domain of E glycoprotein. Virology 1992; 191(2):921-931.
55. Gibbons DL, Vaney MC, Roussel A et al. Conformational change and protein-protein interactions of the fusion protein of Semliki Forest virus. Nature 2004; 427(6972):320-325.
56. Bressanelli S, Stiasny K, Allison SL et al. Structure of a flavivirus envelope glycoprotein in its low-pH-induced membrane fusion conformation. EMBO J 2004; 23(4):728-738.
57. Cecilia D, Gould EA. Nucleotide changes responsible for loss of neuroinvasiveness in Japanese encephalitis virus neutralization-resistant mutants. Virology 1991; 181(1):70-77.
58. Hasegawa H, Yoshida M, Shiosaka T et al. Mutations in the envelope protein of Japanese encephalitis virus affect entry into cultured cells and virulence in mice. Virology 1992; 191(1):158-165.
59. Lee E, Weir RC, Dalgarno L. Changes in the dengue virus major envelope protein on passaging and their localization on the three-dimensional structure of the protein. Virology 1997; 232(2):281-290.
60. Beasley DW, Aaskov JG. Epitopes on the dengue 1 virus envelope protein recognized by neutralizing IgM monoclonal antibodies. Virology 2001; 279(2):447-458.
61. Hurrelbrink RJ, McMinn PC. Attenuation of Murray Valley encephalitis virus by site-directed mutagenesis of the hinge and putative receptor-binding regions of the envelope protein. J Virol 2001; 75(16):7692-7702.
62. Monath TP, Arroyo J, Levenbook I et al. Single mutation in the flavivirus envelope protein hinge region increases neurovirulence for mice and monkeys but decreases viscerotropism for monkeys: Relevance to development and safety testing of live, attenuated vaccines. J Virol 2002; 76(4):1932-1943.
63. Dubovskii PV, Li H, Takahashi S et al. Structure of an analog of fusion peptide from hemagglutinin. Protein Sci 2000; 9(4):786-798.
64. Han X, Bushweller JH, Cafiso DS et al. Membrane structure and fusion-triggering conformational change of the fusion domain from influenza hemagglutinin. Nat Struct Biol 2001; 8(8):715-720.
65. Tamm LK, Han X, Li Y et al. Structure and function of membrane fusion peptides. Biopolymers 2002; 66(4):249-260.
66. Ruigrok RW, Aitken A, Calder LJ et al. Studies on the structure of the influenza virus haemagglutinin at the pH of membrane fusion. J Gen Virol 1988; 69(Pt 11):2785-2795.
67. Stiasny K, Allison SL, Schalich J et al. Membrane interactions of the tick-borne encephalitis virus fusion protein E at low pH. J Virol 2002; 76(8):3784-3790.
68. Chan DC, Kim PS. HIV entry and its inhibition. Cell 1998; 93(5):681-684.
69. Danieli T, Pelletier SL, Henis YI et al. Membrane fusion mediated by the influenza virus hemagglutinin requires the concerted action of at least three hemagglutinin trimers. J Cell Biol 1996; 133(3):559-569.
70. Blumenthal R, Sarkar DP, Durell S et al. Dilation of the influenza hemagglutinin fusion pore revealed by the kinetics of individual cell-cell fusion events. J Cell Biol 1996; 135(1):63-71.
71. Razinkov VI, Melikyan GB, Cohen FS. Hemifusion between cells expressing hemagglutinin of influenza virus and planar membranes can precede the formation of fusion pores that subsequently fully enlarge. Biophys J 1999; 77(6):3144-3151.
72. Kemble GW, Danieli T, White JM. Lipid-anchored influenza hemagglutinin promotes hemifusion, not complete fusion. Cell 1994; 76(2):383-391.
73. Melikyan GB, White JM, Cohen FS. GPI-anchored influenza hemagglutinin induces hemifusion to both red blood cell and planar bilayer membranes. J Cell Biol 1995; 131(3):679-691.
74. Nussler F, Clague MJ, Herrmann A. Meta-stability of the hemifusion intermediate induced by glycosylphosphatidylinositol-anchored influenza hemagglutinin. Biophys J 1997; 73(5):2280-2291.
75. Armstrong RT, Kushnir AS, White JM. The transmembrane domain of influenza hemagglutinin exhibits a stringent length requirement to support the hemifusion to fusion transition. J Cell Biol 2000; 151(2):425-437.
76. Saifee O, Wei L, Nonet ML. The Caenorhabditis elegans unc-64 locus encodes a syntaxin that interacts genetically with synaptobrevin. Mol Biol Cell 1998; 9(6):1235-1252.
77. McNew JA, Weber T, Parlati F et al. Close is not enough: SNARE-dependent membrane fusion requires an active mechanism that transduces force to membrane anchors. J Cell Biol 2000; 150(1):105-117.
78. West JT, Johnston PB, Dubay SR et al. Mutations within the putative membrane-spanning domain of the simian immunodeficiency virus transmembrane glycoprotein define the minimal requirements for fusion, incorporation, and infectivity. J Virol 2001; 75(20):9601-9612.
79. Dutch RE, Lamb RA. Deletion of the cytoplasmic tail of the fusion protein of the paramyxovirus simian virus 5 affects fusion pore enlargement. J Virol 2001; 75(11):5363-5369.
80. Melikyan GB, Markosyan RM, Brener SA et al. Role of the cytoplasmic tail of ecotropic moloney murine leukemia virus Env protein in fusion pore formation. J Virol 2000; 74(1):447-455.
81. Melikyan GB, Jin H, Lamb RA et al. The role of the cytoplasmic tail region of influenza virus hemagglutinin in formation and growth of fusion pores. Virology 1997; 235(1):118-128.

82. Bagai S, Lamb RA. Truncation of the COOH-terminal region of the paramyxovirus SV5 fusion protein leads to hemifusion but not complete fusion. J Cell Biol 1996; 135(1):73-84.
83. Januszeski MM, Cannon PM, Chen D et al. Functional analysis of the cytoplasmic tail of Moloney murine leukemia virus envelope protein. J Virol 1997; 71(5):3613-3619.
84. Nieva JL, Bron R, Corver J et al. Membrane fusion of Semliki Forest virus requires sphingolipids in the target membrane. EMBO J 1994; 13(12):2797-2804.
85. Vashishtha M, Phalen T, Marquardt MT et al. A single point mutation controls the cholesterol dependence of Semliki Forest virus entry and exit. J Cell Biol 1998; 140(1):91-99.
86. Chanel-Vos C, Kielian M. A conserved histidine in the ij loop of the Semliki Forest virus E1 protein plays an important role in membrane fusion. J Virol 2004; 78(24):13543-13552.
87. Chatterjee PK, Eng CH, Kielian M. Novel mutations that control the sphingolipid and cholesterol dependence of the Semliki Forest virus fusion protein. J Virol 2002; 76(24):12712-12722.
88. Stiasny K, Koessl C, Heinz FX. Involvement of lipids in different steps of the flavivirus fusion mechanism. J Virol 2003; 77(14):7856-7862.
89. Burke DS, Monath TP. Flaviviruses. In: Knipe DM, Howley PM, eds. Fields Virology. 4th ed. Philadelphia: Lippincott Williams and Wilkins, 2001:1043-1125.
90. Smith TJ, Kremer MJ, Luo M et al. The site of attachment in human rhinovirus 14 for antiviral agents that inhibit uncoating. Science 1986; 233(4770):1286-1293.
91. Baldwin CE, Sanders RW, Berkhout B. Inhibiting HIV-1 entry with fusion inhibitors. Curr Med Chem 2003; 10(17):1633-1642.
92. Kilby JM, Hopkins S, Venetta TM et al. Potent suppression of HIV-1 replication in humans by T-20, a peptide inhibitor of gp41-mediated virus entry. Nat Med 1998; 4(11):1302-1307.
93. Kanai R, Kar K, Anthony K et al. Crystal structure of west nile virus envelope glycoprotein reveals viral surface epitopes. J Virol 2006; 80(22):11000-11008.
94. Nayak V, Dessau M, Kucera K et al. Crystal structure of dengue virus type 1 envelope protein in the postfusion conformation and its implications for membrane fusion. J Virol 2009; 83(9):4338-4344.
95. Voss JE, Vaney MC, Duquerroy S et al. Glycoprotein organization of Chikungunya virus particles revealed by X-ray crystallography. Nature 2010; 468(7324):709-712.
96. Li L, Jose J, Xiang Y et al. Structural changes of envelope proteins during alphavirus fusion. Nature 2010; 468(7324):705-708.
97. Cockburn JJ, Navarro Sanchez ME et al. Structural insights into the neutralization mechanism of a higher primate antibody against dengue virus. EMBO J 2011; 31(3):767-779.
98. Luca VC, AbiMansour J, Nelson CA, Fremont DH. Crystal structure of the Japanese encephalitis virus envelope protein. J Virol 2012; 86(4):2337-2346.

CHAPTER 9

ENTRY OF RHABDOVIRUSES INTO ANIMAL CELLS

Andrew D. Regan and Gary R. Whittaker*

Department of Microbiology and Immunology, Cornell University, Ithaca, New York, USA
**Corresponding Author: Gary Whittaker—Email: grw7@cornell.edu*

Abstract: Entry is the first step in the infectious life cycle of a virus. In the case of rhabdoviruses, entry is facilitated exclusively by the envelope glycoprotein G and its interactions with the host cell. For vesicular stomatitis virus (VSV), attachment to the cell surface was thought to be facilitated by interactions with the lipid phosphatidylserine, however recent work suggests that it is in fact initiated by recognition of proteinaeous receptors. Clathrin-mediated endocytosis delivers the virions into endosomes where they have been proposed to traffic to multi-vesicular bodies. There, the viral envelope fuses with internal vesicles in a process mediated by glycoprotein G in a pH- and phosphatidylserine-dependent manner. A clear mechanistic understanding of glycoprotein G mediated fusion has yet to be obtained, however current data suggests that it is likely facilitated by events distinct from Class I or Class II fusion proteins of other viruses. Rhabdoviruses are also notable in that their fusion protein exists in a reversible pH-dependent equilibrium, which prevents irreversible preactivation during assembly, and may prove to be relevant in the mediation of cell-to-cell fusion - an alternate form of viral spread.

INTRODUCTION

The Rhabdovirus family comprises hundreds of viruses with a wide variety of hosts, comprising vertebrates, invertebrates and plants, which give rise to various diseases.[1] Virions are 100-430 nm long and 45-100 nm in diameter, with those infecting vertebrates possessing a bullet-shaped morphology. The virus consists of an envelope containing a single glycoprotein (G), a matrix protein (M) and a helical nucleocapsid of nonsegmented, negative-sense RNA and nucleocapsid (N) protein, together with the polymerase (L) and phosphoprotein (P). Here, we focus on the entry mechanism of two mammalian

Viral Entry into Host Cells, edited by Stefan Pöhlmann and Graham Simmons.

Rhabdoviruses, vesicular stomatitis virus (VSV) the type species of the *Vesiculovirus* genus within the *Rhabdoviridae*, and Rabies virus (RABV) the type species of the *Lyssavirus* genus, as well describing some features of viral hematopoietic necrosis virus (VHSV) a member of the *Novirhabdovirus* genus that infects aquatic hosts. The entry pathway of VSV is summarized in Figure 1.

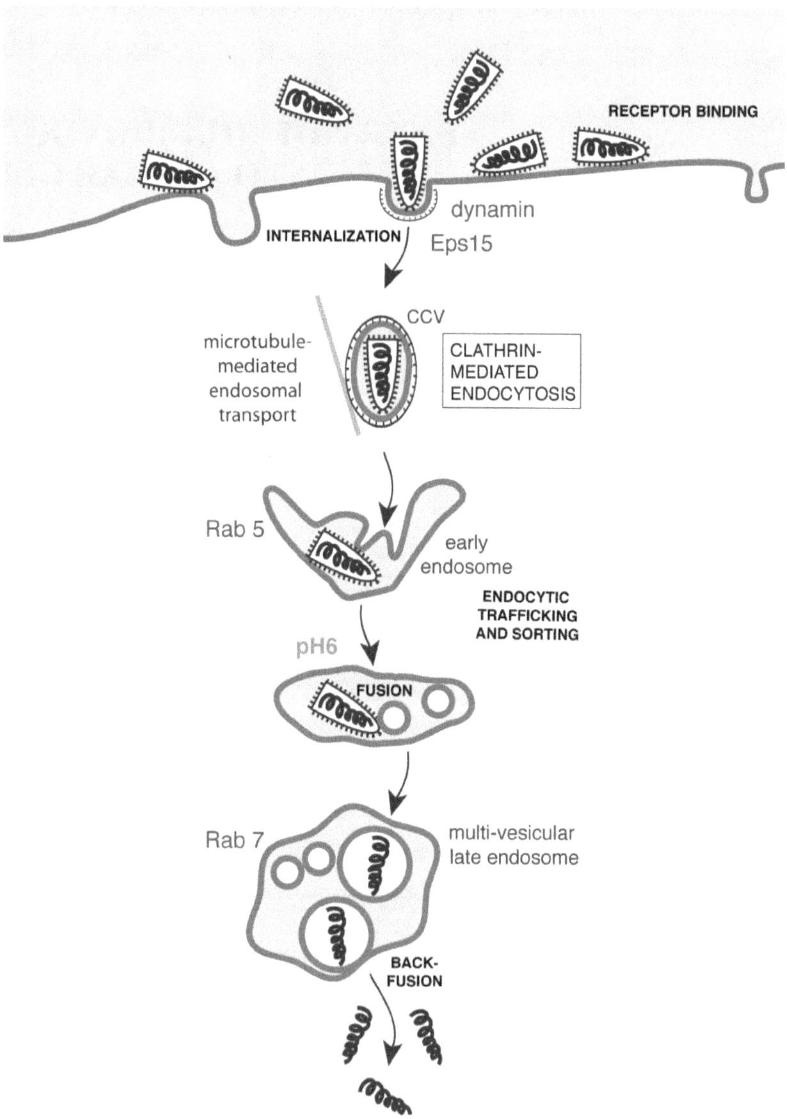

Figure 1. Route of Rhabdovirus entry into host cells. Following receptor binding, Rhabdoviruses are first internalized by clathrin-dependent endocytosis. Virions are trafficked in a microtubule-mediated manner through early endosomes, where low pH-dependent fusion is proposed to occur with internal vesicles of newly forming multi-vesicular endosomes. Back-fusion of these internal vesicles with the limiting late endosomal membrane then allows release of the virus core into the cytoplasm for replication.

VIRAL ATTACHMENT

Attachment to the cell surface represents the first step in entry of any virus, and for Rhabdoviruses this process is carried out by the envelope glycoprotein G. In recent years is has become apparent that the process of Rhabdovirus attachment is far more complex than initially reported. For some members of the Rhabdovirus family (e.g., RABV and VHSV) proteinaceous receptors have been identified,[2,3] while for others (e.g., the prototype VSV) the host cell receptor remains elusive.

Glycoprotein G

The Rhabdovirus glycoprotein G allows both viral attachment[4,5] and fusion.[6-8] Glycoprotein G is a type I transmembrane protein which varies in length between different Rhabdoviruses (VSV Indiana—511aa; RABV- 524aa; VHSV—507aa) and forms approximately 400 homotrimeric units on the surface of the viral envelope.[9] Each unit consists of noncovalently associated polypeptides which exist in a dynamic equilibrium between monomeric and trimeric states.[10-17] Among characterized isolates, glycoprotein G has been shown to contain between two and six potential N-linked glycosylation sites and twelve to sixteen conserved cysteine residues, which likely form internal disulfide bonds of structural significance.[9] No X-ray crystal structure for glycoprotein G has yet been reported, however electron microscopy shows that mature trimers project roughly 8nm from the surface of the viral envelope and consist of a globular head supported by a stalk region.[12]

Host Cell Receptors

The cellular receptors utilized by Rhabdoviruses have proven difficult to positively identify, in part due to the wide spectrum of cell tropism demonstrated by these viruses in vitro. Initial work with the prototype VSV confirmed the presence of at least one specific saturable binding site on the surface of Vero cells, suggesting the presence of a specific receptor.[18] Inhibition of infection was observed when VSV particles were preincubated with Vero cell membrane extracts, presumably by saturating receptor binding domains.[18-21] Inhibition was relieved upon preincubation of Vero cell membrane extracts with phospholipase C, but not by preincubation with neuraminidase, trypsin or heat,[22] leading to the assumption that the receptor was not proteinaceous, but in fact a phospholipid. Subsequently phosphatidylserine (PS) was shown to specifically inhibit VSV infection;[22] hydrophobic domains capable of binding PS, termed p2 peptides, were found in all Rhabdovirus isolates;[23,24] and a second 19 amino acid domain in VSV capable of binding tightly to PS was also identified.[25] Although these data strongly suggested that PS was the host cell receptor for VSV—and perhaps a general entry factor for all Rhabdoviruses—the sum of evidence was strictly indirect. Also complicating this model was the fact that in vivo most or all PS specifically localizes to the inner leaflet of the plasma membrane[26,27]—seemingly inaccessible and incapable of facilitating attachment.

In order to verify whether PS was in fact the host cell receptor for VSV, cell surface exposed PS was quantified by flow cytometry for a wide range of cell types permissive to VSV infection, and binding was shown to be independent of the amount of PS present.[28] In addition, saturating exposed PS with annexin V prior to incubation with VSV particles did not affect viral binding.[28] Although this demonstrates that PS is not the host cell attachment factor for VSV, interactions between glycoprotein G and PS are strong and will be discussed later as they seem to play a downstream role in viral fusion[29]—a fact that likely affected initial characterization.

Although the VSV receptor remains undetermined, proteinaceous receptors for other Rhabdoviruses such as RABV and VHSV have been identified.[2,3] The current model of RABV entry proposes that virions are concentrated at neuromuscular junctions by binding to nicotinic acetylcholine receptor (nAChR).[30-34] This increases the chance of the virion binding to the neural cell adhesion molecule (NCAM) at the presynaptic membrane—which likely acts to facilitate internalization.[34,35] In the case of VHSV, antibodies directed towards fibronectin protected fish cells from infection, and fibronectin was shown to bind VHSV virions, strongly suggesting a role for this protein in viral entry.[2] Although a truly definitive understanding of host cell receptors has not yet been demonstrated for any Rhabdovirus, it is clear that proteinaceous receptors are utilized, that receptors vary between strains, and that attachment and internalization may involve a sequential series of events.

Endocytosis

As virus fusion is a pH-dependent process, Rhabdoviruses must be internalized into acidic endosomes for productive infection. Early studies of VSV entry relied heavily on morphological studies by electron microscopy. In some of these studies, the majority of incoming viruses were shown to be present in pits and vesicles with electron-dense coats, implying a dominant role for clathrin-mediated endocytosis; however noncoated vesicles were also observed.[36,37] In contrast, other investigators, also using electron microscopy, showed a preponderance of viruses in large noncoated vesicles, which were possibly macropinosomes.[38] More recently, the use of a dominant negative inhibitor of clathrin-mediated endocytosis (Eps15Δ95/295), combined with pharmacological approaches and knock-down of clathrin heavy chain using RNAi technology, has clearly shown that clathrin-mediated endocytosis is the predominant route of entry.[39] Additional information on the route of entry has come from studies where VSV has been used as model virus, or as a control virus in pseudotyping experiments. Such work has shown a role for COP-mediated endocytic trafficking, and endosomal Rab proteins,[40,41] and high-throughput RNAi screens have shown that VSV endocytosis is highly regulated by specific kinase families.[42] Additionally, VSV is used extensively as a model for studying trafficking in polarized cells, and such studies have found that the virus utilizes a specific route of endocytosis through the basolateral surface of polarized epithelial cells.[43,44]

MEMBRANE FUSION

Glycoprotein G carries out membrane fusion in a low pH-dependent manner,[45-49] and involves dramatic structural reorganization,[50,51] although a precise mechanistic understanding of this process remains unclear. The optimal pH of fusion for VSV and RABV occurs within a narrow range around 6.0,[6,50,52] but fusion activity may still be observed outside this range. One of the more interesting characteristics of the glycoprotein G low-pH induced conformational changes is that it is fully reversible,[53] in contrast to well characterized class I proteins of other viruses.[54]

Localization of Fusion

Based on the pH trigger for fusion, as well as existing data on endocytic trafficking, it was initially believed that VSV fuses from early endosomes.[49] However, recent work using live-cell imaging of individual viruses, combined with biochemical studies of virus penetration, has led to the novel concept that the release of the virus into the cytosol is actually a two-step event.[55] Virus-cell fusion occurs early in the endocytic pathway, based of the effects of nocodazole in dequenching assays of virus-cell fusion. Spatially however, this fusion event occurs with the internal vesicles present within sorting endosomes (which are proposed to be in excess to the limiting membrane facing the cytosol). Subsequent endosome trafficking is then followed by "back-fusion" of the VSV-containing internal vesicles with the limiting endosomal membrane. VSV infection is dependent on the molecular components involved in multivesicular endosome biogenesis,[56] however inhibiting the formation of the internal vesicles of these so-called multivesicular bodies (MVB) with the PI3-kinase inhibitor wortmannin actually led to a reproducible increase in virus entry; presumably in this case direct fusion with the limiting membrane occurred. It is presently unclear what the role of the two-step fusion process actually is. One suggestion is that this allows exposure of the virus core to cellular chaperones present within the internal MVB vesicles, which may have a role in virus uncoating and genome release.[57]

Glycoprotein G Fusion Domain

Class I viral fusion proteins show extensive α-helical structure and undergo dramatic conformational changes which result in the exposure of a hydrophobic fusion peptide that facilitates membrane fusion between two lipid bilayers.[54,58] Although glycoprotein G appears to form a conventional "spike" which is perpendicular to the viral envelope and thus shows certain features of a class I fusion protein, other features are notably absent including proteolytic cleavage and the presence of coiled coils.[59] Likewise, it shows only limited features of class II fusion protein, which typically lie flat to the vial envelope and are comprised predominantly of β-sheet.[60] Initial attempts to understand glycoprotein G mediated membrane fusion utilized hydrophobic photolabeling and demonstrated that VSV and RABV G was able to interact with the host cell membrane in response to low pH, and that residues 59-221 of VSV G and 103-179 of RV G were in close proximity to the membrane during this process.[61] Analysis of VSV G demonstrated that mutation of another highly conserved region, between residues 118-139, abolished fusion activity or modified the pH of fusion activation.[59,62,63] Other studies have also demonstrated that altering region 395 to 418 for VSV,[64] and 392 to 396 in RABV,[65] have an influence on glycoprotein G-mediated fusion. In addition, double mutants within region 118-139 and region 395-418 of VSV show an additive inhibition of fusion activity.[66] Overall, the generation of a definitive model for Rhabdovirus fusion has produced conflicting data that highlight distinct and separate regions of glycoprotein G as being important for facilitating this process. It seems likely that when a clear model has been established, current theories will be found to overlap, with multiple regions working in concert to facilitate fusion. Although the region between amino acids 118 and 139 is generally thought to represent a candidate internal fusion peptide,[58] data supporting this is limited and the possibility exists that existing data concerning this domain may be explained by modulation of fusion activity due to structural requirements elsewhere in the protein.

Recent studies have in fact identified region 145-164 of VSV G, termed the p2-like peptide, as being a pivotal domain in facilitating glycoprotein G mediated membrane fusion.[29,67] Although initially described as heptad-repeat sequences,[23,24] these differ from the heptad repeats found in class I fusion proteins of other viruses because they are not predicted to form coiled coils. As described earlier, the p2 peptide was originally identified in VHSV as a PS binding region;[23] p2-like peptides were subsequently found in all Rhabdovirus isolates.[24,68] In addition to binding PS, the p2 peptide has also been shown to mediate membrane fusion in a low pH- and PS-dependent manner with kinetics identical to that of Rhabdovirus particles,[67] and to insert into the membrane during fusion.[68] Also supporting this model is the finding that the low pH conformational changes of VSV G and membrane fusion have been shown to directly correlate to the PS content of the target membrane.[69] Further studies revealed that the p2-like domain of VSV was in fact capable of binding PS at acidic and neutral pH, the latter being an event specifically facilitated by electrostatic interactions between PS and two histidine residues within this region.[29] These histidines, which are conserved in the p2-peptides of all Rhabdoviruses, are proposed to become protonated when they are brought into extremely close proximity with the membrane surface, an event likely requiring a separate receptor interaction.[29] Substitution or modification of these histidines abolishes fusion activity[67] suggesting that the neutral pH interaction is required, and potentially represents the initial interaction between VSV G and the target membrane during fusion.

In addition to representing the putative fusion domain, the p2 domain of Rhabdoviruses may also be involved in the translocation of PS across the membrane. As previously mentioned, any model involving PS interactions is intrinsically flawed without an explanation of how PS, which specifically localizes to the inner leaflet of the lipid bilayer, is made available to bind glycoprotein G. Recent studies focused on VHSV G (where the p2 domain was first identified) have shown that protein fragments containing the p2 domain and its flanking region (termed p9) are capable of inducing fusion and translocating PS from the inner to the outer membrane leaflet.[70] This activity required a low pH reorganization of the fragment into a predominantly β-sheet structure and was reversible upon pH neutralization.[70] Region p9 alone does not bind to phospholipids within the membrane and therefore may represent the region responsible for facilitating close contact with the membrane through protein-receptor interactions. Homologous mechanisms likely exist within VSV and RABV, although this remains to be shown.

Further confounding these models is the finding that the membrane proximal domain (amino acids 421 to 461), along with the transmembrane domain itself, have also been shown to be essential for glycoprotein G-mediated membrane fusion.[71-75] In fact, synthetic peptides representing the transmembrane domain of VSV glycoprotein G are also capable of facilitating membrane fusion in a Ca^{2+}-dependent manner.[73]

pH-Dependent Reversibility

Low-pH induced reorganization of the prototypical class I fusion protein influenza HA results in a more favorable energy state, an event which is coupled to the energetically unfavorable event of membrane fusion.[76] Within the constraints of this model it would be expected that a fusion protein could not return to its original energy state after reorganization has occurred—a hypothesis that has been experimentally verified.[76] As discussed, Rhabdovirus glycoprotein G also mediates membrane fusion through a dramatic pH-dependent structural reorganization, however the mechanism appears to be quite distinct from that of HA and other class I viral fusion proteins.

The glycoprotein G of Rhabdoviruses is proposed to exist in three forms: the native state (N), which is observed above pH 7.0 and that which is displayed on the viral envelope after budding; the activated state (A), which is optimal near pH 6.0, involves dramatic conformation reorganization and is capable of facilitating membrane fusion. Below the optimal pH of fusion is the inactivated state (I), which appears structurally similar to the activated state but in which no fusion activity is detectable.[53] Whereas the prototypical class I fusion protein HA remains permanently inactive after the undergoing similar low-pH induced structural reorganizations,[76] the conformational states of glycoprotein G are fully reversible and occur in a pH-dependent manner.[53]

Considering that there is believed to be only one viral fusion event required for entry, the evolution of a reversible fusion protein must be required for a separate event. It is possible that glycoprotein G undergoes reversible folding in order to prevent irreversible preactivation as it travels through the acidic environment of the Golgi, a process which is facilitated by the activation prerequisite of proteolytic cleavage in other class I fusion proteins such as influenza HA.[76] Although not normally considered to be a syncytial virus, cell-cell fusion can be facilitated by VSV at apparently neutral pH in certain polarized cells, which may more closely represent in vivo conditions.[77] Interestingly a number of other low-pH dependent viruses, including coronaviruses and flaviviruses,[78-80] also do not follow the influenza model of low pH inactivation, suggesting a possible common strategy for fusion activation. For Rhabdoviruses, cell-cell fusion may turn out to be a relevant form of viral spread and an important strategy for evading the host immune response.

CONCLUSION

Rhabdovirus entry is a complex process—far more so than initially described -and warrants further study, especially considering the increasing effort to use Rhabdovirus and glycoprotein G pseudotyped vectors for use in human gene therapy.[81] Proteinaceous receptors await identification, which will allow an understanding of their relationship to the viral entry process, especially regarding the role of multivesicular bodies and late endosomes. The process of glycoprotein G-mediated fusion itself remains controversial and fundamentally unclear. Ultimately, the acquisition of a definitive model for Rhabdovirus fusion and virus entry will be dependent on the solving of an X-ray crystal structure for glycoprotein G.

NOTE ADDED IN PROOF

Since the completion of this chapter, the crystal structure of the VSV (Indiana) G protein has been solved.[82] An ectodomain was produced by limited proteolysis of virions at pH 6.25 using thermolysin, yielding G_{th} (residues 1-410 of VSV G), which was then crystallized at pH 7. The structure was solved at 2.4Å resolution, and showed a novel structure for a viral fusion protein. The overall structure of G_{th} is an inverted cone and is apparently in a post-fusion form, consisting of four domains that have features of both class I and class II viral fusion proteins. Domains I and III are predominantly b-sheet, with domain III having a pleckstrin homology (PH) domain implicated in lipid binding and signaling events. Domain II is termed the trimerization domain and comprises a six-helix bundle reminiscent of the core of the post-fusion form of a class I fusion protein. Domain IV (the so-called fusion domain) has an extended b-sheet structure that has extensive similarity to class II fusion proteins, with four hydrophobic residues on the tip of the domain forming a bipartite "fusion patch" comprising residues W^{72}, Y^{73}, Y^{116} and A^{117}. The p2 domain previously implicated in PS binding and fusion, lies within domain IV—although the contribution of histidine residues, previously proposed to be central to fusion activation, is currently unclear. Also, G_{th} is missing the membrane-proximal region previously reported to be essential for fusion. To account for the reversibility of the conformational changes, it is proposed that a large number of acidic amino acids are brought together in the six-helix bundle, with their deprotonation at higher pH inducing strong repulsive forces that destabilize the core trimer and allow reversion to a pre-fusion state. Despite the relative lack of sequence similarity of the G protein, the basic overall topology of the G protein is likely to be conserved across the *Rhabdoviridae*. The VSV G protein clearly represents a novel structural class of viral membrane fusion protein, and perhaps the most remarkable feature is the finding that it shares the same overall domain structure as glycoprotein B of herpes simplex virus (HSV), indicating that these very different viruses may have common evolutionary origins.

REFERENCES

1. Fauquet CM, Mayo MA, Maniloff J et al, eds. Virus Taxonomy. San Diego: Elsevier Academic Press, 2005.
2. Bearzotti M, Delmas B, Lamoureux A et al. Fish rhabdovirus cell entry is mediated by fibronectin. J Virol 1999; 73(9):7703-7709.
3. Lafon M. Rabies virus receptors. J Neurovirol 2005; 11(1):82-87.
4. Wunner WH, Reagan KJ, Koprowski H. Characterization of saturable binding sites for rabies virus. J Virol 1984; 50(3):691-697.
5. Altstiel LD, Landsberger FR. Lipid-protein interactions between the surface glycoprotein of vesicular stomatitis virus and the lipid bilayer. Virology 1981; 115(1):1-9.
6. Gaudin Y, Ruigrok RW, Knossow M et al. Low-pH conformational changes of rabies virus glycoprotein and their role in membrane fusion. J Virol 1993; 67(3):1365-1372.
7. Konieczko EM, Whitaker-Dowling PA, Widnell CC. Membrane fusion as a determinant of the infectibility of cells by vesicular stomatitis virus. Virology 1994; 199(1):200-211.
8. Lecocq-Xhonneux F, Thiry M, Dheur I et al. A recombinant viral haemorrhagic septicaemia virus glycoprotein expressed in insect cells induces protective immunity in rainbow trout. J Gen Virol 1994; 75(Pt 7):1579-1587.
9. Coll JM. The glycoprotein G of rhabdoviruses. Arch Virol 1995; 140(5):827-851.
10. Balch WE, Elliott MM, Keller DS. ATP-coupled transport of vesicular stomatitis virus G protein between the endoplasmic reticulum and the Golgi. J Biol Chem 1986; 261(31):14681-14689.
11. Doms RW, Keller DS, Helenius A et al. Role for adenosine triphosphate in regulating the assembly and transport of vesicular stomatitis virus G protein trimers. J Cell Biol 1987; 105(5):1957-1969.
12. Gaudin Y, Ruigrok RW, Tuffereau C et al. Rabies virus glycoprotein is a trimer. Virology 1992; 187(2):627-632.
13. Kreis TE, Lodish HF. Oligomerization is essential for transport of vesicular stomatitis viral glycoprotein to the cell surface. Cell 1986; 46(6):929-937.
14. Lyles DS, Varela VA, Parce JW. Dynamic nature of the quaternary structure of the vesicular stomatitis virus envelope glycoprotein. Biochemistry 1990; 29(10):2442-2449.
15. Whitt MA, Buonocore L, Prehaud C et al. Membrane fusion activity, oligomerization, and assembly of the rabies virus glycoprotein. Virology 1991; 185(2):681-688.
16. Wilcox MD, McKenzie MO, Parce JW et al. Subunit interactions of vesicular stomatitis virus envelope glycoprotein influenced by detergent micelles and lipid bilayers. Biochemistry 1992; 31(43):10458-10464.
17. Zagouras P, Ruusala A, Rose JK. Dissociation and reassociation of oligomeric viral glycoprotein subunits in the endoplasmic reticulum. J Virol 1991; 65(4):1976-1984.
18. Schlegel R, Willingham MC, Pastan IH. Saturable binding sites for vesicular stomatitis virus on the surface of Vero cells. J Virol 1982; 43(3):871-875.

19. Conti C, Hauttecoeur B, Morelec MJ et al. Inhibition of rabies virus infection by a soluble membrane fraction from the rat central nervous system. Arch Virol 1988; 98(1-2):73-86.
20. Conti C, Mastromarino P, Ciuffarella MG et al. Characterization of rat brain cellular membrane components acting as receptors for vesicular stomatitis virus. Brief report. Arch Virol 1988; 99(3-4):261-269.
21. Mastromarino P, Conti C, Goldoni P et al. Characterization of membrane components of the erythrocyte involved in vesicular stomatitis virus attachment and fusion at acidic pH. J Gen Virol 1987; 68(Pt 9):2359-2369.
22. Schlegel R, Tralka TS, Willingham MC et al. Inhibition of VSV binding and infectivity by phosphatidylserine: Is phosphatidylserine a VSV-binding site? Cell 1983; 32(2):639-646.
23. Coll JM. Heptad-repeat sequences in the glycoprotein of rhabdoviruses. Virus Genes 1995; 10(2):107-114.
24. Coll JM. Synthetic peptides from the heptad repeats of the glycoproteins of rabies, vesicular stomatitis and fish rhabdoviruses bind phosphatidylserine. Arch Virol 1997; 142(10):2089-2097.
25. Hall MP, Burson KK, Huestis WH. Interactions of a vesicular stomatitis virus G protein fragment with phosphatidylserine: NMR and fluorescence studies. Biochim Biophys Acta 1998; 1415(1):101-113.
26. Morrot G, Herve P, Zachowski A et al. Aminophospholipid translocase of human erythrocytes: Phospholipid substrate specificity and effect of cholesterol. Biochemistry 1989; 28(8):3456-3462.
27. Zachowski A, Favre E, Cribier S et al. Outside-inside translocation of aminophospholipids in the human erythrocyte membrane is mediated by a specific enzyme. Biochemistry 1986; 25(9):2585-2590.
28. Coil DA, Miller AD. Phosphatidylserine is not the cell surface receptor for vesicular stomatitis virus. J Virol 2004; 78(20):10920-10926.
29. Carneiro FA, Lapido-Loureiro PA, Cordo SM et al. Probing the interaction between vesicular stomatitis virus and phosphatidylserine. Eur Biophys 2006; 35(2):145-154.
30. Burrage TG, Tignor GH, Smith AL. Rabies virus binding at neuromuscular junctions. Virus Res 1985; 2(3):273-289.
31. Castellanos JE, Castaneda DR, Velandia AE et al. Partial inhibition of the in vitro infection of adult mouse dorsal root ganglion neurons by rabies virus using nicotinic antagonists. Neurosci Lett 1997; 229(3):198-200.
32. Gastka M, Horvath J, Lentz TL. Rabies virus binding to the nicotinic acetylcholine receptor alpha subunit demonstrated by virus overlay protein binding assay. J Gen Virol 1996; 77(Pt 10):2437-2440.
33. Lentz TL, Benson RJ, Klimowicz D et al. Binding of rabies virus to purified Torpedo acetylcholine receptor. Brain Res 1986; 387(3):211-219.
34. Lentz TL, Burrage TG, Smith AL et al. Is the acetylcholine receptor a rabies virus receptor? Science 1982; 215(4529):182-184.
35. Lewis P, Fu Y, Lentz T. Rabies virus entry at the neuromuscular junction in nerve-muscle cocultures. Muscle Nerve 2000; 23(5):720-730.
36. Superti F, Seganti L, Ruggeri FM et al. Entry pathway of vesicular stomatitis virus into different host cells. J Gen Virol 1987; 68(Pt 2):387-399.
37. Matlin KS, Reggio H, Helenius A et al. Pathway of vesicular stomatitis virus leading to infection. J Mol Biol 1982; 156:609-631.
38. Cernescu C, Constantinescu SN, Popescu LM. Electron microscopic observations of vesicular stomatitis virus particles penetration in human fibroblasts. Rev Roum Virol 1990; 41:93-96.
39. Sun X, Yau VK, Briggs BJ et al. Role of clathrin-mediated endocytosis during vesicular stomatitis virus entry into host cells. Virology 2005; 338(1):53-60.
40. Kolokoltsov AA, Fleming EH, Davey RA. Venezuelan equine encephalitis virus entry mechanism requires late endosome formation and resists cell membrane cholesterol depletion. Virology 2006; 347(2):333-342.
41. Daro E, Sheff D, Gomez M et al. Inhibition of endosome function in CHO cells bearing a temperature-sensitive defect in the coatomer (COPI) component epsilon-COP. J Cell Biol 1997; 139(7):1747-1759.
42. Pelkmans L, Fava E, Grabner H et al. Genome-wide analysis of human kinases in clathrin- and caveolae/raft-mediated endocytosis. Nature 2005; 436(7047):78-86.
43. Fuller S, von Bonsdorff CH, Simons K. Vesicular stomatitis virus infects and matures only through the basolateral surface of the polarized epithelial cell line, MDCK. Cell 1984; 38(1):65-77.
44. Gottlieb TA, Ivanov IE, Adesnik M et al. Actin microfilaments play a critical role in endocytosis at the apical but not the basolateral surface of polarized epithelial cells. J Cell Biol 1993; 120(3):695-710.
45. Blumenthal R, Bali-Puri A, Walter A et al. pH-dependent fusion of vesicular stomatitis virus with Vero cells. Measurement by dequenching of octadecyl rhodamine fluorescence. J Biol Chem 1987; 262(28):13614-13619.
46. Florkiewicz RZ, Rose JK. A cell line expressing vesicular stomatitis virus glycoprotein fuses at low pH. Science 1984; 225(4663):721-723.
47. Hernandez LD, Hoffman LR, Wolfsberg TG et al. Virus-cell and cell-cell fusion. Annu Rev Cell Dev Biol 1996; 12:627-661.
48. Matlin KS, Reggio H, Helenius A et al. Pathway of vesicular stomatitis virus entry leading to infection. J Mol Biol 1982; 156(3):609-631.

49. White J, Matlin K, Helenius A. Cell fusion by Semliki Forest, influenza, and vesicular stomatitis viruses. J Cell Biol 1981; 89(3):674-679.
50. Carneiro FA, Ferradosa AS, Da Poian AT. Low pH-induced conformational changes in vesicular stomatitis virus glycoprotein involve dramatic structure reorganization. J Biol Chem 2001; 276(1):62-67.
51. Pak CC, Puri A, Blumenthal R. Conformational changes and fusion activity of vesicular stomatitis virus glycoprotein: [125I]iodonaphthyl azide photolabeling studies in biological membranes. Biochemistry 1997; 36(29):8890-8896.
52. Gaudin Y, Tuffereau C, Durrer P et al. Biological function of the low-pH, fusion-inactive conformation of rabies virus glycoprotein (G): G is transported in a fusion-inactive state-like conformation. J Virol 1995; 69(9):5528-5534.
53. Gaudin Y. Reversibility in fusion protein conformational changes. The intriguing case of rhabdovirus-induced membrane fusion. Subcell Biochem 2000; 34:379-408.
54. Colman PM, Lawrence MC. The structural biology of type I viral membrane fusion. Nat Rev Mol Cell Biol 2003; 4(4):309-319.
55. Le Blanc I, Luyet PP, Pons V et al. Endosome-to-cytosol transport of viral nucleocapsids. Nat Cell Biol 2005; 7(7):653-664.
56. Matsuo H, Chevallier J, Mayran N et al. Role of LBPA and Alix in multivesicular liposome formation and endosome organization. Science 2004; 303(5657):531-534.
57. Uchil P, Mothes W. Viral entry: A detour through multivesicular bodies. Nat Cell Biol 2005; 7(7):641-642.
58. Earp LJ, Delos SE, Park HE et al. The many mechanisms of viral membrane fusion proteins. Curr Top Microbiol Immunol 2005; 285:25-66.
59. Zhang L, Ghosh HP. Characterization of the putative fusogenic domain in vesicular stomatitis virus glycoprotein G. J Virol 1994; 68(4):2186-2193.
60. Kielian M. Class II virus membrane fusion proteins. Virology 2006; 344(1):38-47.
61. Durrer P, Gaudin Y, Ruigrok RW et al. Photolabeling identifies a putative fusion domain in the envelope glycoprotein of rabies and vesicular stomatitis viruses. J Biol Chem 1995; 270(29):17575-17581.
62. Fredericksen BL, Whitt MA. Vesicular stomatitis virus glycoprotein mutations that affect membrane fusion activity and abolish virus infectivity. J Virol 1995; 69(3):1435-1443.
63. Li Y, Drone C, Sat E et al. Mutational analysis of the vesicular stomatitis virus glycoprotein G for membrane fusion domains. J Virol 1993; 67(7):4070-4077.
64. Shokralla S, He Y, Wanas E et al. Mutations in a carboxy-terminal region of vesicular stomatitis virus glycoprotein G that affect membrane fusion activity. Virology 1998; 75:39-50.
65. Gaudin Y, Raux H, Flamand A et al. Identification of amino acids controlling the low-pH-induced conformational change of rabies virus glycoprotein. J Virol 1996; 70:7371-7378.
66. Shokralla S, Chernish R, Ghosh HP. Effect of double-site mutations of vesicular stomatitis virus glycoprotein G on membrane fusion activity. Virology 1999; 256:119-129.
67. Carneiro FA, Stauffer F, Lima CS et al. Membrane fusion induced by vesicular stomatitis virus depends on histidine protonation. J Biol Chem 2003; 278(16):13789-13794.
68. Nunez E, Fernandez AM, Estepa A et al. Phospholipid interactions of a peptide from the fusion-related domain of the glycoprotein of VHSV, a fish rhabdovirus. Virology 1998; 243(2):322-330.
69. Carneiro FA, Bianconi ML, Weissmuller G et al. Membrane recognition by vesicular stomatits virus involves enthalpy-driven protein-lipid interactions. J Virol 2002; 76:3756-3764.
70. Estepa AM, Rocha AI, Mas V et al. A protein G fragment from the salmonid viral hemorrhagic septicemia rhabdovirus induces cell-to-cell fusion and membrane phosphatidylserine translocation at low pH. J Biol Chem 2001; 276(49):46268-46275.
71. Jeetendra E, Ghosh K, Odell D et al. The membrane-proximal region of vesicular stomatitis virus glycoprotein G ectodomain is critical for fusion and virus infectivity. J Virol 2003; 77(23):12807-12818.
72. Jeetendra E, Robison CS, Albritton LM et al. The membrane-proximal domain of vesicular stomatitis virus G protein functions as a membrane fusion potentiator and can induce hemifusion. J Virol 2002; 76(23):12300-12311.
73. Langosch D, Brosig B, Pipkorn R. Peptide mimics of the vesicular stomatitis virus G-protein transmembrane segment drive membrane fusion in vitro. J Biol Chem 2001; 276(34):32016-32021.
74. Cleverley DZ, Lenard J. The transmembrane domain in viral fusion: Essential role for a conserved glycine residue in vesicular stomatitis virus G protein. Proc Natl Acad Sci USA 1998; 95(7):3425-3430.
75. Odell D, Wanas E, Yan J et al. Influence of membrane anchoring and cytoplasmic domains on the fusogenic activity of vesicular stomatitis virus glycoprotein G. J Virol 1997; 71(10):7996-8000.
76. Skehel JJ, Wiley DC. Receptor binding and membrane fusion in virus entry: The influenza hemagglutinin. Annu Rev Biochem 2000; 69:531-569.
77. Roberts PC, Kipperman T, Compans RW. Vesicular stomatitis virus G protein acquires pH-independent fusion activity during transport in a polarized endometrial cell line. J Virol 1999; 73(12):10447-10457.

78. Simmons G, Reeves JD, Rennekamp AJ et al. Characterization of severe acute respiratory syndrome-associated coronavirus (SARS-CoV) spike glycoprotein-mediated viral entry. Proc Natl Acad Sci USA 2004; 101(12):4240-4245.
79. Tscherne DM, Jones CT, Evans MJ et al. Time- and temperature-dependent activation of hepatitis C virus for low-pH-triggered entry. J Virol 2006; 80(4):1734-1741.
80. Chu VC, McElroy LJ, Chu V et al. The avian coronavirus infectious bronchitis virus undergoes direct low-pH-dependent fusion activation during entry into host cells. J Virol 2006; 80(7):3180-3188.
81. Finke S, Conzelmann KK. Recombinant rhabdoviruses: Vectors for vaccine development and gene therapy. Curr Top Microbiol Immunol 2005; 292:165-200.
82. Roche S, Bressanelli S, Rey FA, Gaudin Y. Crystal structure of the low-pH form of the vesicular stomatitis virus glycoprotein G. Science 2006; 313(5784):187-191.

CHAPTER 10

ENTRY OF HERPESVIRUSES INTO CELLS: The Enigma Variations

Claude Krummenacher,*[,1] Andrea Carfí,[2] Roselyn J. Eisenberg[1] and Gary H. Cohen[3]

[1]*Department of Pathobiology, School of Veterinary Medicine, University of Pennsylvania, Philadelphia, Pennsylvania, USA;* [2]*Protein Biochemistry, Novartis Vaccines and Diagnostics, Cambridge, Massachusetts, USA;* [3]*Department of Microbiology, School of Dental Medicine, University of Pennsylvania, Philadelphia, Pennsylvania, USA.*

**Corresponding Author: Claude Krummenacher—Email: ckrummen@vet.upenn.edu*

Abstract: The entry of herpesviruses into their target cells is complex at many levels. Virus entry proceeds by a succession of interactions between viral envelope glycoproteins and molecules on the cell membrane. The process is divided into distinct steps: attachment to the cell surface, interaction with a specific entry receptor, internalization of the particle (optional and cell specific), and membrane fusion. Several viral envelope glycoproteins are involved in one or several of these steps. The most conserved entry glycoproteins in the herpesvirus family (gB, gH/gL) are involved in membrane fusion. Around this functional core, herpesviruses have a variety of receptor binding glycoproteins, which interact with cell surface proteins often from different families. This interaction activates and controls the actual fusion machinery. Interactions with cellular receptors and between viral glycoproteins have to be tightly coordinated and regulated to guarantee successful entry. Although additional entry receptors for herpesviruses continue to be identified, the molecular interactions between viral glycoproteins remain mostly enigmatic. This chapter will review our current understanding of the molecular interactions that occur during herpesvirus entry from attachment to fusion. Particular emphasis will be placed on structure-based representation of receptor binding as a trigger of fusion during herpes simplex virus entry.

Viral Entry into Host Cells, edited by Stefan Pöhlmann and Graham Simmons.

INTRODUCTION

Herpesviruses (HV) constitute a large family of pathogens infecting hosts as different as humans and other mammals, birds, reptiles, amphibians, fish and even oysters.[1] This diversity of hosts indicates that these viruses have developed successful and efficient ways to enter different kind of cells. Despite this remarkable range of targets, some features of the entry machinery are conserved amongst all herpesviruses. Yet, there are many "variations on a theme" around this common mechanism. As in Sir Edward Elgar's "*Enigma Variations*", the mystery of HV entry resides in the original but hidden theme. In this chapter, we will illustrate common aspects of the entry machinery and describe the best-characterized variations developed by these complex viruses. We will focus primarily on molecular aspects of the cell entry mechanisms of human HV (Fig. 1).

Three subfamilies of herpesviruses (α-HV, β-HV and γ-HV) are distinguished by genomic organization, sequence homologies and replicative cycles.[1] Tropism towards a given cell type or tissue (i.e., neurons for α-HV, lymphocytes for γ-HV) is partially determined by the ability of the virus to achieve entry into defined targets. So far, humans are the primary hosts for three α-HV: HSV-1 and 2 (herpes simplex virus type 1 and 2) and VZV (varicella-zoster virus); three β-HV: CMV (cytomegalovirus), HHV-6 and HHV-7 (human herpesvirus 6 and 7); and two γ-HV: EBV (Epstein-Barr virus) and HHV-8 (Kaposi sarcoma herpesvirus).

The large number of variations precludes us from reviewing the whole prodigious body of work contributed by many laboratories over the last decades and many relevant publications are not directly cited here. Previous reviews provide general background on HV biology[1,2] and entry[3-7] and we apologize to colleagues whose original references have not been cited.

STEPS OF HERPESVIRUS ENTRY

All herpesvirus particles consist of an icosahedral capsid which contains the linear DNA genome and is surrounded by layers of tegument proteins.[8] The particle is enveloped by a lipid bilayer acquired from the host cell during egress. The envelope contains a dozen or more viral glycoproteins, several of which play a role in entry (Fig. 1). Ultimately, the virus is considered to have entered a cell when its capsid is released in the cytoplasm after the viral envelope has fused with a cell membrane. The process is divided into distinct steps: attachment to the cell surface, interaction with a specific entry receptor, internalization of the particle, and membrane fusion. In some cells, fusion requires prior internalization, while in others it does not. This complex process necessitates several viral envelope glycoproteins with distinct roles in one or several of these steps.

The attachment glycoproteins vary amongst HV, as do their cell surface ligands. In most cases the attachment protein binds to ubiquitous glycosaminoglycans such as heparan sulfate. These glycoproteins are not essential for entry but this initial interaction tethers the particle to the cell surface, thereby favoring an encounter with a less available entry receptor.

Attachment to the cell surface is not sufficient and virus entry requires an interaction with a specific entry receptor. As with most viruses, usage of a definite entry receptor fulfills two important roles: first, it specifies the types of target cells the virus can infect and second, it ensures that the fusion machinery is activated only when virions are in close proximity to

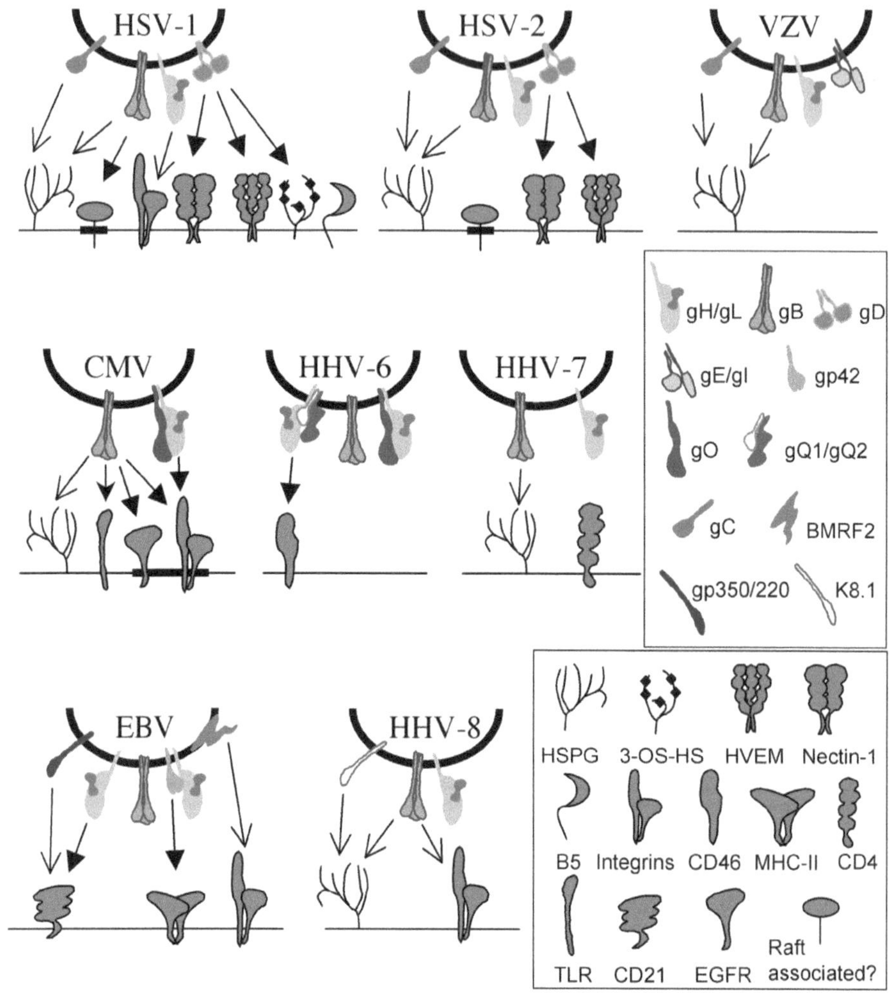

Figure 1. Viral entry glycoproteins interact with cell surface molecules. Attachment interactions are indicated with open arrows and interaction with entry receptors are indicated as filled arrows. The thin horizontal lines below the virions represent the cellular membrane and its thickened portions indicate involvement of lipid rafts. All of the indicated interactions do not necessarily take place on the same cells or at the same time for entry to occur.

the cell membrane. Several receptor-binding glycoproteins have been identified in various HV and these often bind to receptors from different families of cell surface proteins.

Although the three glycoproteins thought to mediate fusion (gB and gH/gL) are conserved, their regulation remains an enigma. Only indirect clues have thus far emerged to unveil the central theme of HV fusion. Some suggest that gH is the fusion protein based on the presence of putative heptad repeats and hydrophobic segments.[9-11] Others indicate that fusion is mostly mediated by gB, because some gB mutants have enhanced fusogenic activities.[12-17] However, these proteins do not share sequence similarities with other known viral fusion proteins.

ENTRY GLYCOPROTEINS: SPECIFICITIES OF HERPESVIRUSES SUBFAMILIES

Alpha-Herpesviruses

These neurotropic viruses primarily infect epithelial and neuronal cells of their hosts and establish latency in neurons.[2] In cell culture, HSV enters most cell lines of vertebrate origin. Initial attachment is mediated primarily by glycoprotein C (gC) binding to heparan sulfate and chondroitin sulfate proteoglycans (HSPG and CSPG).[18,19] In the absence of gC, which is not an essential glycoprotein, gB provides this function.[20,21] Purified forms of both glycoproteins bind to HSPG on cells and to soluble heparin with relatively high affinity.[22-24] Deletion or modification of heparin binding domains of either gB or gC has no significant effect on entry.[25]

In contrast to β-HV and γ-HV, most α-HV envelopes contain glycoprotein D (gD). This glycoprotein is essential for entry[26] and is the primary receptor binding protein for HSV-1 and HSV-2. Since gD does not share any structural characteristics with known viral fusion proteins,[27] its interaction with a cellular receptor is thought to activate the fusion process. HSV and various animal α-HV use members of the nectin family as receptors for gD.[6,28] Human nectin-1 is used by HSV-1, HSV-2, porcine pseudorabies virus (PRV) and bovine herpesvirus 1 (BHV-1).[29,30] Nectin-2 is used by PRV and some mutant forms of HSV-1 and 2[31,32] while human nectin-like-5/CD155, the poliovirus receptor, is used by PRV and BHV-1.[29] Nectins are conserved proteins and it is likely that some of them are receptors for other animal α-HV in their respective hosts.[33-35] In the case of HSV, two other unrelated receptors are able to mediate virus entry by binding to gD. First, HVEM (herpesvirus entry mediator), a member of the tumor necrosis factor (TNF) receptor family is used by wild type (wt) HSV.[36] Second, a form of heparan sulfate (HS) that is specifically modified by 3-O-sulfotransferases (3-O-STs) binds HSV-1 gD and allows virus entry.[37] Most HSV-1 and HSV-2 laboratory strains use both nectin-1 and HVEM and all the tested isolates use both receptors despite variations in serotype and origins of the virus (i.e., oral and genital lesions, encephalitis, disseminated infection of neonates).[38] This suggests that HVEM and nectin-1 are both involved in HSV infection in humans. One should note that these unrelated molecules bind HSV gD independently and do not act as coreceptors during entry. It is therefore fascinating that gD binds to structurally different receptors and yet activates a conserved fusion machinery (see below for more details).

VZV diverges from other α-HV because it lacks a gD gene.[3] Thus gD is not necessarily a factor in the neurotropism of α-HV. In the case of VZV, gE plays an essential role in virus-cell fusion[39] and may be the receptor binding protein of this virus. In contrast, HSV gE is not required for virus entry, although it greatly enhances virus spread.[40-42] In this case, gE, mostly found associated with gI, targets newly made HSV particles to specific regions of the cell (e.g., axons, junctions).[43-45] Interestingly, gD and gE genes are thought to have evolved by duplication and divergence[46] and the glycoproteins share several features. For instance, HSV gE and gD play redundant roles in virion envelopment.[47] Furthermore VZV gE and HSV gE, like HSV gD, accumulate at junctions between epithelial cells[48-51] suggesting an interaction of gE with a receptor involved in cell adhesion.[42,52] Coincidentally, VZV particles are highly cell-associated and spread involves direct contact of uninfected cells with infected cells rather than with free virus.

The core fusion machinery of α-HV is composed of gB and gH/gL. For HSV, gD, gB and gH/gL are necessary and sufficient to promote cell-cell fusion in the presence of an entry receptor.[14,53-55] The gB ectodomain is a homotrimer[56] while gH and gL form a noncovalently linked heterodimer.[57] Although gL lacks a transmembrane region it is required for proper folding, processing and transport of the membrane anchored gH.[57-59] Its presence in the virion envelope is always in association with gH. Several mutations causing syncytial phenotypes map to the gB cytoplasmic tail. This hyperactive membrane fusion activity gives rise to large multinucleated cells at the site of infection in vivo and in cell culture.[14,15] Moreover, recent data suggest that HSV gB binds to the surface of cells that lack HS,[60] and gH/gL binds directly to integrin αVβ3,[61] but the role and/or identity of cell receptors for components of the HSV fusion machinery remain speculative.

Beta-Herpesviruses

The prototypical human β-HV genome (HCMV) has genes for a large number of glycoproteins, but not for gD. The major envelope constituent, gB, mediates attachment by interacting with HSPG and it is also essential for fusion.[62] Interestingly, CMV gB is cleaved,[63] and has the surface unit (SU)-transmembrane unit (TM) topology of receptor-binding/fusion proteins found on other types of viruses (e.g., retroviruses, orthomyxoviruses).[64,65] CMV gB binds to EGFR (epidermal growth factor receptor) or to various integrins, via a disintegrin-like domain, to promote entry.[66,67] Additionally gH interacts with integrin αVβ3.[68] Altogether these interactions lead to formation of a complex including EGFR and integrins in lipid rafts.[68] Furthermore, CMV gB interacts with toll-like receptor 2 (TLR2) which might indirectly influence entry.[7] To complicate matters further, gH/gL coexpression in the absence of gB is sufficient to cause fusion of several cell types while gB alone can induce syncytia formation in U373 glioblastoma cells.[69,70] Adding to the "variations", CMV gH and gL are associated with a heavily glycosylated viral glycoprotein, gO.[71] There are no known cellular ligands for CMV gO. In HHV-6, gO and gQ (expressed as gQ1 and gQ2,)[72] associate with gH/gL in a mutually exclusive way,[73,74] and only gH/gL/gQ interacts with cellular CD46, a candidate receptor for HHV-6.[75-77] The third component of the β-HV gH/gL complex has no defined role in fusion, but might be important in determining cell tropism. The last human β-HV, HHV-7, likely attaches to the cell surface through the interaction of gB with HS.[78] Finally, CD4 appears to play a role in T-cell infection by HHV-7, although its involvement as a receptor during attachment or fusion is unclear.[79]

Gamma-Herpesviruses

The attachment of EBV to the cell surface is mediated by glycoprotein gp350/220, which binds the complement receptor 2 (CR2, CD21) in lieu of HSPG.[4,80] Additionally, the EBV BMRF2 protein binds integrins (α5β1) during infection of polarized cells.[81] As a receptor binding protein, EBV gp42 interacts with MHC class II to mediate entry into B cells.[82,83] The structure of gp42 is unlike that of any known ligand for MHC-II[84] and gp42 uses its flexible N-terminus to interact with gH/gL.[85] Binding of MHC-II to gp42, in association with gH/gL, is necessary for EBV entry into B lymphocytes, but not epithelial cells.[86] The corresponding requirements for entry of EBV into epithelial cells are not known but gH might be directly involved, possibly by binding CR2/CD21.[87,88] This gp42-based system provides an elegant mechanism for regulation of

EBV's tropism; virions produced in B cells contain little gp42 (because it is tied up with MHC II) and do not require gp42 to infect epithelia, whereas virions released from epithelial cells carry more gp42 to efficiently target B cells.[89,90]

HHV-8 attachment is probably mediated by gB and K8.1 (a positional homolog of EBV gp350/220) both of which can bind HSPG.[91,92] Both gB and the gH/gL complex are necessary and sufficient for membrane fusion of this virus.[93] HHV-8 gB also interacts with integrins $\alpha3\beta1$[94,95] although, like in the case of EBV, it is not clear whether the interactions with integrins play an active role in fusion or an accessory role in attachment.

An elegant study by Kaleeba and Berger identified the cysteine transporter xCT as an entry receptor for HHV-8. Expression of xCT in nonpermissive cells allows HHV-8 entry and cell fusion. The level of expression of xCT correlates with cell susceptibility to HHV-8.[96]

ENTRY PATHWAYS: FUSION AT THE PLASMA MEMBRANE AND ENDOCYTOSIS

Herpesviruses follow different entry routes according to the cell type they infect. For instance, CMV enters fibroblasts by direct fusion with the plasma membrane but is endocytosed in endothelial and retinal epithelial cells.[97,98] In contrast, EBV enters epithelial cells by direct fusion but is endocytosed in normal B lymphocytes,[99,100] and HHV-8 is endocytosed in fibroblasts.[101] Although HSV can fuse directly with the plasma membrane, we now know it is endocytosed in most cells types.[102-104]

The case of HSV is exemplary. For a long time its entry was thought to be independent of a low pH environment.[105] This idea was reinforced by electron micrographs of a related porcine α-HV, PRV, which clearly revealed direct fusion of the virus envelope with the plasma membrane of pig cells.[106] Nicola et al[104] revisited the effect of pH on HSV entry and found that endosomal acidification is, in fact, required for entry into several cell lines that had not previously been examined in this context. They found that epithelial cells and keratinocytes are more likely to require endosomal acidification than cells of neuronal lineage.[102] HSV entry via a low pH independent endocytosis pathway is indeed as widespread as low pH dependent endocytosis.[102,103] In contrast entry into African green monkey Vero cells, widely used in laboratories studying HSV, occurs by fusion at the cell surface,[104] and thus far, this seems to be the only identified cell line where this mechanism is used by HSV. The three pathways of HSV entry require the same glycoproteins (gB, gH/gL, gD) as well as a gD receptor.[102,103] Given that the same viral fusion machinery appears to be used, it is unclear why, in some cells, entry occurs only in a low pH-environment. Mouse melanoma B78H1 cells lack functional gD receptors and virus is not endocytosed. However, when these cells are engineered to express HVEM or nectin-1, entry occurs via a pH independent endosomal route and internalization is blocked by neutralizing anti-gD antibodies.[103] In contrast receptor-negative Chinese hamster ovary (CHO) cells internalize HSV in the absence of a gD receptor but this endocytic event leads to virus degradation since proper fusion cannot occur.[104] The reason why the virus enters different cells by different pathways remains to be elucidated at the molecular and cellular levels.

TRIGGERING THE FUSION MACHINERY: HSV gD BINDING TO RECEPTORS

HSV-1 gD is a type I membrane anchored protein with an ectodomain of 316 residues that is organized around a core immunoglobulin (Ig) fold (aa 55-185).[27] Structurally, the core is related to cellular V-type Ig folds (i.e., CD58), but its atypical disulfide pattern precluded sequence-based prediction of the Ig fold.[27,107] We speculate that the virus appropriated this cellular structure as a backbone for the glycoprotein since most receptor binding activities and conformational changes involve the N-and C-terminal extensions that wrap around the Ig core.[27,108]

Four functional regions were defined by linker insertion mutagenesis of gD prior to solution of its structure.[109] These discrete regions (FR1: amino acids 27-43; FR2: 126-131; FR3: 225-246 and FR4: 277-300) fold together to form a structural domain where most of the action occurs.[27] FR1 within the N-terminal extension encompasses a segment critical to HVEM and nectin-1 binding.[27,110] The N-terminus rests on a long helix ($\alpha 3$), corresponding to FR3, which in turn is supported by FR2, the only functional region identified within the Ig fold. FR4 at the C-terminus of the ectodomain folds back around the Ig-core and interacts with FR1 at the N-terminus and with the $\alpha 3$ helix (i.e., FR3) in a manner that is crucial for triggering fusion[108] (see below).

The Various gD Receptors

HSV gD binds at least three different receptors: HVEM, nectin-1 and 3-OS-HS.[29,36,37] Any one of these receptors can be used for entry, as well as for virus spread from cell to cell and for cell-cell fusion in a virus-free assay.[111-113] Remarkably, despite their obvious structural differences, the three receptors all bind to gD with similar affinity.[37,114-116] In the remaining section, we will give an overview of the specificity of each receptor and describe a model for receptor-mediated activation of gD based on recent structural data.

HVEM

This member of the TNF receptor family was the first to be identified as a receptor for wt HSV-1.[36] The structural details of the HVEM/gD interaction, solved by X-ray crystallography,[27] show that HVEM contacts gD via residues within its first and second cysteine-rich domains (CRD1, CRD2).[117,118] The prominent tyrosine 23 in CRD1 plays an essential role by forming multiple hydrogen bonds with 3 amino acids of gD. On the other side of the interface, the amino acids on gD that contact HVEM are all located on the two arms of an N-terminal hairpin (aa 1-32)[119] (Fig. 2C). A thorough analysis of the 14 contact residues on gD showed that three regions were critical. The first comprises a central pocket into which the side chain of HVEM Tyr23 protrudes. The second is a linear β-strand (aa 27 to 29) that creates a β-sheet with a β-strand from HVEM CRD1 (aa 35 to 37). Interestingly, the structure of HVEM bound to one of its natural ligands, BTLA (B- and T-cell attenuator) shows a similar interface.[120] Despite having unrelated amino acid sequences and overall structures, BTLA and gD both form an intermolecular β-sheet with HVEM. Another HVEM ligand, named LIGHT,[121] binds to the opposite side of HVEM and has opposite regulatory effects to BTLA on lymphocyte activation.[122]

Nectin-1

This cell adhesion molecule belongs to the family of nectins characterized by three Ig-folds in their extracellular domain.[123] Dimeric nectin-1 trans-interacts with its ligand on an adjacent cell to mediate cell aggregation and to organize intercellular junctions.[124] The ligands for nectin-1 are: nectin-1 itself (homophilic) or nectin-3 or -4 (heterophilic).[124] The binding site for gD is located on the most distal Ig-like domain (V-domain) of nectin-1.[125,126] Epitope mapping and mutagenesis approaches indicate that the contact surface for gD consists of residues on the predicted β-strands C', C'' and D as well as on the connecting loops.[127,128] When the V-domain alone is expressed as a soluble protein, it binds to gD with the same affinity as the full ectodomain.[126] However, this domain needs to be correctly presented and folded for the receptor to be functional on cells[129,130] and shows reduced activity as an entry receptor when presented alone.[125]

On the gD side, the binding domain for nectin-1 appears to be topologically distinct from the binding domain for HVEM.[110] The N-terminal HVEM binding loop (aa 7-32) can be deleted from gD without significantly affecting nectin-1 usage.[131] Residues important for nectin-1 binding are located on and downstream of the α3 helix and on a region closer to the N-terminus (Fig. 2). The most critical residue identified so far is the exposed Tyr38.[110] A point mutation of this residue (Y38A) significantly reduces binding to nectin-1 and a double mutation forming an N-terminal disulfide bridge (Y38C-A3C) completely ablates it. Because the gD Ig core is wrapped by N- and C-terminal extensions that are involved in binding to nectin-1, it seems unlikely that gD mimics the Ig-to-Ig binding of a natural nectin-1 ligand. Nevertheless, the binding site for gD does overlap a functional site involved in cell adhesion since soluble gD blocks nectin-1 mediated cell aggregation.[132,133] Soluble gD also disrupts cell aggregates, suggesting that the viral glycoprotein competes with the homophilic ligand of nectin-1.[132] This is crucial since the nectin-1 ligand that is normally engaged at cell junctions must be displaced to allow for virus binding to this receptor.[134]

3-OS Modified Heparan Sulfate

In addition to protein receptors, gD binds to HS molecules decorated with specific sulfate groups by 3-O-sulfotransferases.[37] The sequential activity of these sulfotransferases generates highly specific sulfate patterns on the carbohydrate backbone of HS. Patterns generated by 3-O-ST isoforms 2 to 6 can be recognized by HSV-1 gD.[37,135-138] Two negatively charged pockets on gD could be receptacles for specific sulfate moieties of HS,[27] but the mode of action and usage of 3-OS-HS for HSV entry in cells and spread in the host is not yet understood.[113,139]

Receptor-Mediated Activation of Viral Entry

The initial structural determination of gD bound to HVEM[27] followed by that of the full ectodomain of unliganded gD[108] have been critical for our understanding of the receptor-mediated activation of gD during entry (Fig. 2). The wealth of information gathered from previous antibody mapping studies and scores of gD mutants was nicely explained when laid on the structural data. The first structure of gD showed that three of the four functional regions combine to form the HVEM binding site.[27] However, functional

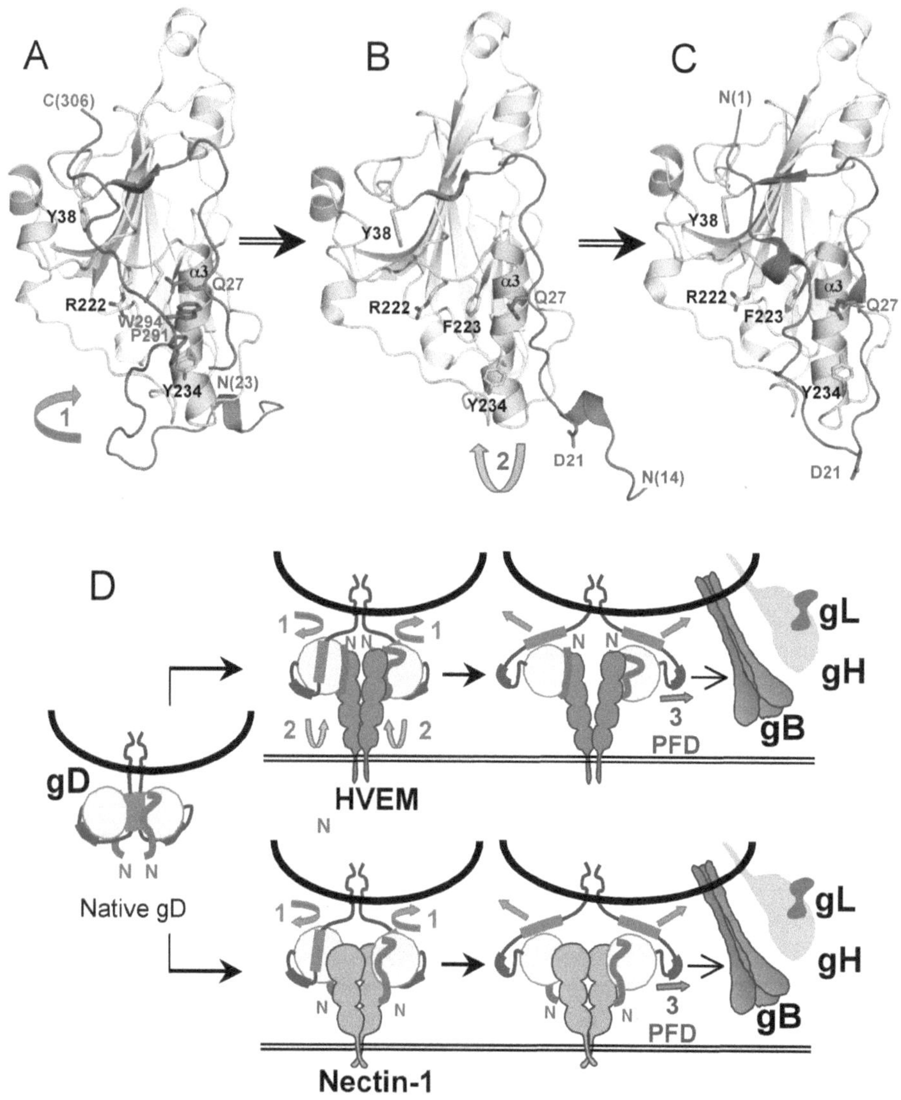

Figure 2. Structures of the gD ectodomain. Three independent structures of the gD ectodomain are presented to illustrate the conformations adopted by gD before and after it binds HVEM. A) gD306t$_{307cys}$. Full ectodomain in prereceptor binding state.[108] The C-terminus folds back around the Ig core and runs close to the N-terminal strand. B) gD285t. Unbound form of the ectodomain truncated at position 285.[27] The C-terminus is absent and the N-terminus is extended. In this intermediate conformation, the putative nectin-1 binding site is exposed and the N-terminus can fold back to create the HVEM binding hairpin. C) gD285t bound to HVEMt. Same form of gD as in panel B when bound to HVEM.[27] The N-terminus of gD forms a hairpin that contacts the receptor. HVEM has been omitted to better illustrate the binding site. D) Model of gD activation upon binding to HVEM and nectin-1. Numbered arrows indicate the temporal order of conformational changes that occur upon receptor binding. Arrows 1 and 2 represent conformational changes detailed in panels A-C. Conformational changes affecting the pro-fusion domain (PFD) are hypothetical.

region 4 (FR4), located close to the C-terminus of the ectodomain, was not resolved in this structure. Although FR4 is not directly involved in gD binding to either HVEM or nectin-1, it significantly affects the interactions. Insertions and substitutions in this region lead to forms of gD that are not functional in entry or cell-cell fusion.[109,140-142] In fact, these mutations or even a complete deletion of FR4 result in a 50-100 fold increase in affinity between gD and either nectin-1 or HVEM.[126,141,143] This increased affinity is due to a faster rate of complex formation, not to an increased stability of the complex. In addition, certain mutations in either the N and C-terminus of gD prevent binding of the monoclonal antibody AP7 suggesting that the C-and N-termini of gD are close to each other.[109,144] This led to the speculation that the C-terminus was somehow "in the way" of the receptor binding sites.

Recent structural data not only confirmed this prediction but also explained the role of FR4 in gD function.[108] The structure of the C-terminus of the gD ectodomain was solved by engineering a cysteine at the C-terminus of the ectodomain ($gD306t_{307cys}$).[108] The resulting disulfide-linked dimer stabilized the entire C-terminus, which is otherwise too flexible to allow crystallization of the complete ectodomain.[27] $gD306t_{307cys}$ has an extensive dimer interface that largely involves the C-terminal region. Because gD can be crosslinked as a dimer on the virion envelope[145] we speculated that dimerization may play a role in folding of the C-terminal residues in the native virion protein. However, because the $gD306t_{307cys}$ dimer was created with a disulfide bond, one cannot be certain that the position of the protomers relative to each other in this dimer represents exactly the dimeric form of gD in virions. It is hypothesized that the extra disulfide bond mimics the role of the missing transmembrane region and cytoplasmic tail, and in turn stabilizes the C-terminus. In support of this hypothesis, functional and immunological data show that $gD306t_{307cys}$ has characteristics similar to native gD on the viral envelope. Importantly, the C-terminus of $gD306t_{307cys}$ was trapped in its prereceptor-binding "native" conformation.

In unliganded gD, the N-terminus is extended and flexible but folds into a hairpin to bind HVEM[27] (Fig. 2B,C). A comparison of structures shows that the last 18 residues of unbound $gD306t_{307cys}$ occupy the same space as the 16 N-terminal amino acids of gD bound to HVEM[108](Fig. 2A,C). Thus, the formation of the N-terminal HVEM-binding hairpin requires the displacement of the gD C-terminus. The atomic details of the gD/nectin-1 interaction are currently unknown, but mutagenesis studies point to specific gD residues that are likely involved.[110,131,142,146,147] Three of these residues (Tyr38, Arg222 and Phe223) are masked by the C-terminus in $gD306t_{307cys}$ while others (e.g., His39, Gly218, Asp215, Leu220, Pro221) are buried in the dimer interface.[108] This arrangement suggests that binding of nectin-1 also necessitates displacement of the C-terminus; furthermore this event would result in loosening of the native dimer. To confirm that the C-terminus must be displaced upon receptor binding, a double cysteine mutant was engineered to lock the C-terminus in the position observed in $gD306t_{307cys}$ by introducing an intramolecular disulfide bond (Cys37-Cys302). Formation of the correct disulfide bonds was confirmed by X-ray crystallography. In addition, this form of gD bound the AP7 antibody, confirming that the C-terminus was correctly folded near the N-terminus. Despite adopting the correct conformation, the mutant with a locked C-terminus failed to bind HVEM or nectin-1. Moreover it was essentially not functional in HSV entry.[108] Altogether, these observations explain how two unrelated receptors bind to different sites on gD and yet generate a similar conformational change to trigger membrane fusion.

Although the C-terminus must be flexible enough to allow for receptor binding, flexibility alone is not sufficient for function. A prominent feature of this region is the

A

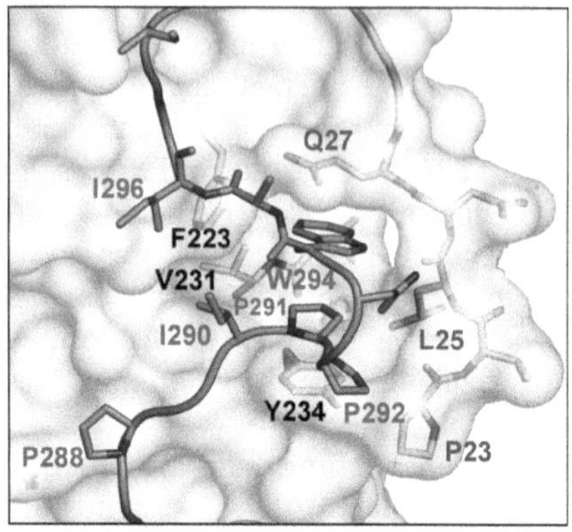

B

		277 … 319
HSV-1 (17)	*(human)*	ALLEDPVGTVAPQI**PPNW**HIPSIQDAATPYHPPATPNNMGLIA..
HSV-2 (HG52)	*(human)*	ALLEDPAGTVSSQI**PPNW**HIPSIQDVA-PHHAPAAPSNPGLII..
CeHV-1 (B virus)	*(cercop.)*	TPGDEPAPAVAAQL**PPNW**HVPEASDVTIQGPAPAPSGHTGAVV..
CeHV-2	*(cercop.)*	PPADGSLPAVSAQL**PPNW**HVPEADDVTIQGPAPTPADHTGAIV..
HVPapio-2	*(baboon)*	PPADGSPPAVSAQL**PPNW**HVPEADDVTIQGPVPAPAGHTGAIV..
PRV (Kaplan)	*(swine)*	ETPHRPFAPPAVV**PSGW**PQPAEPFQPRTPAAPGVSRHRSVIVG..
BHV-1 (Cooper)	*(bovine)*	KPGPSPDADR**PEGW**PSLEAITHPPPAPATPAAPDAVPVSVGIG..
BHV-5 (SV507/99)	*(bovine)*	PGPGGPGADVDR**PEGW**PSLEAITRPPPTPATPAPPTALPVGIG..
CpHV-1	*(caprine)*	EHAPGGRADASR**PEGW**PSLEDITRAPDPPTTPTLPPAAPLGVR..
EHV-1 (Kentucky A)	*(equine)*	NQ**PFKW**PSIKHLVPRLDEVDEVIEPVTKPPKTSKSNSTFVGIS..
FeHV	*(feline)*	SSG**PNW**IDIENYTPKNNVPIIISDDDVPTAPPKGMNNQSVVIP..
CaHV	*(canine)*	EIKSKVSSEKEIVNLEDYFKNTNVIDTTTPTPINNTPKNITVG..
PhoHV	*(phocine)*	AEEADEESKKLEEYLKISDVNLIDGNIETQLPINNSKTNITIA..
MaHV-1	*(kangaroo)*	VPALAATVVPKV**PPNW**HAPKIYEFPHTNPNPDSKTPWGTIAGG..
MaHV-2	*(kangaroo)*	EHEMSILAATVVPRV**PPNW**HAPKIYDFPQSNPTPATNPPLGII..

Figure 3. Conservation of the PxxW triggering motif in mammalian and marsupial α-HV gD. A) Residues W294 and Pro291 insert in a pocket on the gD surface to anchor the C-terminus of the ectodomain (sticks) in its native configuration. B) Alignment of membrane proximal regions of α-HV gD. The virus strains and hosts are indicated and the conserved PxxW site is highlighted in the sequence. The N-glycosylation consensus sites, which are found mostly in gD without PxxW motifs, are underlined. The beginning of the transmembrane sequence is boxed (left). Residue numbers correspond to the HSV-1 gD open reading frame.

anchor-like insertion of Pro291 and Trp294 (PxxW motif) into a pocket on the gD surface (Fig. 3). Mutational analysis confirmed the critical role of Trp294 suggested by its structural positioning[108] (Fig. 3) and its location in the middle of the previously defined FR4.[109] A mutation of Trp294 to alanine destabilizes the structure of the C-terminus thus exposing the nectin-1 binding site and favoring formation of the HVEM-binding loop. This mutant (W294A) has a higher affinity for both receptors similar to that observed for forms of gD lacking residues downstream of 285.[108] Importantly, a gD-null virus complemented with this mutated gD is significantly impaired in entry suggesting that an excessively flexible C-terminus is harmful for gD function. This suggests that displacement of the C-terminus must occur concomitantly with receptor binding to activate the fusion machinery. In the context of virus entry this mechanism would prevent premature activation of the fusion machinery by allowing the requisite conformational changes only when the virion is in close contact with the target cell.

A similar activation system is likely to be shared by many α-HV that encode a gD protein with the same anchoring sequence PxxW near the transmembrane region (Fig. 3). However the means by which the activation signal is transmitted to gB and/or gH/gL is not yet clear. The structure/function data suggest two possibilities. The first is that the conformational change in the C-terminus leads to presentation of a gD domain that is then able to interact with the fusion protein(s). Two recent studies[148,149] identified a linear proline rich domain (aa 260-285), upstream of the C-terminal anchor, which is important for gD function after receptor binding. In all X-ray structures of gD this segment is partially unresolved, suggesting it is highly flexible.[27,108] It is hypothesized that this region, named the pro-fusion domain (PFD) contacts other essential glycoproteins, possibly after refolding to a more ordered structure.[148] Because it is flexible, one can also envisage that the PFD acts as a hinge to present another part of gD to activate the fusion protein.[149] Indeed analogous proline rich regions connect functional domains of unrelated viral envelope proteins.[65,150] The second possibility (not necessarily exclusive of the first) for activating the fusion machinery is that opening of the C-terminus modifies the fusion machinery to allow stabilization of a fusion intermediate. This hypothesis relies on additional cues for full activation, which might explain why, in some cells, endocytosis and acidic pH are required for fusion.[151] Both models imply that this conformational change in gD only takes place when the virus is apposed to the cell surface. Premature conformational changes would be deleterious and this idea is supported by the nonfunctional phenotype of gDW294A with its highly flexible C-terminus. Thus, HSV gD is not merely a receptor-binding protein but is a finely tuned device that controls the activation of the viral fusion machinery.

CONCLUSION AND FUTURE CHALLENGES

The identification of cellular receptors and the determination of the structure of glycoproteins bound to receptors marked critical achievements in understanding how HV entry is initiated. Recent structural advances have augmented the structural model developed here. Solving the structure of HSV gD bound to nectin-1[152] and EBV gp42 bound to HLA[153] reinforced the concept of receptor-mediated activation of fusion through conformational changes. Furthermore, the determination of the structures of gB[154,155] and gHgL[156-158] significantly improved our understanding of fusion and its regulation. It is clear that the conserved gB is the effector of fusion for HV.[159] In contrast, gHgL show no structural homology with known fusion proteins and is considered a regulator of gB activity.[160] These structures provide highly informative, albeit static, views of the HV entry glycoproteins which will help further functional studies to elucidate their coordinated mechanism of action. One of the future challenges is to identify how the viral glycoproteins interact with each other, and in which order, to regulate and promote fusion.

Many challenges also reside on the host side. Receptors, and possibly co-receptors, for many HV have yet to be identified. There is increasing evidence that both gB and gH/gL interact with cell surface molecules that facilitate the entry process in a way that is often unclear. As additional essential and accessory receptors are being identified, it is important to determine their mechanistic role but also the part they play in determining in vivo tropism and spread. Though many details of the theme underlying entry of herpesviruses have been uncovered, many variations still remain enigmatic.

ACKNOWLEDGMENTS

We are grateful to the members of the Cohen and Eisenberg laboratory for their critical reading of this chapter. We thank Dr. E. Heldwein for allowing us to cite unpublished data. C. Krummenacher is supported by NIH grants AI097171 and DE022137, R.J. Eisenberg is supported by NIH grants AI056045 and AI076231, and G.H. Cohen is supported by NIH grant AI18289.

REFERENCES

1. Roizman B, Pellett PE. The family herpesviridae: A brief introduction. In: Knipe DM, Howley PM, eds. Fields Virology. 4th ed. Philadelphia: Lippincott, Willams and Wilkins, 2001:2381-2398.
2. Mettenleiter TC. Pathogenesis of neurotropic herpesviruses: Role of viral glycoproteins in neuroinvasion and transneuronal spread. Virus Res 2003; 92(2):197-206.
3. Cole NL, Grose C. Membrane fusion mediated by herpesvirus glycoproteins: The paradigm of varicella-zoster virus. Rev Med Virol 2003; 13(4):207-222.
4. Speck P, Haan KM, Longnecker R. Epstein-Barr virus entry into cells. Virology 2000; 277(1):1-5.
5. Spear PG, Longnecker R. Herpesvirus entry: An update. J Virol Methods 2003; 77(19):10179-10185.
6. Campadelli-Fiume G, Cocchi F, Menotti L et al. The novel receptors that mediate the entry of herpes simplex viruses and animal alphaherpesviruses into cells. Rev Med Virol 2000; 10(5):305-319.
7. Compton T. Receptors and immune sensors: The complex entry path of human cytomegalovirus. Trends Cell Biol 2004; 14(1):5-8.
8. Grunewald K, Desai P, Winkler DC et al. Three-dimensional structure of herpes simplex virus from cryo-electron tomography. Science 2003; 302:1396-1398.
9. Gianni T, Martelli PL, Casadio R et al. The ectodomain of herpes simplex virus glycoprotein H contains a membrane alpha-helix with attributes of an internal fusion peptide, positionally conserved in the herpesviridae family. J Virol 2005; 79(5):2931-2940.
10. Lopper M, Compton T. Disulfide bond configuration of human cytomegalovirus glycoprotein B. J Virol 2002; 76(12):6073-6082.
11. Galdiero S, Falanga A, Vitiello M et al. Fusogenic domains in herpes simplex virus type 1 glycoprotein H. J Biol Chem 2005; 280:28632-28643.
12. Foster TP, Melancon JM, Kousoulas KG. An alpha-Helical domain within the carboxyl terminus of Herpes simplex virus type 1 (HSV-1) glycoprotein B (gB) is associated with cell fusion and resistance to heparin inhibition of cell fusion. Virology 2001; 287(1):18-29.
13. Bzik DJ, Fox BA, DeLuca NA et al. Nucleotide sequence of a region of the herpes simplex virus type 1 gB glycoprotein gene: Mutations affecting rate of virus entry and cell fusion. Virology 1984; 137:185-190.
14. Gage PJ, Levine M, Glorioso JC. Syncytium-inducing mutations localize to two discrete regions within the cytoplasmic domain of herpes simplex virus type 1 glycoprotein B. J Virol 1993; 67:2191-2201.
15. Engel JP, Boyer EP, Goodman JL. Two novel single amino acid syncytial mutations in the carboxy-terminus of glycoprotein B of herpes simplex virus type 1 confer a unique pathogenic phenotype. Virology 1993; 192:112-120.
16. Ruell N, Zago A, Spear PG. Alanine substitution of conserved residues in the cytoplasmic tail of herpes simplex virus gB can enhance or abolish cell fusion activity and viral entry. Virology 2006; 344:17-24.
17. Navarro D, Paz P, Tugizov S et al. Glycoprotein B of human cytomegalovirus promotes virion penetration into cells, transmission of infection from cell to cell, and fusion of infected cells. Virology 1993; 197:143-158.
18. Herold BC, WuDunn D, Soltys N et al. Glycoprotein C of herpes simplex virus type 1 plays a principal role in the adsorption of virus to cells and in infectivity. J Virol 1991; 65:1090-1098.
19. Shieh MT, WuDunn D, Montgomery RI et al. Cell surface receptors for herpes simplex virus are heparan sulfate proteoglycans. J Cell Biol 1992; 116:1273-1281.
20. WuDunn D, Spear PG. Initial interaction of herpes simplex virus with cells is binding to heparan sulfate. J Virol 1989; 63:52-58.
21. Cheshenko N, Herold BC. Glycoprotein B plays a predominant role in mediating herpes simplex virus type 2 attachment and is required for entry and cell-to-cell spread. J Gen Virol 2002; 83(9):2247-2255.
22. Tal-Singer R, Peng C, Ponce de Leon M et al. Interaction of herpes simplex virus glycoprotein gC with mammalian cell surface molecules. J Virol 1995; 69(7):4471-4483.
23. Rux AH, Lou H, Lambris JD et al. Kinetic analysis of glycoprotein C of herpes simplex virus types 1 and 2 binding to heparin, heparan sulfate, and complement component C3b. Virology 2002; 294(2):324-332.

24. Williams RK, Straus SE. Specificity and affinity of binding of herpes simplex virus type 2 glycoprotein B to glycosaminoglycans. J Virol 1997; 71:1375-1380.
25. Laquerre S, Argnani R, Anderson DB et al. Heparan sulfate proteoglycan binding by herpes simplex virus type 1 glycoproteins B and C, which differ in their contributions to virus attachment, penetration, and cell-to-cell spread. J Virol 1998; 72:6119-6130.
26. Ligas MW, Johnson DC. A herpes simplex virus mutant in which glycoprotein D sequences are replaced by β-galactosidase sequences binds to but is unable to penetrate into cells. J Virol 1988; 62:1486-1494.
27. Carfi A, Willis SH, Whitbeck JC et al. Herpes simplex virus glycoprotein D bound to the human receptor HveA. Molecular Cell 2001; 8(1):169-179.
28. Spear PG, Eisenberg RJ, Cohen GH. Three classes of cell surface receptors for alphaherpesvirus entry. Virology 2000; 275:1-8.
29. Geraghty RJ, Krummenacher C, Eisenberg RJ et al. Entry of alphaherpesviruses mediated by poliovirus receptor related protein 1 and poliovirus receptor. Science 1998; 280:1618-1620.
30. Cocchi F, Menotti L, Mirandola P et al. The ectodomain of a novel member of the immunoglobulin subfamily related to the poliovirus receptor has the attribute of a bona fide receptor for herpes simplex virus types 1 and 2 in human cells. J Virol 1998; 72(12):9992-10002.
31. Warner MS, Martinez W, Geraghty RJ et al. A cell surface protein with herpesvirus entry activity (HveB) confers susceptibility to infection by herpes simplex virus type 2, mutants of herpes simplex virus type 1 and pseudorabies virus. Virology 1998; 246:179-189.
32. Lopez M, Cocchi F, Menotti L et al. Nectin2alpha (PRR2alpha or HveB) and nectin2delta are low-efficiency mediators for entry of herpes simplex virus mutants carrying the Leu25Pro substitution in glycoprotein D. J Virol 2000; 74(3):1267-1274.
33. Milne RSB, Connolly SA, Krummenacher C et al. Porcine HveC, a member of the highly conserved HveC/nectin 1 family, is a functional alphaherpesvirus receptor. Virology 2001; 281:315-328.
34. Connolly SA, Whitbeck JC, Rux AH et al. Glycoprotein D homologues in herpes simplex virus type 1, pseudorabies virus, and bovine herpes virus type 1 bind directly to human HveC (nectin-1) with different affinities. Virology 2001; 280:7-18.
35. Shukla D, DalCanto MC, Rowe CL et al. Striking similarity of murine nectin-1α to human nectin-1α (HveC) in sequence and activity as a gD receptor for alphaherpesvirus entry. J Virol 2000; 74:11773-11781.
36. Montgomery RI, Warner MS, Lum BJ et al. Herpes simplex virus-1 entry into cells mediated by a novel member of the TNF/NGF receptor family. Cell 1996; 87:427-436.
37. Shukla D, Liu J, Blaiklock P et al. A novel role for 3-O-sulfated heparan sulfate in herpes simplex virus 1 entry. Cell 1999; 99:13-22.
38. Krummenacher C, Baribaud F, Ponce De Leon M et al. Comparative usage of herpesvirus entry mediator A and nectin-1 by laboratory strains and clinical isolates of herpes simplex virus. Virology 2004; 322(2):286-299.
39. Mallory S, Sommer M, Arvin AM. Mutational analysis of the role of glycoprotein I in varicella-zoster virus replication and its effects on glycoprotein E conformation and trafficking. J Virol 1997; 71(11):8279-8288.
40. Saldanha CE, Lubinski J, Martin C et al. Herpes simplex virus type 1 glycoprotein E domains involved in virus spread and disease. J Virol 2000; 74(15):6712-6719.
41. Dingwell KS, Brunetti CR, Hendricks RL et al. Herpes simplex virus glycoproteins E and I facilitate cell-to-cell spread of virus in vivo and across junctions of cultured cells. J Virol 1994; 68:834-845.
42. Polcicova K, Goldsmith K, Rainish BL et al. The extracellular domain of herpes simplex virus gE is indispensable for efficient cell-to-cell spread: Evidence for gE/gI receptors. J Virol 2005; 79(18):11990-12001.
43. Johnson DC, Webb M, Wisner TW et al. Herpes simplex virus gE/gI sorts nascent virions to epithelial cell junctions, promoting virus spread. J Virol 2001; 75(2):821-833.
44. McMillan TN, Johnson DC. Cytoplasmic domain of herpes simplex virus gE causes accumulation in the trans-Golgi network, a site of virus envelopment and sorting of virions to cell junctions. J Virol 2001; 75(4):1928-1940.
45. Wang F, Tang W, McGraw HM et al. Herpes simplex virus type 1 glycoprotein e is required for axonal localization of capsid, tegument, and membrane glycoproteins. J Virol 2005; 79(21):13362-13372.
46. McGeoch DJ. Evolutionary relationships of virion glycoprotein genes in the S regions of alphaherpesvirus genomes. J Gen Virol 1990; 71:2361-2367.
47. Farnsworth A, Goldsmith K, Johnson DC. Herpes simplex virus glycoproteins gD and gE/gI serve essential but redundant functions during acquisition of the virion envelope in the cytoplasm. J Virol 2003; 77(15):8481-8494.
48. Wisner T, Brunetti C, Dingwell K et al. The extracellular domain of herpes simplex virus gE is sufficient for accumulation at cell junctions but not for cell-to-cell spread. J Virol 2000; 74(5):2278-2287.
49. Krummenacher C, Baribaud I, Eisenberg RJ et al. Cellular localization of nectin-1 and glycoprotein D during herpes simplex virus infection. J Virol 2003; 77(16):8985-8999.
50. Norrild B, Virtanen I, Lehto VP et al. Accumulation of herpes simplex virus type 1 glycoprotein D in adhesion areas of infected cells. J Gen Virol 1983; 64(11):2499-2503.

51. Mo C, Schneeberger EE, Arvin AM. Glycoprotein E of varicella-zoster virus enhances cell-cell contact in polarized epithelial cells. J Virol 2000; 74(23):11377-11387.
52. Dingwell KS, Johnson DC. The herpes simplex virus gE-gI complex facilitates cell-to-cell spread and binds to components of cell junctions. J Virol 1998; 72(11):8933-8942.
53. Turner A, Bruun B, Minson T et al. Glycoproteins gB, gD, and gHgL of herpes simplex virus type 1 are necessary and sufficient to mediate membrane fusion in a Cos cell transfection system. J Virol 1998; 72:873-875.
54. Pertel PE, Fridberg A, Parish ML et al. Cell fusion induced by herpes simplex virus glycoproteins gB, gD, and gH-gL requires a gD receptor but not necessarily heparan sulfate. Virology 2001; 279(1):313-324.
55. Muggeridge MI. Characterization of cell-cell fusion mediated by herpes simplex virus 2 glycoproteins gB, gD, gH and gL in transfected cells. J Gen Virol 2000; 81(8):2017-2027.
56. Heldwein EE, Lou H, Whitbeck JC et al. Crystal structure of the ectodomain of HSV-1 glycoprotein B reveals an elongated trimer. Turku, Finland: 30th International Herpesvirus Workshop, 2005.
57. Hutchinson L, Browne H, Wargent V et al. A novel herpes simplex virus glycoprotein, gL, forms a complex with glycoprotein H (gH) and affects normal folding and surface expression of gH. J Virol 1992; 66:2240-2250.
58. Peng T, Ponce de Leon M, Novotny MJ et al. Structural and antigenic analysis of a truncated form of the herpes simplex virus glycoprotein gH-gL complex. J Virol 1998; 72(7):6092-6103.
59. Duus KM, Hatfield C, Grose C. Cell surface expression and fusion by the varicella-zoster virus gH:gL glycoprotein complex: Analysis by laser scanning confocal microscopy. Virology 1995; 210(2):429-440.
60. Bender FC, Whitbeck JC, Lou H et al. Herpes simplex virus glycoprotein B binds to cell surfaces independently of heparan sulfate and blocks virus entry. J Virol 2005; 79(18):11588-11597.
61. Parry C, Bell S, Minson T et al. Herpes simplex virus type 1 glycoprotein H binds to alphavbeta3 integrins. J Gen Virol 2005; 86(1):7-10.
62. Jones TR, Lee SW, Johann SV et al. Specific inhibition of human cytomegalovirus glycoprotein B-mediated fusion by a novel thiourea small molecule. J Virol 2004; 78(3):1289-1300.
63. Britt WJ, Vugler LG. Processing of the gp55-116 envelope glycoprotein complex (gB) of human cytomegalovirus. J Virol 1989; 63:403-410.
64. Colman PM, Lawrence MC. The structural biology of type I viral membrane fusion. Nat Rev Mol Cell Biol 2003; 4(4):309-319.
65. Kielian M, Rey FA. Virus membrane-fusion proteins: More than one way to make a hairpin. Nat Rev Microbiol 2006; 4(1):67-76.
66. Feire AL, Koss H, Compton T. Cellular integrins function as entry receptors for human cytomegalovirus via a highly conserved disintegrin-like domain. Proc Natl Acad Sci USA 2004; 101(43):15470-15475.
67. Wang X, Huong SM, Chiu ML et al. Epidermal growth factor receptor is a cellular receptor for human cytomegalovirus. Nature 2003; 424:456-461.
68. Wang X, Huang DY, Huong SM et al. Integrin alphavbeta3 is a coreceptor for human cytomegalovirus. Nat Med 2005; 11(5):515-521.
69. Tugizov S, Navarro D, Paz P et al. Function of human cytomegalovirus glycoprotein B: Syncytium formation in cells constitutively expressing gB is blocked by virus-neutralizing antibodies. Virology 1994; 201(2):263-276.
70. Kinzler ER, Compton T. Characterization of human cytomegalovirus glycoprotein-induced cell-cell fusion. J Virol 2005; 79(12):7827-7837.
71. Huber MT, Compton T. The human cytomegalovirus UL74 gene encodes the third component of the glycoprotein H-glycoprotein L-containing envelope complex. J Virol 1998; 72(10):8191-8197.
72. Akkapaiboon P, Mori Y, Sadaoka T et al. Intracellular processing of human herpesvirus 6 glycoproteins Q1 and Q2 into tetrameric complexes expressed on the viral envelope. J Virol 2004; 78(15):7969-7983.
73. Mori Y, Akkapaiboon P, Yang X et al. The human herpesvirus 6 U100 gene product is the third component of the gH-gL glycoprotein complex on the viral envelope. J Virol 2003; 77(4):2452-2458.
74. Mori Y, Akkapaiboon P, Yonemoto S et al. Discovery of a second form of tripartite complex containing gH-gL of human herpesvirus 6 and observations on CD46. J Virol 2004; 78(9):4609-4616.
75. Mori Y, Yang X, Akkapaiboon P et al. Human herpesvirus 6 variant A glycoprotein H-glycoprotein L-glycoprotein Q complex associates with human CD46. J Virol 2003; 77(8):4992-4999.
76. Santoro F, Greenstone HL, Insinga A et al. Interaction of glycoprotein H of human herpesvirus 6 with the cellular receptor CD46. J Biol Chem 2003; 278(28):25964-25969.
77. Santoro F, Kennedy PE, Locatelli G et al. CD46 is a cellular receptor for human herpesvirus 6. Cell 1999; 99(7):817-827.
78. Secchiero P, Sun D, De Vico AL et al. Role of the extracellular domain of human herpesvirus 7 glycoprotein B in virus binding to cell surface heparan sulfate proteoglycans. J Virol 1997; 71(6):4571-4580.
79. Lusso P, Secchiero P, Crowley RW et al. CD4 is a critical component of the receptor for human herpesvirus 7: Interference with human immunodeficiency virus. Proc Natl Acad Sci USA 1994; 91:3872-3876.

80. Nemerow GR, Wolfert R, McNaughton ME et al. Identification and characterization of the Epstein-Barr virus receptor on human B lymphocytes and its relationship to the C3d complement receptor (CR2). J Virol 1985; 55(2):347-351.
81. Tugizov SM, Berline JW, Palefsky JM. Epstein-Barr virus infection of polarized tongue and nasopharyngeal epithelial cells. Nat Med 2003; 9(3):307-314.
82. Li Q, Spriggs MK, Kovats S et al. Epstein-Barr virus uses HLA class II as a cofactor for infection of B lymphocytes. J Virol 1997; 71(6):4657-4662.
83. Haan KM, Kwok WW, Longnecker R et al. Epstein-Barr virus entry utilizing HLA-DP or HLA-DQ as a coreceptor. J Virol 2000; 74(5):2451-2444.
84. Mullen MM, Haan KM, Longnecker R et al. Structure of the Epstein-Barr virus gp42 protein bound to the MHC class II receptor HLA-DR1. Mol Cell 2002; 9(2):375-385.
85. Wang X, Hutt-Fletcher LM. Epstein-Barr virus lacking glycoprotein gp42 can bind to B cells but is not able to infect. J Virol 1998; 72:158-163.
86. Li Q, Turk SM, Hutt-Fletcher LM. The Epstein-Barr virus (EBV) BZLF2 gene product associates with the gH and gL homologs of EBV and carries an epitope critical to infection of B cells but not of epithelial cells. J Virol 1995; 69(7):3987-3994.
87. Borza CM, Morgan AJ, Turk SM et al. Use of gHgL for attachment of Epstein-Barr virus to epithelial cells compromises infection. J Virol 2004; 78(10):5007-5014.
88. Molesworth SJ, Lake CM, Borza CM et al. Epstein-Barr virus gH is essential for penetration of B cells but also plays a role in attachment of virus to epithelial cells. J Virol 2000; 74(14):6324-6332.
89. Wang X, Kenyon WJ, Li Q et al. Epstein-Barr virus uses different complexes of glycoproteins gH and gL to infect B lymphocytes and epithelial cells. J Virol 1998; 72(7):5552-5558.
90. Borza CM, Hutt-Fletcher LM. Alternate replication in B cells and epithelial cells switches tropism of Epstein-Barr virus. Nat Med 2002; 8(6):594-599.
91. Wang FZ, Akula SM, Pramod NP et al. Human herpesvirus 8 envelope glycoprotein K8.1A interaction with the target cells involves heparan sulfate. J Virol 2001; 75(16):7517-7527.
92. Birkmann A, Mahr K, Ensser A et al. Cell surface heparan sulfate is a receptor for human herpesvirus 8 and interacts with envelope glycoprotein K8.1. J Virol 2001; 75(23):11583-11593.
93. Pertel PE. Human herpesvirus 8 glycoprotein B (gB), gH, and gL can mediate cell fusion. J Virol 2002; 76(9):4390-4400.
94. Wang FZ, Akula SM, Sharma-Walia N et al. Human herpesvirus 8 envelope glycoprotein B mediates cell adhesion via its RGD sequence. J Virol 2003; 77(5):3131-3147.
95. Akula SM, Pramod NP, Wang FZ et al. Integrin alpha3beta1 (CD 49c/29) is a cellular receptor for Kaposi's sarcoma-associated herpesvirus (KSHV/HHV-8) entry into the target cells. Cell 2002; 108(3):407-419.
96. Kaleeba JAR, Berger EA. Kaposi's sarcoma-associated herpesvirus fusion-entry receptor: Cystine transporter xCT. Science 2006; 311:1921-1924.
97. Ryckman BJ, Jarvis MA, Drummond DD et al. Human cytomegalovirus entry into epithelial and endothelial cells depends on genes UL128 to UL150 and occurs by endocytosis and low-pH fusion. J Virol 2006; 80(2):710-722.
98. Compton T, Nepomuceno RR, Nowlin DM. Human cytomegalovirus penetrates host cells by pH-independent fusion at the cell surface. Virology 1992; 191(1):387-395.
99. Miller N, Hutt-Fletcher LM. Epstein-Barr virus enters B cells and epithelial cells by different routes. J Virol 1992; 66(6):3409-3414.
100. Tanner J, Weis J, Fearon D et al. Epstein-Barr virus gp350/220 binding to the B lymphocyte C3d receptor mediates adsorption, capping, and endocytosis. Cell 1987; 50:203-213.
101. Akula SM, Naranatt PP, Walia NS et al. Kaposi's sarcoma-associated herpesvirus (human herpesvirus 8) infection of human fibroblast cells occurs through endocytosis. J Virol 2003; 77(14):7978-7990.
102. Nicola AV, Straus SE. Cellular and viral requirements for rapid endocytic entry of herpes simplex virus. J Virol 2004; 78(14):7508-7517.
103. Milne RS, Nicola AV, Whitbeck JC et al. Glycoprotein D receptor-dependent, low-pH-independent endocytic entry of herpes simplex virus type 1. J Virol 2005; 79(11):6655-6663.
104. Nicola AV, McEvoy AM, Straus SE. Roles for endocytosis and low pH in herpes simplex virus entry into HeLa and Chinese hamster ovary cells. J Virol 2003; 77(9):5324-5332.
105. Wittels M, Spear PG. Penetration of cells by herpes simplex virus does not require a low pH-dependent endocytic pathway. Virus Res 1990; 18:271-290.
106. Granzow H, Weiland F, Jons A et al. Ultrastructural analysis of the replication cycle of pseudorabies virus in cell culture: A reassessment. J Virol 1997; 71(3):2072-2082.
107. Long D, Wilcox WC, Abrams WR et al. Disulfide bond structure of glycoprotein D of herpes simplex virus types 1 and 2. J Virol 1992; 66:6668-6685.
108. Krummenacher C, Supekar VM, Whitbeck JC et al. Structure of HSV gD in the prereceptor binding conformation reveals a mechanism for receptor-mediated activation of virus entry. EMBO J 2005; 24(23):4144-4153.

109. Chiang HY, Cohen GH, Eisenberg RJ. Identification of functional regions of herpes simplex virus glycoprotein gD by using linker-insertion mutagenesis. J Virol 1994; 68:2529-2543.
110. Connolly SA, Landsburg DJ, Carfi A et al. Potential nectin-1 binding site on herpes simplex virus glycoprotein D. Journal of Virology 2005; 79(2):1282-1295.
111. Terry-Allison T, Montgomery RI, Warner MS et al. Contributions of gD receptors and glycosaminoglycan sulfation to cell fusion mediated by herpes simplex virus 1. Virus Res 2001; 74(1-2):39-45.
112. Cocchi F, Menotti L, Dubreuil P et al. Cell-to-cell spread of wild-type herpes simplex virus type 1, but not of syncytial strains, is mediated by the immunoglobulin-like receptors that mediate virion entry, nectin1 (PRR1/HveC/HIgR) and nectin2 (PRR2/HveB). J Virol 2000; 74(8):3909-3917.
113. Shukla D, Spear PG. Herpesviruses and heparan sulfate: An intimate relationship in aid of viral entry. J Clin Invest 2001; 108(4):503-510.
114. Whitbeck JC, Peng C, Lou H et al. Glycoprotein D of herpes simplex virus (HSV) binds directly to HVEM, a member of the TNFR superfamily and a mediator of HSV entry. J Virol 1997; 71(8):6083-6093.
115. Krummenacher C, Nicola AV, Whitbeck JC et al. Herpes simplex virus glycoprotein D can bind to poliovirus receptor-related protein 1 or herpesvirus entry mediator, two structurally unrelated mediators of virus entry. J Virol 1998; 72:7064-7074.
116. Liu J, Shriver Z, Pope RM et al. Characterization of a heparan sulfate octasaccharide that binds to herpes simplex virus type 1 glycoprotein D. J Biol Chem 2002; 277(36):33456-33467.
117. Whitbeck JC, Connolly SA, Willis SH et al. Localization of the gD-binding region of the human herpes simplex virus receptor, HveA. J Virol 2001; 75:171-180.
118. Connolly SA, Landsburg DJ, Carfi A et al. Structure-based analysis of the herpes simplex virus glycoprotein D binding site present on herpesvirus entry mediator HveA(HVEM). J Virol 2002; 76:10894-10904.
119. Connolly SA, Landsburg DJ, Carfi A et al. Structure-based mutagenesis of herpes simplex virus glycoprotein D defines three critical regions at the gD/HveA interface. J Virol 2003; 77(14):8127-8140.
120. Compaan DM, Gonzalez LC, Tom I et al. Attenuating lymphocyte activity: The crystal structure of the BTLA-HVEM complex. J Biol Chem 2005; 280(47):39553-39561.
121. Mauri DN, Ebner R, Kochel KD et al. LIGHT, a new member of the TNF superfamily, and lymphotoxin (LT) α are ligands for herpesvirus entry mediator (HVEM). Immunity 1998; 8:21-30.
122. Croft M. The evolving crosstalk between costimulatory and coinhibitory receptors: HVEM-BTLA. Trends Immunol 2005; 26(6):292-294.
123. Sakisaka T, Takai Y. Biology and pathology of nectins and nectin-like molecules. Curr Opin Cell Biol 2004; 16(5):513-521.
124. Takai Y, Irie K, Shimizu K et al. Nectins and nectin-like molecules: Roles in cell adhesion, migration, and polarization. Cancer Sci 2003; 94(8):655-667.
125. Cocchi F, Lopez M, Menotti L et al. The V domain of herpesvirus Ig-like receptor (HIgR) contains a major functional region in herpes simplex virus-1 entry into cells and interacts physically with the viral glycoprotein D. Proc Natl Acad Sci USA 1998; 95:15700-15705.
126. Krummenacher C, Rux AH, Whitbeck JC et al. The first immunoglobulin-like domain of HveC is sufficient to bind herpes simplex virus gD with full affinity while the third domain is involved in oligomerization of HveC. J Virol 1999; 73:8127-8137.
127. Krummenacher C, Baribaud I, Ponce de Leon M et al. Localization of a binding site for herpes simplex virus glycoprotein D on the herpesvirus entry mediator C by using anti-receptor monoclonal antibodies. J Virol 2000; 74:10863-10872.
128. Martinez WM, Spear PG. Amino acid substitutions in the V domain of nectin-1 (HveC) that impair entry activity for herpes simplex virus types 1 and 2 but not for Pseudorabies virus or bovine herpesvirus 1. J Virol 2002; 76(14):7255-7262.
129. Geraghty RJ, Fridberg A, Krummenacher C et al. Use of chimeric nectin-1(HveC)-related receptors to demonstrate that ability to bind alphaherpesvirus gD is not necessarily sufficient for viral entry. Virology 2001; 285:366-375.
130. Struyf F, Plate AE, Spear PG. Deletion of the second immunoglobulin-like domain of nectin-1 alters its intracellular processing and localization and ability to mediate entry of herpes simplex virus. J Virol 2005; 79(6):3841-3845.
131. Manoj S, Jogger CR, Myscofski D et al. Mutations in herpes simplex virus glycoprotein D that prevent cell entry via nectins and alter cell tropism. Proc Natl Acad Sci USA 2004; 101(34):12414-12421.
132. Krummenacher C, Baribaud I, Sanzo JF et al. Effects of herpes simplex virus on structure and function of nectin- 1/HveC. J Virol 2002; 76(5):2424-2433.
133. Sakisaka T, Taniguchi T, Nakanishi H et al. Requirement of interaction of nectin-1alpha/HveC with afadin for efficient cell-cell spread of herpes simplex virus type 1. J Virol 2001; 75(10):4734-4743.
134. Yoon M, Spear PG. Disruption of adherens junctions liberates nectin-1 to serve as receptor for herpes simplex virus and pseudorabies virus entry. J Virol 2002; 76(14):7203-7208.

135. O'Donnell CD, Tiwari V, Oh MJ et al. A role for heparan sulfate 3-O-sulfotransferase isoform 2 in herpes simplex virus type 1 entry and spread. Virology 2006; 346(2):452-459.
136. Tiwari V, O'Donnell CD, Oh MJ et al. A role for 3-O-sulfotransferase isoform-4 in assisting HSV-1 entry and spread. Biochem Biophys Res Commun 2005; 338(2):930-937.
137. Xu D, Tiwari V, Xia G et al. Characterization of heparan sulphate 3-O-sulphotransferase isoform 6 and its role in assisting the entry of herpes simplex virus type 1. Biochem J 2005; 385(2):451-459.
138. Xia G, Chen J, Tiwari V et al. Heparan sulfate 3-O-sulfotransferase isoform 5 generates both an antithrombin-binding site and an entry receptor for herpes simplex virus, type 1. J Biol Chem 2002; 277(40):37912-37919.
139. Akhtar J, Shukla D. Viral entry mechanisms: cellular and viral mediators of herpes simplex virus entry. FEBS J 2009; 276(24):7228-7236.
140. Milne RSB, Hanna SL, Rux AH et al. Function of herpes simplex virus type 1 gD mutants with different receptor-binding affinities in virus entry and fusion. J Virol 2003; 77(16):8962-8972.
141. Rux AH, Willis SH, Nicola AV et al. Functional region IV of glycoprotein D from herpes simplex virus modulates glycoprotein binding to the herpes virus entry mediator. J Virol 1998; 72:7091-7098.
142. Jogger CR, Montgomery RI, Spear PG. Effects of linker-insertion mutations in herpes simplex virus 1 gD on glycoprotein-induced fusion with cells expressing HVEM or nectin-1. Virology 2004; 318(1):318-326.
143. Willis SH, Rux AH, Peng C et al. Examination of the kinetics of herpes simplex virus glycoprotein D binding to the herpesvirus entry mediator, using surface plasmon resonance. J Virol 1998; 72(7):5937-5947.
144. Minson AC, Hodgman TC, Digard P et al. An analysis of the biological properties of monoclonal antibodies against glycoprotein D of herpes simplex virus and identification of amino acid substitutions that confer resistance to neutralization. J Gen Virol 1986; 67:1001-1013.
145. Handler CG, Eisenberg RJ, Cohen GH. Oligomeric structure of glycoproteins in herpes simplex virus type 1. J Virol 1996; 70:6067-6075.
146. Yoon M, Zago A, Shukla D et al. Mutations in the N termini of herpes simplex virus type 1 and 2 gDs alter functional interactions with the entry/fusion receptors HVEM, Nectin-2, and 3-O-sulfated heparan sulfate but not with Nectin-1. J Virol 2003; 77(17):9221-9231.
147. Whitbeck JC, Muggeridge MI, Rux A et al. The major neutralizing antigenic site on herpes simplex virus glycoprotein D overlaps a receptor-binding domain. J Virol 1999; 73:9879-9890.
148. Fusco D, Forghieri C, Campadelli-Fiume G. The pro-fusion domain of herpes simplex virus glycoprotein D (gD) interacts with the gD N terminus and is displaced by soluble forms of viral receptors. Proc Natl Acad Sci USA 2005; 102(26):9323-9328.
149. Zago A, Jogger CR, Spear PG. Use of herpes simplex virus and pseudorabies virus chimeric glycoprotein D molecules to identify regions critical for membrane fusion. Proc Natl Acad Sci USA 2004; 101(50):17498-17503.
150. Lavillette D, Maurice M, Roche C et al. A proline-rich motif downstream of the receptor binding domain modulates conformation and fusogenicity of murine retroviral envelopes. J Virol 1998; 72(12):9955-9965.
151. Whitbeck JC, Zuo Y, Milne RSB et al. Stable association of herpes simplex virus with target membrane is triggered by low pH in the presence of the gD receptor HVEM. J Virol 2006; 80:3773-3780.
152. Di Giovine P, Settembre EC, Bhargava AK et al. Structure of herpes simplex virus glycoprotein D bound to the human receptor nectin-1. PLoS Pathog 2011; 7:e1002277.
153. Kirschner AN, Sorem J, Longnecker R et al. Structure of Epstein-Barr virus glycoprotein 42 suggests a mechanism for triggering receptor-activated virus entry. Structure 2009; 17:223-233.
154. Backovic M, Longnecker R, Jardetzky TS. Structure of a trimeric variant of the Epstein-Barr virus glycoprotein B. Proc Natl Acad Sci USA 2009; 106:2880-2885.
155. Heldwein EE, Lou H, Bender FC et al. Crystal structure of glycoprotein B from herpes simplex virus 1. Science 2006; 313:217-220.
156. Chowdary TK, Cairns TM, Atanasiu D et al. Crystal structure of the conserved herpesvirus fusion regulator complex gH-gL. Nat Struct Mol Biol 2010; 17:882-8.
157. Backovic M, DuBois RM, Cockburn JJ et al. Structure of a core fragment of glycoprotein H from pseudorabies virus in complex with antibody. Proc Natl Acad Sci USA 2010; 107:22635-22640.
158. Matsuura H, Kirschner AN, Longnecker R et al. Crystal structure of the Epstein-Barr virus (EBV) glycoprotein H/glycoprotein L (gH/gL) complex. Proc Natl Acad Sci USA 2010; 107:22641-22646.
159. Heldwein EE, Krummenacher C. Entry of herpesviruses into mammalian cells. Cell Mol Life Sci 2008; 65:1653-1668.
160. Eisenberg RJ, Atanasiu D, Cairns TM et al. Herpes virus fusion and entry: a story with many characters. Viruses 2012; 4:800-832.

INDEX

Symbols

β1 integrin 13, 42, 46-48, 63, 86

A

Apoptosis 42, 43, 51-54, 56-59, 62-64, 135
Attachment 1-18, 25-29, 33, 37, 42-48, 52, 55, 62, 64, 74, 83-89, 95, 98-101, 103-105, 108, 110-114, 116, 118-120, 133, 137, 138, 151, 152, 154, 159, 167, 169, 170, 178-183
Attachment glycoprotein 95, 98-101, 103, 105, 108, 110-114, 116, 118, 120, 179
Avian sarcoma and leukosis virus (ASLV) 128-132, 141

B

Binding 1, 2, 4, 6-18, 27, 28, 33, 34, 36, 44-46, 51-53, 55, 58, 60, 61, 63, 72, 74-78, 80, 84-88, 90, 95, 100-104, 106, 109, 111-118, 120, 130-132, 134, 137-141, 151, 152, 154, 158, 159, 161, 162, 168-170, 172, 174, 178, 180-182, 184-189

C

Carbohydrate 1-3, 12-14, 16-18, 42, 44, 48, 52, 62, 84-88, 101, 104, 185
Cathepsin B (CTSB) 48, 49, 89, 90
Cathepsin L (CTSL) 48, 49, 89, 90, 106
Caveolae 30, 31, 34, 36, 76, 88
Caveolin 31, 32, 34-37, 76, 88
CCR5 6, 7, 10, 13, 139-141
CD4 5-10, 13-15, 18, 132, 135, 137-141, 182
CD46 103, 182
Charge 2-4, 6, 7, 10, 109
Class I fusion protein 129, 130, 140, 141, 150, 151, 154, 156, 160, 162, 171-174
Class II fusion protein 150-157, 159-162, 167, 171, 174
Clathrin 30-34, 42, 47, 72, 73, 76, 77, 88, 89, 151, 158, 167, 168, 170
Conformational change 2, 10, 17, 25-28, 32, 33, 47, 48, 51, 78, 79, 89, 95, 101, 106-120, 131, 137, 139-141, 158, 160, 161, 170-172, 174, 184, 186-189
Coreceptor 6, 7, 11-13, 27, 62, 75, 76, 132, 137, 139-141, 154, 181
Coxsackievirus 25, 27, 32-36, 45
C-type lectin 13, 14, 86-88
CXCR4 6, 10, 13, 139-141

D

DC-SIGN 5, 13-15, 87, 90, 138, 154

E

Ebola 5, 15, 83, 84, 130, 151
Endocytosis 30-32, 36, 37, 42, 46-48, 72, 73, 76, 77, 80, 88, 89, 104, 131, 134, 158, 167, 168, 170, 183, 189
Entry 1-4, 6, 9, 11-14, 16-18, 24-26, 30-32, 34-37, 42, 43, 46-48, 51, 63, 72, 73, 76-80, 83-90, 95, 96, 98, 103, 104, 110, 111, 120, 128-132, 134-141, 150, 151, 161, 162, 167-171, 173, 178-185, 187-189
Envelope 6, 9, 15, 72, 78, 79, 86, 96, 98, 99, 105, 106, 109, 110, 128, 129, 131, 132, 136, 137, 141, 150, 151, 154, 156, 158, 162, 167, 169, 171, 173, 178, 179, 181-183, 187, 189
Envelope protein 86, 128, 129, 131, 132, 136, 151, 154, 156, 158, 162, 189
Ephrin-B2 96, 102, 104

F

Filovirus 83-90, 130, 151
Foamy virus (FV) 128, 129, 131-135
Foot and mouth disease virus (FMDV) 5, 8, 24, 25, 27, 32, 33
Fusion 7-9, 13, 31, 58, 72, 73, 76-80, 83, 84, 86, 88-90, 95, 98-120, 128-132, 134, 137, 139-141, 150-162, 167-174, 178-184, 186-189
Fusion glycoprotein 95, 98-100, 105, 106, 110-114, 118, 120
Fusion loop 150, 152-160, 162
Fusion pore 78, 117, 141, 159, 160
Fusion protein (F) 9, 11, 78, 84, 89, 95, 98, 99, 101, 102, 104-120, 129, 130, 140, 141, 150-162, 167, 171-174, 180-182, 189

G

Glycoprotein (G) 2-4, 6, 13, 15, 17, 18, 57, 73-75, 77, 83-85, 87-89, 95, 96, 98-114, 116-120, 129-134, 139, 167, 169-174, 178-185, 189

H

Hemagglutinin (HA) 15, 16, 74, 75, 78-80, 98, 101, 103, 105, 107, 109, 110, 131, 151, 156, 158, 160, 162, 172, 173
Hemifusion 78, 159, 160
Heptad repeat (HR) 45, 84, 99, 105-109, 130, 140, 172, 180
Herpesvirus (HV) 178-183, 188, 189
Human immunodeficiency virus (HIV) 2, 5-7, 9, 10, 13-15, 18, 78, 83, 87, 89, 105, 106, 109, 128-141, 151, 154, 161, 162

I

Influenza 2, 5, 15, 16, 31, 58, 72-79, 83, 89, 100, 101, 103, 105-107, 109, 110, 131, 151, 156, 158, 160, 162, 172, 173
Inhibitor 11, 15, 17, 31-34, 36, 42, 45, 47, 48, 52, 53, 55-57, 62, 87-90, 116, 128, 131, 136, 140, 141, 161, 170, 171
Integrin 5, 8, 11, 12, 17, 27, 28, 33, 34, 42, 46-48, 63, 86, 104, 182, 183
Interferon (IFN) 57-62, 64
Intracellular signal 36, 54

J

Junctional adhesion molecule-A (JAM-A) 42, 44-46, 48, 52-55, 62, 63

L

Lectin 10, 13-15, 84, 86-88, 137, 138, 154

M

Membrane fusion 72, 78, 83, 84, 88-90, 95, 98, 103, 105, 107, 109-111, 113, 114, 117, 119, 120, 128, 130-132, 137, 140, 141, 150-152, 154, 156, 158, 160-162, 170-174, 178, 179, 182, 183, 187
Mitogen-activated protein kinase (MAPK) 56, 57

N

Nectin 103, 181, 183-189
Neuraminidase 16, 17, 52, 96, 98, 100, 101, 103, 105, 116, 169
Nuclear factor-κB (NF-κB) 42, 43, 54-59, 62-64
Nuclear import 72, 79

P

Paramyxovirus 78, 95-98, 100, 101, 103, 105-114, 116-120, 130, 151
Pathogenesis 4, 15, 35, 43, 62, 64, 77, 79, 128, 129, 141
pH-Dependence 72, 73, 78, 86, 88, 89, 167, 168, 170, 172, 173, 183
Phosphatidylserine (PS) 167, 169, 170, 172, 174
Picornavirus 11, 24-28, 31, 37
Poliovirus 24-30, 32, 33, 36, 37, 181
Proteoglycan 2-4, 7, 33, 137, 181
Proteolysis 48, 49, 84, 90, 173

R

Rabies 5, 168
Receptor 1-18, 24-34, 37, 42, 44-48, 52-54, 57, 59, 61-64, 72-78, 83-90, 95, 96, 98, 100, 101, 103, 104, 111-120, 128, 130-132, 134, 137-141, 151, 152, 154, 158, 159, 167-170, 172, 173, 178-189
Reovirus 42-64, 90
Retrovirus 7, 45, 78, 105, 106, 128, 129, 131-133, 135, 141, 151, 182
Rhinovirus 5, 10, 24-29, 33, 34

S

Sialic acid 5, 12, 16-18, 44-46, 48, 52, 54, 72-76, 80, 101-103, 114-116, 120
Signaling lymphocyte activation molecule (SLAM) 103, 117
Structure 2-4, 6, 10, 16, 17, 25, 26, 29, 30, 32, 45-47, 49-51, 74, 78, 87, 88, 95, 100-104, 106-110, 113, 114, 116-118, 120, 128, 133, 134, 137, 139, 141, 150-158, 160-162, 169, 171-174, 178, 182, 184-189

T

Transmembrane anchor 150, 151, 153-156, 159-161
Tyro3 receptor tyrosine kinase 86

V

Vesicular stomatitis virus (VSV) 87, 167-174

FSC
www.fsc.org
MIX
Papier | Fördert gute Waldnutzung
FSC® C083411

Kolibri 360 GmbH
Ferdinand-Jühlke-Straße 7
99095 Erfurt, Deutschland
produktsicherheit@kolibri360.de